# PLANT-BASED BIOACTIVE COMPOUNDS AND FOOD INGREDIENTS

*Encapsulation, Functional, and Safety Aspects*

*Innovations in Plant Science for Better Health: From Soil to Fork*

# PLANT-BASED BIOACTIVE COMPOUNDS AND FOOD INGREDIENTS

*Encapsulation, Functional, and Safety Aspects*

*Edited by*

**Junaid Ahmad Malik, PhD**

**Megh R. Goyal, PhD**

**Preeti Birwal, PhD**

**Ritesh B. Watharkar, PhD**

AAP | APPLE ACADEMIC PRESS

First edition published 2024

**Apple Academic Press Inc.**
1265 Goldenrod Circle, NE,
Palm Bay, FL 32905 USA

760 Laurentian Drive, Unit 19,
Burlington, ON L7N 0A4, CANADA

**CRC Press**
2385 NW Executive Center Drive,
Suite 320, Boca Raton FL 33431

4 Park Square, Milton Park,
Abingdon, Oxon, OX14 4RN UK



**Library and Archives Canada Cataloguing in Publication**

Title: Plant-based bioactive compounds and food ingredients : encapsulation, functional, and safety aspects / edited by Junaid Ahmad Malik, Megh R. Goyal, Preeti Birwal, Ritesh B. Watharkar
Names: Malik, Junaid Ahmad, 1987- editor. | Goyal, Megh R., editor. | Birwal, Preeti, editor. | Watharkar, Ritesh B., editor.
Series: Innovations in plant science for better health.
Description: First edition. | Series statement: Innovations in plant science for better health: from soil to fork | Includes bibliographical references and index.
Identifiers: Canadiana (print) 20230226507 | Canadiana (ebook) 20230226663 | ISBN 9781774912911 (hardcover) | ISBN 9781774912928 (softcover) | ISBN 9781003372226 (ebook)
Subjects: LCSH: Plant bioactive compounds. | LCSH: Microencapsulation. | LCSH: Functional foods. | LCSH: Veganism—Health aspects. | LCSH: Food—Safety measures.
Classification: LCC QK898.B54 P53 2023 | DDC 572/.2—dc23

**Library of Congress Cataloging-in-Publication Data**

CIP data on file with US Library of Congress

ISBN: 978-1-77491-291-1 (hbk)
ISBN: 978-1-77491-292-8 (pbk)
ISBN: 978-1-00337-222-6 (ebk)

# ABOUT THE BOOK SERIES: INNOVATIONS IN PLANT SCIENCE FOR BETTER HEALTH: FROM SOIL TO FORK

**SERIES EDITOR-IN-CHIEF:**

**Hafiz Ansar Rasul Suleria, PhD**
School of Agriculture and Food, Faculty of Veterinary and Agricultural Sciences, The University of Melbourne, Parkville, Victoria, Australia
email: h.suleria@hotmail.com

This book series offers academia, engineers, technologists, and users from different disciplines information to gain knowledge on the breadth and depth of this multifaceted field. The volumes explore the fields of phytochemistry along with its potential and extraction techniques. The volumes discuss the therapeutic perspectives of biochemical compounds in plants and animal and marine sources in an interdisciplinary manner because the field requires knowledge of many areas, including agricultural, food, and chemical engineering; manufacturing technology along with applications from diverse fields like chemistry; herbal drug technology; microbiology; animal husbandry; and food science; etc. There is an urgent need to explore and investigate the innovations, current shortcomings, and future challenges in this growing area of research.

We welcome book proposals on the following specialty areas (but not limited to):

- Food and function
- Nutritional composition of different foods materials
- Functional and nutraceutical perspectives of foods
- Extraction of bioactive molecules
- Phytomedicinal properties of different plants
- Traditional plants used as functional foods
- Phytopharmacology of plants used in metabolic disorders
- Importance of spices and medicinal and functional foods
- Natural products chemistry
- Food processing waste/byproducts: management and utilization
- Herbals as potential bioavailability enhancer and herbal cosmetics

- Phytopharmaceuticals for the delivery of bioactives
- Alternative and complementary medicines
- Ethnopharmacology and ethnomedicine
- Marine phytochemistry
- Marine microbial chemistry
- Other related includes nuts, seed spices, wild flora, etc.

# ABOUT THE BOOK SERIES EDITOR-IN-CHIEF

**Hafiz Ansar Rasul Suleria, PhD,** is a Senior Lecturer (ARC DECRA Fellow) in the School of Agriculture and Food at the University of Melbourne, Australia. He was formerly the Alfred Deakin Research Fellow at Deakin University, Victoria, Australia. He did his postdoctoral fellowship at the Department of Food, Nutrition, Dietetic and Health at Kansas State University, USA. He is also an Honorary Fellow of the Diamantina Institute, Faculty of Medicine, The University of Queensland (UQ), Australia. Dr. Suleria has been awarded an International Postgraduate Research Scholarship and the Australian Postgraduate Award for his PhD research at the UQ School of Medicine and the Translational Research Institute in collaboration with the Commonwealth and Scientific and Industrial Research Organization, Australia.

Dr. Suleria formerly worked as a lecturer in the Department of Food Sciences, Government College University Faisalabad, Pakistan. He also worked as a research associate in the PAK-US Joint Project funded by the Higher Education Commission, Pakistan, and Department of State, USA, with collaboration of the University of Massachusetts, USA, and the National Institute of Food Science and Technology, University of Agriculture Faisalabad, Pakistan.

Dr. Suleria has published more than 80 peer-reviewed scientific papers in professional journals and has co-edited several books. He is also in collaboration with more than ten universities where he is working as a co-supervisor/special member for PhD and postgraduate students. His major research focus is on food nutrition, particularly in screening of bioactive molecules, including isolation, purification, and characterization using various cutting-edge techniques from different plants, marine, and animal sources; in vitro, in vivo bioactivities; and cell culture and animal modeling.

## BOOKS IN THE SERIES:

- **Assessment of Medicinal Plants for Human Health: Phytochemistry, Disease Management, and Novel Applications**
  Editors: Megh R. Goyal, PhD, and Durgesh Nandini Chauhan, MPharm
- **Bioactive Compounds from Multifarious Natural Foods for Human Health: Foods and Medicinal Plants**
  Editors: Hafiz Ansar Rasul Suleria, PhD, Megh R. Goyal, PhD, and Huma Bader Ul Ain, PhD

- **Bioactive Compounds from Plant Origin: Extraction, Applications, and Potential Health Benefits**
  Editors: Hafiz Ansar Rasul Suleria, PhD, and Colin Barrow, PhD
- **Bioactive Compounds of Medicinal Plants: Properties and Potential for Human Health**
  Editors: Megh R. Goyal, PhD, and Ademola O. Ayeleso, PhD
- **Cereals and Cereal-Based Foods: Functional Benefits and Technological Advances for Nutrition and Healthcare**
  Editors: Megh Goyal, PhD, Kamaljit Kaur, PhD, and Jaspreet Kaur, PhD
- **Curative and Preventive Properties of Medicinal Plants: Research on Disease Management and Animal Model Studies**
  Editors: Megh R. Goyal, PhD, Junaid Ahmad Malik, PhD, and Ademola Olabode Ayeleso, PhD
- **Health Benefits of Secondary Phytocompounds from Plant and Marine Sources**
  Editors: Hafiz Ansar Rasul Suleria, PhD, and Megh Goyal, PhD
- **Herbs, Spices, and Medicinal Plants for Human Gastrointestinal Disorders: Health Benefits and Safety**
  Editors: Megh R. Goyal, PhD, Preeti Birwal, PhD, and Durgesh Nandini Chauhan, MPharm
- **Himalayan Medicinal Plants for the Treatment of Depression: A Source of Rich Antidepressant Agents**
  Editors: Abdur Rauf, PhD, Prabhakar Semwal, PhD,and Hafiz Ansar Rasul Suleria, PhD
- **Human Health Benefits of Plant Bioactive Compounds: Potentials and Prospects**
  Editors: Megh R. Goyal, PhD, and Hafiz Ansar Rasul Suleria, PhD
- **Phytochemical and Pharmacological Investigation of the Family Rutaceae**
  Editors: Abdur Rauf, PhD, Hafiz Ansar Rasul Suleria, PhD, and Syed Muhammad Mukarram Shah, PhD
- **Phytochemicals from Medicinal Plants: Scope, Applications, and Potential Health Claims**
  Editors: Hafiz Ansar Rasul Suleria, PhD, Megh R. Goyal, PhD, and Masood Sadiq Butt, PhD
- **Plant Secondary Metabolites for Human Health: Extraction of Bioactive Compounds**
  Editors: Megh R. Goyal, PhD, P. P. Joy, and Hafiz Ansar Rasul Suleria, PhD
- **Plant- and Marine- Based Phytochemicals for Human Health: Attributes, Potential, and Use**
  Editors: Megh R. Goyal, PhD, and Durgesh Nandini Chauhan, MPharm

- **Plant-Based Bioactive Compounds and Food Ingredients: Encapsulation, Functional, and Safety Aspects**
  Editors: Junaid Ahmad Malik, PhD, Megh R. Goyal, PhD, Preeti Birwal, PhD, and Ritesh B. Watharkar, PhD
- **Plant-Based Functional Foods and Phytochemicals: From Traditional Knowledge to Present Innovation**
  Editors: Megh R. Goyal, PhD, Arijit Nath, PhD, and Hafiz Ansar Rasul Suleria, PhD
- **The Functional Foods: Nutrient and Health Benefits**
  Editors: Megh R. Goyal, PhD, Junaid Ahmad Malik, PhD, and Anu Kumari
- **The Role of Phytoconstitutents in Health Care: Biocompounds in Medicinal Plants**
  Editors: Megh R. Goyal, PhD, Hafiz Ansar Rasul Suleria, PhD, and Ramasamy Harikrishnan, PhD
- **The Therapeutic Properties of Medicinal Plants: Health-Rejuvenating Bioactive Compounds of Native Flora**
  Editors: Megh R. Goyal, PhD, PE, Hafiz Ansar Rasul Suleria, PhD, Ademola Olabode Ayeleso, PhD, T. Jesse Joel, and Sujogya Kumar Panda

# ABOUT THE EDITORS

**Junaid Ahmad Malik, PhD**
*Lecturer, Department of Zoology,*
*Government Degree College, Bijbehara,*
*Kashmir, Jammu and Kashmir, India*

Junaid Ahmad Malik, PhD, is a Lecturer with the Department of Zoology at Government Degree College, Bijbehara, Kashmir (J&K), India, and is actively involved with teaching and research activities. He has more than eight years of research experience. His areas of interest are ecology, soil macrofauna, wildlife biology, conservation biology, etc. Dr. Malik has published over 20 research articles and technical papers in international peer-reviewed journals and has authored and edited books, book chapters, and more than 10 popular editorial articles. He also serves as an editor and reviewer of several journals. He has participated in several state, national, and international conferences, seminars, workshops, and symposia and has more than 20 conference papers to his credit. He is a life member of the Society for Bioinformatics and Biological Sciences.

Dr. Malik received a BSc (2008) in Science from the University of Kashmir, Srinagar, J&K; MSc (2010) in Zoology from Barkatullah University, Bhopal, Madhya Pradesh, India; and PhD (2015) in Zoology from the same university. He completed his BEd program in 2017 at the University of Kashmir, Srinagar, J&K.

**Megh R. Goyal, PhD, PE**
*Retired Professor in Agricultural and Biomedical Engineering, University of Puerto Rico, Mayaguez Campus; Senior Acquisitions Editor, Biomedical Engineering and Agricultural Science, Apple Academic Press, Inc.*

Megh R. Goyal, PhD, PE, is a Retired Professor of Agricultural and Biomedical Engineering from the General Engineering Department at the College of Engineering at the University of Puerto Rico–Mayaguez Campus (UPRM). He is a Senior Acquisitions Editor and Senior Technical Editor-in-Chief for Agricultural and Biomedical Engineering for Apple Academic Press, Inc. During his professional career, he has worked as a Soil Conservation Inspector; Research Assistant at Haryana Agricultural University and Ohio State University; Research Agricultural Engineer/Professor at the Department of Agricultural Engineering of UPRM; and Professor of Agricultural and Biomedical Engineering in the General Engineering Department of UPRM. He spent a one-year sabbatical leave in 2002–2003 at the Biomedical Engineering Department of Florida International University, Miami, USA.

Dr. Goyal was the first agricultural engineer to receive the professional license in agricultural engineering from the College of Engineers and Surveyors of Puerto Rico. In 2005, he was proclaimed the "Father of Irrigation Engineering in Puerto Rico for the Twentieth Century" by the American Society of Agricultural and Biological Engineers, Puerto Rico Section, for his pioneering work on micro irrigation, evapotranspiration, agroclimatology, and soil and water engineering.

During his professional career of 52 years, he has received many awards, including Scientist of the Year, Membership Grand Prize for the American Society of Agricultural Engineers Campaign, Felix Castro Rodriguez Academic Excellence Award, Man of Drip Irrigation by the Mayor of Municipalities of Mayaguez/Caguas/Ponce and Senate/Secretary of Agriculture of ELA, Puerto Rico, and many others. He has been recognized as one of the experts "who rendered meritorious service for the development of [the] irrigation sector in India" by the Water Technology Centre of Tamil Nadu Agricultural University in Coimbatore, India, and ASABE who bestowed on him the 2018 Netafim Microirrigation Award.

Dr. Goyal has authored more than 200 journal articles and edited more than 120 books. AAP has published many of his books, including *Management of Drip/Trickle or Micro Irrigation; Evapotranspiration: Principles and Applications for Water Management;* ten-volume set on *Research Advances in Sustainable Micro Irrigation.* He has also authored the textbooks *Elements of Agroclimatology* (Spanish) by UNISARC, Colombia, and two *Bibliographies on Drip Irrigation.* Dr. Goyal has also developed several book series with AAP, including Innovations in Agricultural & Biological Engineering, Innovations and Challenges in Micro Irrigation; and Innovations in Plant Science for Better Health: From Soil to Fork.

Dr. Goyal received his BSc degree in Engineering from Punjab Agricultural University, Ludhiana, India, and his MSc and PhD degrees from the Ohio State University, Columbus, Ohio, USA. He also earned a Master of Divinity degree from the Puerto Rico Evangelical Seminary, Hato Rey, Puerto Rico, USA.

**Preeti Birwal, PhD**
*Scientist (Processing and Food Engineering), Department of Processing and Food Engineering, College of Agricultural Engineering and Technology, Punjab Agricultural University, Ludhiana, Punjab, India*

Preeti Birwal, PhD, is working as a Scientist (processing and food engineering) in the Department of Processing and Food Engineering at the College of Agricultural Engineering and Technology at Punjab Agricultural University, Ludhiana, Punjab, India. She is currently working in the area of nonthermal food preservation, fermented beverages, food packaging, and technology of millet-based beer. She has served at Jain (Deemed to be University), Bangalore, as a member of the board of examiners and placements. She has participated at several national and international conferences and seminars and has delivered lectures as a resource person on doubling farmers' income through dairy technology training sponsored by the directorate of Extension, Ministry of Agriculture and Farmers Welfare, Government of India. Dr. Birwal has published research papers, an edited book, book chapters, popular articles, conference papers, abstracts, and editorial opinions. She is advising several MTech scholars in food technology and has successfully guided five postgraduate students for their dissertation work. She also serves as an external examiner for various Indian state agricultural universities and as an editor and reviewer of several journals. Dr. Birwal has been named outstanding reviewer of the month by the online journal *Current Research in Nutrition and Food Science.* She has successfully completed AUTOCAD 2D & 3D certification. She is a life member of IDEA. She graduated with a degree in Dairy Technology from ICAR–National Dairy Research Institute, Karnal, India; a master's degree in Food Process Engineering and Management from NIFTEM, Haryana; and PhD (Dairy Engineering) from ICAR–NDRI, Bangalore, India. She has received several fellowships from MHRD, Nestle India, GATE, and UGC-RGNF.

**Ritesh B. Watharkar, PhD**

Ritesh B. Watharkar, PhD, is currently working as an Assistant Professor in the College of Agriculture and Food Business Management at MIT-ADT University, Maharashtra, India.

He received his BTech degree in Food Technology from Dr. Balasaheb Sawant Konkan Agriculture University, Dapoli, Maharashtra, India; his MTech in Food Processing and Engineering degree is from the Karunya University, India; his PhD in Food Engineering and Technology degree from Tezpur University (A Central University), Assam, India.

He was formerly a Senior Research Fellow under a funded project by the Ministry of Food Processing Industries, New Delhi (2013–2015), and Assistant Professor at K.K. Wagh Food Tech College (2011–2013); at Shramshakti College of Food Technology, affiliated by Mahatma Phule Agricultural University (2019–2020).

During his professional career, he received awards such as senior research fellow under the UNESCO/China Great Wall Fellowship Program and worked for one year in the School of Biosystem Engineering and Food Science at Zhejiang University, Hangzhou, China. He was also awarded for best popular article from *Krishidut Magazine,* Maharashtra, India. Currently, he is a life member of the Association of Food Scientists and Technologists, India. He has published more than 10 journal articles and attended more than 10 national and international conferences. He has also published more than 10 popular articles in the *Marathi Literacy Magazine*.

# CONTENTS

# CONTRIBUTORS

**Raksha Agrawal**
Department of Pharmacy, Gyan Ganga Institute of Technology and Sciences, Jabalpur, Madhya Pradesh, India

**Sameer Ahmad**
Vignan's Foundation for Science, Technology & Research (Deemed to be University), Guntur, Andhra Pradesh, India

**Farhana Mehraj Allai**
Department of Food Technology, Islamic University of Science and Technology, Awantipora, India

**Cintia da Silva Araújo**
Department of Food Engineering, Federal University of Espírito Santo, Alegre, Brazil

**K. A. Athmaselvi**
Indian Institute of Food Processing Technology, Pudukkottai Road, Thanjavur, Tamil Nadu, India

**Z. R. Azaz Ahmad Azad**
Department of Postharvest Engineering and Technology, Faculty of Agricultural Sciences, Aligarh Muslim University, Aligarh, Uttar Pradesh, India

**Sanika Bhokarikar**
Department of Food Process Engineering, S.R.M. Institute of Science and Technology, Kancheepuram, Tamil Nadu, India

**Preeti Birwal**
Department of Processing and Food Engineering, College of Agricultural Engineering and Technology, Punjab Agricultural University, Ludhiana, Punjab, India

**Durgesh Nandini Chauhan**
Columbia Institute of Pharmacy, Raipur, Chhattisgarh, India

**Nagendra Singh Chauhan**
Drugs Testing Laboratory Avam Anusandhan Kendra, Raipur, Chhattisgarh, India

**Shreya Chitnavis**
Department of Pharmacy, Gyan Ganga Institute of Technology and Sciences, Jabalpur, Madhya Pradesh, India

**Gargi Ghoshal**
Dr. S. S. Bhatnagar University Institute of Chemical Engineering & Technology, Panjab University, Chandigarh, India

**Megh R. Goyal**
Senior Technical Editor-in-Chief (Agriculture and Biomedical Engineering) for AAP, Retd. Professor in Agric. & Biomedical Eng., University of Puerto Rico-Mayaguez, Mayaguez–Puerto Rico, USA

**Khalid Gul**
Department of Food Process Engineering, National Institute of Technology, Rourkela, India

**Pinar Gumus**
Department of Nutrition and Dietetic, Faculty of Health Sciences, Kilis 7 Aralik University, Turkey

**Rubeka Idrishi**
School of Agro and Rural Technology–Technology Complex, Indian Institute of Technology (IIT), Guwahati, Assam, India

**Anupam Jaiswal**
Department of Pharmacy, Gyan Ganga Institute of Technology and Sciences, Jabalpur, Madhya Pradesh, India

**Lovejot Kaur**
Food Grains & Oilseeds Processing Division, ICAR-Central Institute of Post-Harvest Engineering and Technology, Ludhiana, Punjab, India

**Maninder Kaur**
Department of Processing and Food Engineering, Punjab Agricultural University, Ludhiana, Punjab, India

**Ramandeep Kaur**
Department of Processing and Food Engineering, KVK, Punjab Agricultural University, Moga, Punjab, India

**Arvind Kumar**
Department of Dairy Science and Food Technology, Institute of Agricultural Sciences, Banaras Hindu University, Varanasi, Uttar Pradesh, India

**Leandro Levate Macedo**
Department of Food Science, Federal University of Lavras, Lavras, Brazil

**Dipendra Kumar Mahato**
CASS Food Research Centre, School of Exercise and Nutrition Sciences, Deakin University, Burwood, VIC, Australia

**Junaid Ahmad Malik**
Department of Zoology, Govt. Degree College, Bijbehara, Kashmir, Jammu and Kashmir, India.

**Arshied Manzoor**
Department of Postharvest Engineering and Technology, Faculty of Agricultural Sciences, Aligarh Muslim University, Aligarh, Uttar Pradesh, India

**Madeneni Madhava Naidu**
Department of Spices and Flavor Sciences, CSIR-CFTRI, Mysuru, Karnataka, India;
Academy of Scientific and Innovative Research (AcSIR), Ghaziabad, Uttar Pradesh, India

**Prerna Nath**
Horticulture and crop processing Division, ICAR-Central Institute of Post-Harvest Engineering and Technology, Ludhiana, Punjab, India

**Sadaf Nazir**
Department of Food Technology, Institute of Engineering and Technology, Bundelkhand University, Jhansi, Uttar Pradesh, India

**Stuti Pandey**
Department of Pharmacy, Gyan Ganga Institute of Technology and Sciences, Jabalpur, Madhya Pradesh, India

**Rekha Pathak**
Department of Pharmacy, Gyan Ganga Institute of Technology and Sciences, Jabalpur, Madhya Pradesh, India

**Amrita Poonia**
Department of Dairy Science and Food Technology, Institute of Agricultural Sciences, Banaras Hindu University, Varanasi, Uttar Pradesh, India

**Sanjoy Sadhukhan**
Plant Molecular Biology Laboratory, Department of Botany, Raiganj University, Raiganj, Uttar Dinajpur, West Bengal, India

**Anup Kumar Sarkar**
Department of Botany, Dukhulal Nibaran Chandra College, Aurangabad, Murshidabad, West Bengal, India

**Swati Sethi**
Food Grains & Oilseeds Processing Division, ICAR-Central Institute of Post-Harvest Engineering and Technology, Ludhiana, Punjab, India

**Kamal Shah**
Institute of Pharmaceutical Research, GLA University, Mathura, Uttar Pradesh, India

**Bharti Sharma**
Department of Dairy Science and Food Technology, Institute of Agricultural Sciences, Banaras Hindu University, Varanasi, Uttar Pradesh, India

**Monika Sharma**
Southern Regional Station (SRS), ICAR-National Dairy Research Institute (NDRI), Adugodi, Bengaluru, India

**Pandule Vishal Shrirang**
Hatsun Agro Product Ltd., Palacode, Tamil Nadu, India

**Ramandeep Singh Sidhu**
Mechanical Engineering, Guru Nanak Dev Engineering College, Ludhiana, Punjab, India

**Uma Maheshwari Srinivasa**
Department of Spices and Flavor Sciences, CSIR-CFTRI, Mysuru, Karnataka, India; Academy of Scientific and Innovative Research (AcSIR), Ghaziabad, Uttar Pradesh, India

**Aryasree Sukumar**
Department of Food Process Engineering, SRM Institute of Science and Technology, Chengalpattu, Tamil Nadu, India

**Megha Verma**
Department of Pharmacy, Gyan Ganga Institute of Technology and Sciences, Jabalpur, Madhya Pradesh, India

**Wallaf Costa Vimercati**
Department of Food Science, Federal University of Lavras, Lavras, Brazil

**Ritesh B. Watharkar**
College of Management, Food and Agriculture Management Department, MIT Art, Design and Technology University, Pune, Maharashtra, India

**Deep N. Yadav**
Food Grains & Oilseeds Processing Division, ICAR-Central Institute of Post-Harvest Engineering and Technology, Ludhiana, Punjab, India

**Insha Zahoor**
Department of Postharvest Engineering and Technology, Faculty of Agricultural Sciences, Aligarh Muslim University, Aligarh, Uttar Pradesh, India

# ABBREVIATIONS AND SYMBOLS

| | |
|---|---|
| 6σ | process width |
| *a* | requirement of HCl for sample titration |
| ABTS | 2,2'-azino-bis(3-ethylbenzothiazoline-6-sulfonic acid |
| $A_c$ | area of smallest circums0cribed circle |
| ACAT | acyl CoA cholesteryl acyltransferase |
| ACC | acetyl CoA carboxylase |
| AD | Alzheimer's disease |
| AdipoR2 | adiponectin receptor 2 |
| AFM | atomic force microscopy |
| Ag | silver |
| AGE | advanced glycation end products |
| AGMARK | Agricultural Produce Grading and Marketing Act (India) |
| AIDS | acquired immunodeficiency syndrome |
| Akt | protein kinase B |
| ALA | alfa-linolenic acid |
| ALP | alkaline phosphatase |
| ALT | alanine transaminase |
| AMD | age-related macular degeneration |
| AMPK | AMP-activated kinase |
| AMPK | AMP-activated protein kinase |
| $A_p$ | area of projected body in its rest position |
| AP-1 | activator protein 1 |
| ARA | arachidonic acid |
| ARV | antiretroviral |
| AST | aspartate aminotransferase |
| Au | gold |
| *b* | titration of blank with HCl |
| BAL | bronchoalveolar lavage |
| Bax | Bcl-2-associated X protein |
| BC | basal carcinoma |
| Bcl-2 | B-cell lymphoma 2 |
| B-CPAP | B-cell adaptor for phosphoinositide |
| BD | bulk density |
| BIS | Bureau of Indian Standards |

| | |
|---|---|
| BRC | British Retail Consortium |
| BRCGS | British Retail Consortium Global Standard |
| C | total assessed defects per unit |
| c/EBP)-α (C/EBPs) | CCAAT enhancer-binding protein |
| CA | cellulose acetate |
| CABG | coronary artery bypass grafting |
| CAC | Codex Alimentarius Commission |
| CAT | catalase |
| CCP | critical control points |
| CDSN | corneo desmosin |
| CFAQ | caffeoyl-feruloyl-quinic acids |
| CGA | chlorogenic acids |
| CLA | conjugated linoleic acid |
| CMC | critical micellar concentration |
| CoA | coenzyme A |
| COX-2 | cyclooxygenase-2 |
| $C_p$ | capability index |
| CPZ | capsazepine |
| CQA | caffeoylquinic acids |
| CS | coffee silverskin |
| c-Src | proto-oncogene tyrosine-protein kinase Src |
| Cu | copper |
| CuO | copper oxide |
| CVD | cardiovascular diseases |
| $D_3$/ $D_4$ | constant |
| $D_c$ | diameter of smallest circumscribed circle |
| DHA | docosahexaenoic acid |
| $D_i$ | diameter of largest inscribed circle |
| diCQA | di-caffeoylquinic acids |
| DIT | diet-induced thermogenesis |
| DMBA | 12-dimethylbenz (α) anthracene |
| DOX | doxorubicin |
| DPPH | 2,2-diphenyl-1-picrylhydrazyl |
| DSHEA | Dietary Supplement Health and Education Act |
| DW | dry weight |
| EAC | Ehrlich ascites carcinoma |
| EDTA | ethylenediamine tetraacetic acid |
| EGCG | epigalocatechin-3-gallate |
| EGF | epidermal growth factor |

| | |
|---|---|
| EPA | eicosapentaenoic acid |
| ER | expansion ratio |
| ERK | extracellular-signal-regulated kinase |
| ESRD | end stage renal disease |
| ET | extrusion technology |
| FA | fatty acids |
| FADD | FAS-associated death domain protein |
| FAO | Food and Agriculture Organization |
| FAS | fatty acid synthase |
| FDA | Food and Drug Administration |
| FEMA | Flavor and Extract Manufacturer's Association |
| FFA | free fatty acid |
| FOSHU | food for special dietary uses |
| FQA | feruloyl quinic acids |
| FRAP | ferric-reducing ability |
| FSANZ | Food Standards Australia and New Zealand |
| FSQMS | Food Safety and Quality Management Systems |
| FSSA | Food Safety and Standards Act |
| FSSAI | Food Safety and Standard Authority of India |
| FSSAI | Food Safety and Standards Authority of India |
| GABA | gamma-aminobutyric acid |
| GAE | gallic acid equivalent |
| GAP | good agricultural practices |
| GDP | good distribution practices |
| GE | ginger extract |
| GFR | glomerular filtration rate |
| GFSI | global food safety initiative |
| GGT | gamma-glutamyl transpeptidase |
| GHP | good hygiene practice |
| GI | gastrointestinal |
| GIT | gastrointestinal tract |
| GLA | gamma linolenic acid |
| GLUT | glucose transporters |
| GMP | good manufacturing practice |
| GOT | glutamic-oxaloacetic transaminase |
| GPx | glutathione peroxidase |
| GRP | good retail practices |
| GSH | reduced glutathione |
| GTP | good transport practices |

| | |
|---|---|
| HACCP | hazard analysis critical control point |
| HbA1c | hemoglobin A1c |
| HD | Huntington's disease |
| HDL | high-density lipoprotein |
| HDPP | high-density polypropylene |
| HFD | high fat diet |
| HHP | high hydrostatic pressure |
| HHPE | high hydrostatic pressure extraction |
| HIV | human immunodeficiency virus |
| HMG-CoA | 3-hydroxy-3-methyl-glutaryl-coenzyme A |
| HOMA-β | homeostasis model assessment of β-cell function |
| HPA | human pancreatic amylase |
| HSV | herpes simplex virus |
| HUVECs | human umbilical vein endothelial cells |
| HVED | high voltage electrical discharges |
| IBD | inflammatory bowel disease |
| IFN-γ | interferon-γ |
| IFS | International Food Standard |
| IKKβ kinase | inhibitory kappa B (IκB) kinase β |
| IL | interleukins |
| IL-10 | interleukin 10 |
| IL-1ra | interleukin-1 receptor antagonist |
| IL-1β | interleukin-1-beta |
| IOP | Institute of Packaging |
| ISO | International Organization for Standardization |
| JNK | c-Jun N-terminal kinase |
| k | no. of independent variables |
| KLK | human kallikrein-related peptidase |
| L.A.B. | lactic acid bacteria |
| LA | linoleic acid |
| LC | long chain |
| LCL | lower control limit |
| LC-MS | liquid chromatography-mass spectroscopy |
| LDH | lactate dehydrogenase |
| LDL | low density lipoprotein |
| LDPE | low density polyethylene |
| LUV | large unilamellar vesicles |
| LXR | liver-X-receptors |
| LYC-SLNs | lycopene incorporated solid lipidic nanoparticles |

| | |
|---|---|
| *M* | HCl molarity |
| m | no. of restrictions |
| MAE | microwave-assisted extraction |
| MAPK | mitogen activated protein kinase |
| MC | moisture content |
| MCT | medium chain triglycerides |
| MgO | magnesium oxide |
| MgO | magnesium oxide |
| ml | milliliter |
| ML-1a | myelogenous leukemia 1a |
| MLV | multilamellar vesicles |
| MMP | matrix metalloproteinases |
| MPLSR | modified partial least squares regression |
| MTX | methotrexate |
| MUFA | mono unsaturated fatty acid |
| N | size of lot |
| n | sample size |
| NDDs | neurodegenerative disorders |
| NDEA | nitrosodiethylamine |
| NDO | neurogenic detrusor overactive |
| NET | neuroendocrine tumor |
| NF-κB | nuclear factor kappa-light-chain-enhancer of activated B cells |
| NLC | nanostructure lipid carrier |
| NOEL | non-observed effect level |
| NOS | nitric oxide synthase |
| NP | nanoparticles |
| Nrf2/HO-1 | nuclear factor erythroid 2-related factor 2/heme oxygenase 1 |
| NSMs | nanostructured |
| NTA | nanoparticle tracking analysis |
| O/W | oil in water |
| OC | operating characteristics |
| OHSAS | occupational health and safety management systems |
| oxLDL | oxidized low-density lipoprotein |
| $\bar{p}$ | mean |
| p | sample proportion |
| p300-HAT | histone acetyltransferase p300 |
| PARP | poly ADP-ribose polymerase |

| | |
|---|---|
| PBD | plant-based diet |
| PCL | polycaprolactone |
| PCNA | proliferating cell nuclear antigen |
| pCoQA | p-coumaroyl quinic acids |
| PCR | principal component regression |
| Pd | palladium |
| PD | Parkinson's disease |
| PDCA | Plan, Do, Check and Act |
| PEF | pulsed electric field |
| PEG | polyethylene glycol |
| PET | polyethylene terephthalate |
| PETA | people for the ethical treatment of animals |
| PGC | peroxisome proliferator-activated receptor-γ coactivator |
| PGMS | polyglycerol monostearate |
| PLE | pressurized liquid extraction |
| PLGA | poly (lactic-co-glycolic acid) |
| PLSR | partial least squares regression |
| PMF | 5-hydroxy 6.7.8.4- tetramethoxyflavon |
| PON | paraoxonase |
| PPARs | peroxisome proliferator-activated receptors |
| PRDM16 | PR domain containing 16 |
| Pt | platinum |
| PUFA | poly unsaturated fatty acids |
| QMS | quality management system |
| QTL | quantitative trait loci |
| R | past values average/range |
| RBO | rice bran oil |
| RDA | recommended daily allowance |
| RES | reticulo-endothelial system |
| ROS | reactive oxygen species |
| RPE | retinal pigment epithelial |
| RSD | relative standard deviation |
| *s* | weight of dry sample in grams |
| SAS | synthetic amorphous silver |
| SC | Scientific Committee |
| SCG | spent coffee grounds |
| SCN-DOX | sulfatide containing nanoliposome-doxirubicin |
| SD | solid dispersion |
| SEDDS | self-emulsifying drug delivery system |

| | |
|---|---|
| SFA | saturated fatty acid |
| SFE | supercritical fluid extraction |
| $SiO_2$ | silicon dioxide |
| SME | specific mechanical energy |
| SNEDDS | self-nanoemulsifying drug delivery system |
| SOD | superoxide dismutase |
| SPC | soy phosphatidyl choline/statistical process control |
| SPM | scanning probe microscopy |
| SPS | sanitary and phytosanitary measures |
| SQF | Safe Quality Food Standard |
| SREBP-1 | sterol regulatory element-binding transcription factor 1 |
| SSE | residual sum of squares |
| STAT3 | signal transducer and activator of transcription 3 |
| STPP | sodium tripolyphosphate |
| STZ | streptozotocin |
| SUV | small unilamellar vesicles |
| T1DM | Type 1 diabetes mellitus |
| T2DM | Type 2 diabetes mellitus |
| TAG | triacylglycerides |
| TBA | thiobarbituric acid |
| TBARS | thiobarbituric acid reactive substances |
| TBT | technical barriers to trade |
| TC | Total Cholesterol/Technical Committee |
| TGF-β | transforming growth factor beta 1 |
| Th2 | T-helper cell 2 |
| $TiO_2$ | titanium dioxide |
| TNF-α | tumor necrosis factor- α |
| tNOX | tumor-associated nicotinamide adenine dinucleotide oxidase |
| TQM | total quality management |
| TRP | transient receptor potential |
| TRPV1 | transient receptor potential vanilloid subtype 1 |
| U.K. | United Kingdom |
| U.S.A. | United States of America |
| UAE | ultrasound-assisted extraction |
| UCL | upper control limit |
| UCP | uncoupling protein |
| VEGF | vascular endothelial growth factor |
| VR1 | vanilloid receptor-1 |

| | |
|---|---|
| W/O | water in oil |
| WAI | water absorption index |
| WHO | World Health Organization |
| WPI | whey protein isolate |
| WSI | water solubility index |
| WTO | World Trade Organization |
| $x$ | sample mean |
| $x_1$ | mean of sample 1 |
| $x_2$ | mean of sample 2 |
| $z$ | standard deviation |
| ZnO | zinc oxide |
| $\alpha$ | consumer's risk |
| $\beta$ | producer's risk |
| $\mu$ | population mean |
| $\rho$ | defective fraction/density |
| $\sigma$ | variance |
| $\sigma/\sqrt{n}$ | standard deviation of a population (z) |
| $\omega$-3 | omega-3 |

# PREFACE

The need for plant-based nutritious foods has risen dramatically in recent years, driven by customer needs and demands. Consumers now a days seek nutritious and safe meals to boost immunity and combat illnesses with an emphasis on preserving the majority of natural qualities throughout processing. Global food and nutritional security determine the requirement for functional ingredients/food value in food products. Each food product undergoes some processing before reaching a consumer. Plant-based functional foods include phytochemicals, which have nutraceutical benefits. Clinical trials have shown that this type of diet can lower the risk of chronic illnesses.

As customer interest in natural plant-based healthy food grows, the food industry is likewise looking for healthier goods. Different forms of foods (such as cereals, millets, fruits, and vegetables) have large number of functional elements that are good for our health in terms of physiological impacts and nutritional function. Plant-based products are regarded safe by consumers and industries due to their natural basis. As a result, food processors are focusing on developing plant-based fortified food items that are high in bioactive chemicals to minimize risk of illnesses. As a result, product development in the food processing industry offers fascinating potential and unique challenges.

Food components are encased in the carrier agent using the encapsulation process. This technology is quite beneficial in terms of food safety and shelf stability. Improved stability and processing are two of the most significant reasons for encapsulating plant-based components.

This book tries to show how encapsulation methods and plant-based materials may be used in a variety of ways. The book is divided into four sections. The first section, Encapsulation Technologies, delves into the ideas, advantages, and techniques for encapsulating bioactives-based on cereals, spices, and coffee. It also discusses the role of nanoliposomes for encapsulation of natural foods. The second section, Functional Aspects of Plant-based Foods, discusses the functional aspects of plant-based foods, nutraceutical-based functional food design, diverse functional food items, functional aspects of food, and the role of functional components. The third section, Health Benefits of Plant-Based Foods, covers selected plants (such as cereals, pepper, and coffee) and their chief phytochemicals and bioactive compounds along with their health benefits. The fourth section, Safety Aspects of Functional and Natural Foods, examines the role of functional foods in food safety, as well as an overview of industrial food safety issues and techniques for monitoring food quality and safety.

This book volume on *Plant-Based Bioactive Compounds and Food Ingredients: Encapsulation, Functional, and Safety Aspects* is a wealth of knowledge and a useful reference tool for students, researchers, scientists, processors, and businesses, among others. Through this book volume, they will gain knowledge on encapsulation processes and their applications, functional foods, and food safety issues. This book is significant because it compiles contemporary academic data on advanced cereal innovation, spice-based functional food elements, and their health advantages, and encapsulation technologies for plant-based ingredient creation to produce safe and high-quality food. Aside from a broad understanding of food, the ideas of encapsulation and functional food, as well as their applications in food processing, has been summarized.

Due to the extensive effort and support of all contributing writers' this book turned out to be a wealth of knowledge that went beyond our disclosure and expectations. The chapters have been prepared to include the names of the contributing writers. The editing crew admires the writers' ability, commitment, and endurance. Contributors, correspondents, and the editing team also worked hard to make this book engaging for readers hailing from a variety of food sectors and academic institutions.

We are thankful to the staff of Apple Academic Press Inc. for their devotion, support, and guidance in achieving the project's overall purpose. We also want to express our gratitude to our friends and family for their unwavering support, encouragement, love, and compassion during the preparation and editing process of this book volume.

**—Editors**

# PART I
## ENCAPSULATION TECHNOLOGIES

CHAPTER 1

# ROLE OF NANOLIPOSOMES FOR ENCAPSULATION OF NATURAL FOODS

BHARTI SHARMA, DIPENDRA KUMAR MAHATO, and ARVIND KUMAR

## ABSTRACT

The major challenge today is to obtain a formulation that can exhibit satisfactory bioavailability of the nutraceutical. Nanotechnology appears as a better approach to enhance the solubility, stability, and permeability of the encapsulated material. The small size of the nanoparticles along with their composition offers great research opportunities. Nanoparticles can have different structural formulations, like nanoemulsions, micelles, nanoliposomes, and nanocochelates. The use of liposomes in the pharmaceutical industry for better and targeted drug delivery and in chemotherapy has shown promising results; therefore, the food industry also intends to utilize them for delivery of bioactive components of the food, such as polyphenols, flavor components, fatty acids, and enzymes. Protection of sensitive bioactive molecules in the gastrointestinal system and even when present in external environment, storage stability, and enhanced bioavailability in the body are some of the benefits that are offered by the nanoliposomes. Hence, this chapter will focus on the advantages of nanotechnology, with a brief about different nano-based delivery systems. This will be followed by an introduction to nanoliposomes, their classification, and methods of preparation in detail. With a discussion on the advantages and disadvantages of nanoliposomal technology, this chapter will end up with its application and present status in the market.

Plant-Based Bioactive Compounds and Food Ingredients: Encapsulation, Functional, and Safety Aspects.
Junaid Ahmad Malik, Megh R. Goyal, Preeti Birwal, & Ritesh B. Watharkar (Eds.)

## 1.1 INTRODUCTION

Food not only provides energy but also acts as medicine and it helps in healing. There are certain nutrient and non-nutrient compounds present in food that helps in exerting beneficial effects.[11,14,72] Such foods that provide distinguished health benefits apart from basic nutrition are known as nutraceuticals; and these are found useful in preventing certain diseases like cardiac problems, osteoporosis, improving skin health, and delaying age-related problems.[23,55] It is therefore essential that these compounds get utilized by our body, thus showing more bioavailability to the body. In other words, the proper delivery of these compounds in the body is important.

With nanotechnology, the goal is the delivery of the bioactive components at the proper time and at targeted location in the body. As many compounds have poor water solubility or low permeability in the small intestine, therefore their bioavailability can be improved by increasing the dose which, however, is not the best solution as it can result in several side effects. Nanosize systems allow quicker absorption after oral administration and also allow easy movement between the cells of the gastrointestinal tract. The surface area (SA) increases due to the nanosized structure which further increases the absorption process.[28,55,62] Besides nanotechnology can also be used to improve attributes like flavor, color, texture, and odor of the food. The nanoparticles can be designed from various formulations like nanoemulsions, micelles, nanoliposomes, and nanocochelates. Nanoliposomes are bilayer lipid vesicles with a hydrophilic head and lipophilic fatty acid tail. They are composed of phospholipids and are useful in areas of drug delivery, as a diagnostic agent and in food industries. They are the nanometric version of liposomes.[20,38,53]

Studies also suggest that the liposomes are removed through the reticuloendothelial system (RES) from the blood including the clearance by liver and spleen. Smaller size not only allows enhanced absorption but also prevents clearance from the blood. Moreover, modifying the vesicle surface by incorporation of PEG, polyethylene glycol, increase the stability, bioavailability, and also protection from detection by cells of RES system.[4,29–34] Liposome and nanoliposomes are similar in various aspects such as chemical and structural properties. Nonetheless, nanoliposomes have an added advantage over liposomes. Nanoliposomes have larger SA and have greater potential to improve solubility, enhance bioavailability, better permeability to the cellular membrane, and increase controlled release of the encapsulated compounds.[28,55]

This chapter focuses on advantages, limitations, and formulation methods of nanoliposomes in the food industry. Their applications, current use, and future perspectives have also been considered.

## 1.2 TYPES OF NANO-BASED DELIVERY SYSTEMS

Nanotechnology has huge potential in the food industry. As nanoparticles have greater SA, this leads to enhancement of wettability and dissolution rate of bioactive compounds is found.[13,55–57,65] Based on the methods used to design these nanoparticles, they can be of various forms differing in their structure and physicochemical properties. These include nanoliposomes (10–300 nm), nanocochelates (50–500 nm), micelles <100 nm), nanoemulsions (10–100 nm), lipid nanoparticles (100–200 nm), and cocervates (10–600 nm).[47–51]

### *1.2.1 NANOCOCHELATES*

Different types of nano-based delivery system have different size and characteristics. All of these have different application and limitations. Nanocochelates, for example, can encapsulate both hydrophilic and lipophilic compounds and thus have better stability and protection from degradation than others. However, its method of manufacturing is expensive which limits its use.[62] Both nanocochelates and nanoliposomes are small vesicles that are surrounded by a bilayer of lipids. However, the difference between them lies in their composition, that is, the presence of phosphatidylserine and calcium ions along with cholesterol in nanocochelates. Nanoliposomes, on the other hand, are vesicles consisting of phospholipids and cholesterol only.[54,74] Lipid nanoparticles have also shown stability of the nanoparticles and controlled release of the bioactive compound but its solid lipid core results in crystallization and thereby adding to its disadvantage.

### *1.2.2 COACERVATES*

The most complex of all the delivery systems is coacervates. It is a biopolymer complex that contains two oppositely charged polymers (proteins and polysaccharides) held together via electrostatic forces. When these polymers interact with each other, they result in phase separation of polyelectrolytes in a solution. The next step is the deposition of a shell or coacervate phase around the bioactive material.[7] For example, for the nanoencapsulation of capsaicin, an active component found in pepper, gelatin, and gum acacia was used. A tannin treatment was done afterward and thus resulted in improved stability.[33,84] Coacervates are capable of encapsulating small molecules (lipophilic) like favoring oils but their complex structure also makes them expensive to use.

### *1.2.3 NANOEMULSIONS*

Nanoemulsions are suitable for encapsulation of both amphiphilic and lipophilic components. They are formed by colloidal dispersion of liquid and oil which are stabilized by surfactants or emulsifiers. These can be produced by using techniques like ultrasonication and microfluidization[47,76,78].

### *1.2.4 MICELLES*

Micelles form another nano-based system which is amphiphilic in nature. They are formed by many amphiphilic molecules which when they reach the critical micellar concentration, assemble, and form micelles. Their hydrophilic groups face toward outside while hydrophobic groups are present in the core. As no use of exogenous energy is required during their formation, therefore these molecules are also thermodynamically stable.[45] Previous studies have reported use of micelles in increasing the bioavailability of curcumin, the compound present in turmeric, and is known for its anti-oxidant and anti-inflammatory properties.[67,85]

In order to design these nanoparticles containing the nutraceutical, drug, or any other bioactive compound, biodegradable polymers are used. Encapsulation using polymer has certain advantages as follows:

- Proper release of the compound
- Protection from degradation during processing or when ingested
- Better transport from the cell membrane
- Modified biodistribution in the body

If an increased intracellular delivery is required, carrier molecules can be attached with the particles. As nutraceutical functions as medicine also, therefore it is important to achieve targeted delivery, which becomes feasible with the help of these nanoparticles. Cochemé et al. exhibited the formulation of coenzyme Q10 (a nutraceutical) by a moiety lipophilic triphenyl phosphonium cation, which helps it to deliver at the targeting organelle mitochondria. The cation can be attached directly to coenzyme Q10 or chemically to a nanocarrier.[18]

## 1.3 NANOLIPOSOMES

These are spherical tiny bodies consisting of phospholipids which are arranged in bilayers and form a vesicle. The formation of nanoliposomes was first observed by Bangham et al., when egg lecithin was dispersed in water and arranged its hydrophilic part, that is, the polar part is present on the outside at the surface of the liposome, whereas the hydrophobic part is present in the core.[3,29,77] The inside of the vesicle contains an aqueous solution which makes it fit for entrapment of hydrophilic molecules and the lipid bilayer is suitable for lipophilic molecules.[55]

Nanoliposomes, nanometric versions of liposomes, are bi-layered vesicles but smaller in size. Because of their amphiphilic nature, these nanovesicles can trap all hydrophilic, hydrophobic, and amphiphilic molecules; however, the main problem lies in the manufacturing of such small vesicles.[9,31,53] These are categorized based on the number of bilayers present as follows:

- Multilamellar vesicles (MLV)—As the name indicates, these vesicles are onion-like and contain multiple concentric phospholipid spherical layers which are separated from each other by layers of water (Figure 1.1).
- Unilamellar vesicles—Unlike MLV, these are a single phospholipid bilayer sphere with an aqueous solution inside it. These are further classified into two categories based on their size: small unilamellar vesicles (SUV si10000) and large unilamellar vesicles (LUV si10000).[1,61]

Certain molecules like polymers, antigens, or cholesterol can also be attached to the nanovesicles which can enhance the stability and shelf-life of the bioactive component. They can also help in targeting the nanoliposome where needed. In addition to this, antioxidant compounds like alpha-tocopherol can also be incorporated in the vesicles to prevent the oxidation of phospholipid ingredients.[26,49] For example, the first liposomal drug PEGylated liposomal doxorubicin (Doxil) was approved by the FDA, and it contains PEG grafted on the vesicle in order to escape the RES system and be in the circulatory system for a period of time. Another example is encapsulation of Vitamin D3 to fortify the beverages, using nanoliposomal technology.[15,48] They are very useful in industries like pharmaceuticals, cosmetic, and food because of their encapsulation and targeted delivery property. They are beneficial in targeted delivery or enhancing the bioavailability of the compound as their smaller size allows them to penetrate the biological membrane easily.[44,66]

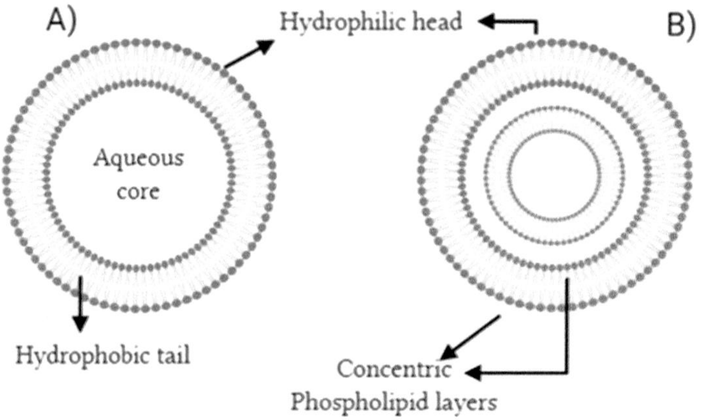


**FIGURE 1.1** Structure of (A) unilamellar liposomes and (B) multilamellar liposomes.

## 1.4 PREPARATION OF NANOLIPOSOME

The formation of nanoliposomes requires high energy input. Thus, methods such as sonication, microfluidization, and extrusion are employed for the purpose.[53,62] The phospholipid layer of the nanoliposome is primarily made by using lecithin, a component derived from egg yolk and soy, which also makes it cheap economically. Fatty acids can also be used for this purpose. It is a type of phospholipid and the concentration of these phospholipids used for preparation determines the formation of the vesicles.[21,25,30]

Gentine et al.[25] found that increasing the concentration of lipid up to 15 mM showed an increase in the diameter of the vesicles. However, if the phospholipid concentration is above 20–25 mM, then the formulation is poor. This is due to the limited solubility of phospholipids in the solvent such as ethanol.

Besides nanoliposomal ingredients, solvent selection is also important for the formulation. There are various solvents like chloroform, propyl acetate, diethyl ether, and acetone which can be employed for the preparation. But factors like the route of administration, method of preparation, dosage, and intended use are to be considered before the selection of solvent.[38,50]

For the preparation of liposomes, organic solvents, like methanol, ethanol, and isopropanol, and ethyl acetate are used. The process of preparation of formulation requires the evaporation of the solvent that has been used in the end but evidences for the presence of trace amount of solvent in the product have been found. They exhibit cytotoxic effects and can cause several problems like modification in the permeability of the membrane, formation of emulsions, protein degeneration, reduced stability, and limited access to nutrients. Therefore, the removal of these solvents along with the evaluation of their residual concentration is necessary.[19,50] There are numerous methods that can be used for the preparation of nanoliposomes.[38,53] The following factors become decisive while selecting the method of preparation:

- Properties of the bioactive compound that needs to be encapsulated (its charge and sensitivity to pH, temperature, and so on).
- Type of solvent that would be used for the suspension.
- Toxicity level in the final product.
- Size and shelf life of the vesicle.

### *1.4.1 SONICATION*

Sonication involves the use of high-intensity sound waves, that is, the waves whose frequency is above the audible hearing range (20 Hz to 20 kHz), and can produce vesicles of nanometer size.[40] Ultrasound waves are classified on the basis of their power intensity as low-intensity and high-intensity waves. Low-intensity ultrasound (<1 W/cm$^2$) can pass through the material without causing any damage therefore used in diagnosing techniques like sonography, whereas high intensity (10–1000 W/cm$^2$) can break down particles and is therefore suitable for application in the formulation of nanoliposomes.[46,60]

The addition of phospholipids in the aqueous solution does not automatically take the shape of bi-layered vesicles; energy is required for their organization. Sonication is used to arrange the lipids in the form of bilayers by providing sufficient energy.[49] The instrument used for the preparation is known as sonicator. There are two types of sonicators: probe sonicator and bath sonicator.

#### 1.4.1.1 PROBE SONICATOR

In this technique, the sonicator's tip is placed in the MLV flask. The sample was then sonicated for 10–15 min resulting in formation of SUV. The advantage of probe tip sonicators is their capability to deliver high-energy input into the suspension. However, on the other hand, this results in overheating sometimes and causes degradation of lipid suspension. In addition to this, the tips used in the sonication tend to release small particles of titanium into the suspension, which can cause toxicity, therefore must be removed by centrifugation before using it.[61] Figure 1.2 shows different parts of sonicator.

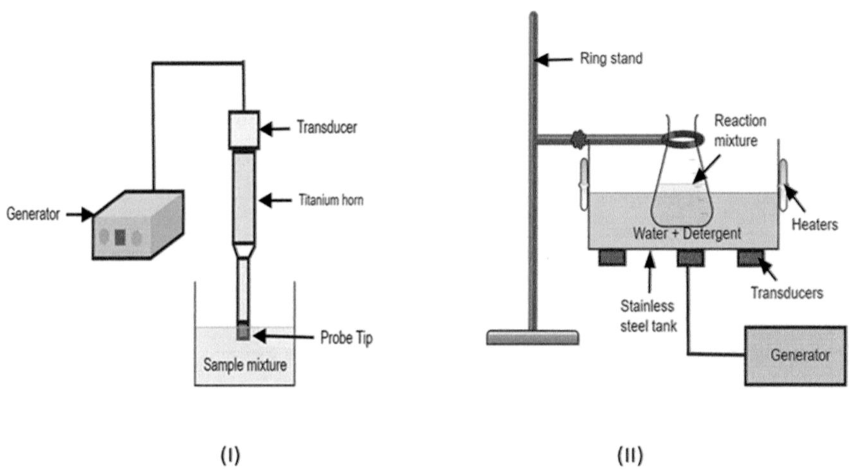


**FIGURE 1.2** Instrument design of (I) probe sonicator and (II) bath sonicator.
*Source:* Concept adapted from Koshani and Jafari.[40,43]

#### 1.4.1.2 BATH SONICATOR

These are the most widely used sonicators and are better than probe sonicators as they can solve the overheating and contamination problem. For the preparation of nanovesicles, the bath sonicator is filled with water which is at room temperature. a few drops of liquid detergent is also added, then the flask containing MLV suspension is placed in the sonicator with the help of a ring stand and a test tube clamp. It should be noted that the level of liquid both inside and outside the flask should be equal. After this, sonication is performed for 20–40 min. Another advantage is that the suspension can be stored in a sterile vessel, thus minimizing the chance of contamination. The longer process time is the limitation of the method.[1,61,64] The flow chart shows the method of preparation of nanoliposomes using sonication (Figure 1.3).

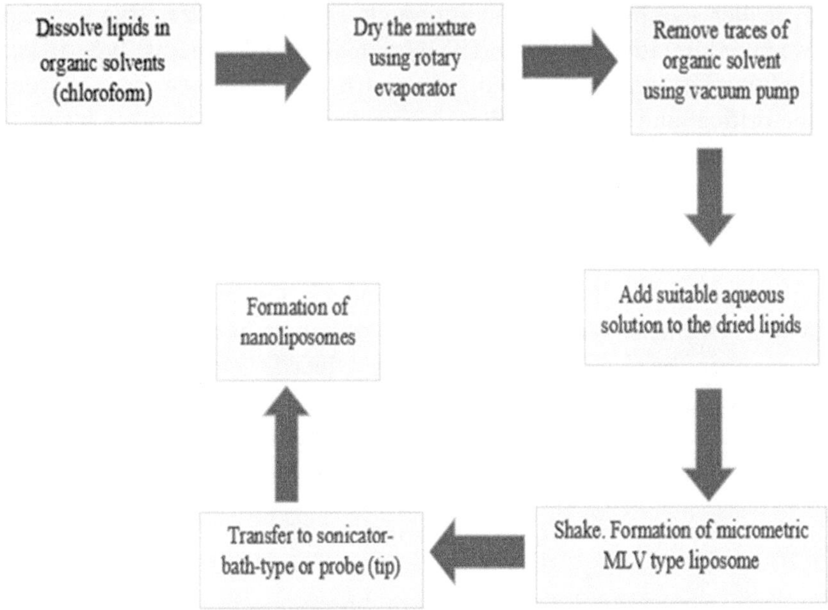


**FIGURE 1.3** Flow chart for the method of preparation of nanoliposomes using sonication.

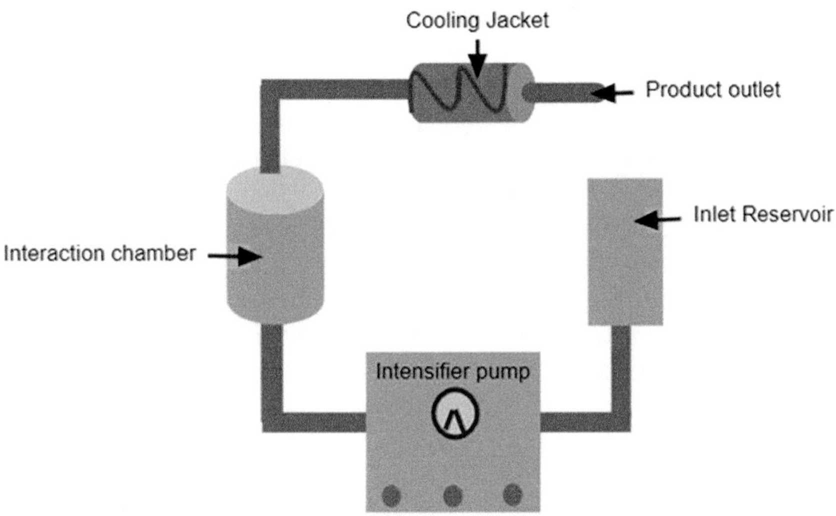


**FIGURE 1.4** Components of the microfluidizer.

*Source:* Concept adapted from Mozafari et al.[53]

### 1.4.2 MICROFLUIDIZATION

The principle of the technique is the use of high pressure to divide the particles into smaller size by passing them through a narrow orifice. This technique does not require toxic solvents. The working of the device involves division of the pressure stream into two parts. Each of which then passes through a fine orifice, and then their flow is directed toward each other inside the microfluidizer. The pressure used can be up to 10,000 psi. In order to use the microfluidizer, the dispersion of nanoliposomal ingredients and their solvent is passed through the inlet reservoir to an intensifier pump that initiates high pressure which leads the product to the interaction chamber with a velocity greater than 400 m/s. It is a Y-shaped chamber where the stream is separated into microchannels, which are of the size of human hair. Within the interaction chamber, the size of the particles gets reduced. The disadvantage of this method includes material loss, contamination, and damage to the structure of the material[34,53,81]. Figure 1.4 shows the different parts of the microfluidizer.

### 1.4.3 HEATING OR MOZAFARI METHOD

The basic mechanism of formation of liposomes is the interaction and arrangement of lipids and water molecules. The arrangement requires input of energy which is provided by heating while stirring in this method. Here, the liposome components are added to an aqueous medium such as distilled water or a buffer and then heated in the presence of glycerol. Stirring is also performed simultaneously for the formation of vesicles.[52] Glycerol is added to increase the stability of the vesicles and is also nontoxic and therefore does not require to be removed from the final product. The process includes hydration of the phospholipids in an aqueous medium for a couple of hours under an inert condition of either nitrogen or argon. After that, the lipid dispersion is mixed with the bioactive component in a heat-resistant flask. Glycerol (3%) is added to a final volume concentration followed by placing the flask on a hot-plate stirrer and mixture is homogenized at 800–1000 rpm until all lipids are dissolved.[21,53,55] The heating temperature usually ranges from 40°C to 120°C depending upon the properties of liposomal ingredient. The temperature should be below the transition temperature of the lipids. For example, in case of cholesterol, liposomes can be prepared at 120°C.[52]

In this technique, nanoliposome formulation can be done without performing filtration or sonication. Another advantage of the method is that it does not need to be dissolved in volatile organic solvents like ether or methanol which are toxic and need to be removed from the product after the formation of vesicles. This is the method by which liposomes can be prepared using a single instrument without involving any toxic solvent.

A modification of the heating method, known as Mozafari method, enables the preparation of liposomes without requirement of prehydration of the liposomal ingredients. They are directly added to a preheated mixture of glycerol and nisin and are heated while stirring. In order to stabilize the formulation after its preparation, it is kept above its transition temperature under inert environment.[49]

## 1.5 VISUALIZATION AND CHARACTERIZATION OF NANOPARTICLES

Different analytical techniques are being used to determine the shape, structure, and size of the vesicles and also for their quantification. The method to be used should give reproducible, fast, and clear results.[21,51,55]

### *1.5.1 SCANNING PROBE MICROSCOPY*

Microscopy-based techniques are availed to determine the morphology or the structure of the nanoparticles. But these techniques require complex sample preparation like staining, labeling, and then fixation which requires more time and leads to alteration in the structure of the particles. However, in SPM, imaging can be done with a simple preparation process. This technique measures the attraction or repulsion of the vesicle between the surface and the tip of the probe.[55,68] Based on the type of interaction involved, there are different microscopy techniques that can be used. Atomic force microscopy (AFM) is one of the most commonly practiced techniques as it does not deform the vesicle by applying shear or lateral force. It, therefore, maintains the morphology of the vesicle. But, when the nanoliposomes get deposited on the substrate that is being used for AFM, a change in shape can be observed. The distortion depends on the fluidity and chemical composition of the vesicle.[2,32]

### *1.5.2 FLOW CYTOMETRY*

In this technique, measurement of each particle is done by analyzing the light scattered by each of them when the cell suspension is passed through the instrument.[82] This technique is widely employed in cell and microbiology to detect and classify the cells one at a time. It can be applied in assessing the diameter and size of a single nanoliposome by measuring the light scattered.[69,82] The nanoliposomes are labeled with fluorescent markers in order to differentiate them from impurities or other foreign particles or noise from the electronic system. The technique is very fast as it has the capability to detect millions of cells in a few seconds. It is used for detection of multilamellar and large unilamellar vesicles in a continuous flow system.[16,21,35] The vesicles are measured at 10° forward scatter and at side scatter of 90°. Flow cytometry is useful for determining the size distribution of the liposomes and is found more beneficial when the solution is not homogenous.[21,82]

### *1.5.3 MASS SPECTROMETRY*

Spectroscopy techniques are widely used for quantification of the particles, that is, to assess the concentration and size of the particles. In this technique, the light at a certain wavelength is projected on the particle solution and then measures the light scattered to determine their size.[10,62] Another technique, liquid chromatography-mass spectroscopy method (LC–MS) is used for determining the concentration and the chemical composition of the separated particles. The method has been used in vivo for the quantification of nanoliposomes by monitoring the drugs that are being encapsulated in the nanoliposomes. As the structure of nanovesicles is fragile and there are chances of the release of drug during the process which can alter the quantification, therefore, it is essential to separate the encapsulated and released drugs.[17,20,22] The separation of free and encapsulated material is usually done by performing solid-phase extraction and using Oasis HLB cartridges.[22,83] Besides, most of the nanoliposomal quantification study has been performed in plasma samples only. It is difficult to measure in the tissues because the quantification process requires homogenous mixture and this can damage the vesicle. The method developed by Su and Liu[77] involves processing of tissues without damaging the nanoliposome. Their method suggests use of ball mill instead of a homogenizer for the sample preparation.

### *1.5.4 NANOPARTICLE TRACKING ANALYSIS*

This method uses light scattering techniques and tracks the position change of single nanoliposome by measuring its movement under Brownian motion. In a particular medium, the lipid vesicle acquires a charge and the magnitude of the attraction/repulsion between those vesicles is measured as zeta potential. The zeta potential of the liposomal particles is measured by inducing electric field across the dispersion.[36,62] This can help in understanding the stability of the vesicle. The technique gives high-resolution results, is cost-effective, and can also be utilized for measuring concentration and distribution along with the size of nanoliposomes. The process includes the injection of the particles in the cell of the instrument where they are irradiated by laser beam on the optical surface through a liquid layer.[20,24,66] The region where the vesicles are present is detected by refraction. The region is then illuminated and seen under the microscope. The movement of the vesicles can be observed by using a charge-coupled device camera. A software is enabled with the instrument that can identify the center of each vesicle and then can determine the particle size.[20,24,63] The limitation of the technique includes its incapability to detect particles greater 1000 nm.[71] A novel technique, known as multispectral advanced nanoparticle tracking analysis (NTA), has been described by Singh et al.,[75] which is advanced than the conventional NTA and can measure liposome samples up to 2000 nm. Thus, the technique is useful in characterization of heterogenous samples where different-sized vesicles are present.

## 1.6 ADVANTAGES AND DISADVANTAGES OF NANOLIPOSOMAL TECHNOLOGY

The small size of the nanoliposomes along with the ability to encapsulate hydro-, lipo-, and amphiphilic compounds are the supremacy of this technology. In addition to this, there are certain other advantages such as

- Easier absorption of the bioactive compound by the body, which thereby increasing the bioavailability of the product.[12,62]
- The size and chemical structure of the nanovesicle allows for controlled release, at the targeted site, of the drug or the bioactive compound. Their small size allows them to cross membrane barriers of the cell and the charge on the surface also mediates their transport.[61]
- Reduced cytotoxicity is another advantage of nanoencapsulation. In a study, Lin et al.[42] reported the problem of dose-limiting

cytotoxicity of a drug, Doxorubicin (DOX), by encapsulating it in the sulfatide-containing nanoliposome (SCN-DOX). Sulfatide is a lipid that can bind to several glycoproteins and is involved in biological processes like cell growth and signal transduction. The experiment used animal models for testing, and it was found that SCN-DOX resulted in four times the lower stacking of the drug in the heart of a rat and thereby reduced cardiotoxicity.

- Drug delivery at a specific target is one of the major advantages of the technique. The drug DOX is used in chemotherapy and its encapsulation not only reduced the cytotoxicity level but also improved its distribution by delivering at the specific targets. The targeted delivery can be explained by the binding of SCN-DOX to a specific glycoprotein, tenascin-C, which is found in higher concentration in the surroundings of tumor cells.[42]
- The encapsulated material is protected from the external environment like pH, enzymes, temperature, and some other chemical changes; this results in enhanced stability of the material provided by its encapsulation.[53,86]
- In the food industry, it is useful in enhancing the pleasant flavor and aroma and also masking the unpleasant ones.[53]

Though the advantages of the nanoliposomal technology show its potential in the food and pharmaceutical industry, yet there are certain limitations which can create an obstacle in exploring its full use. The major disadvantage is the cost of the manufacturing techniques owing to its requirement of high input energy for the formulation.[62] Another limitation is that sometimes the encapsulated material does not get released at the site of delivery and therefore results in reduced efficiency of the drug or bioactive component. Additionally, the nanoliposomal particles are characterized as foreign by the immune system and therefore rapidly cleared out of the circulatory system.[62,79] Moreover, some studies have reported that the nanoliposomes gets accumulated in organs like spleen and liver apart other than the targeted sites. This results in toxicity and shows some side effects like reduction in the activity of macrophages and hand and foot syndrome.[70,77]

## 1.7 APPLICATIONS OF NANOTECHNOLOGY FOR ENCAPSULATION

In the food and related industries, the encapsulation technique is mainly used for preserving and protecting the bioactive components like vitamins, polyphenols, antioxidants, and carotenoids. The objective behind their encapsulation is to increase their shelf life, enhance the bioavailability and protect them from the harsh environment in the GI tract.[71,80] There are certain examples of food products, in which the fatty acids have been encapsulated and positive effect on the quality of the product has been observed. A study by Augustin et al. showed that encapsulation of omega-3 oils in the particle size range of 1000 nm or above resulted in the oxidative stability of emulsions and powder in the buttermilk. In some studies, encapsulation has shown improvement in sensor properties.[5]

The nanoliposomal technology has been reported to accelerate the cheese ripening process.[37] Addition of proteinases to the mix before the isolation of the curd is found to be an effective method to save time. However, the addition of enzymes results in premature proteolysis and cause poor curd consistency and low yield. Moreover, several enzymes are lost in the whey. For proper ripening, a controlled release of the enzymes is required. Therefore, encapsulation is found to improve the stability and also protected the enzymes from the harsh environment. Besides, fortification of food products is a good way to add nutrients and increase their bioavailability. Nanoliposomal technology has shown its potential here too. For example, milk is deficient in iron and ascorbic acid; a study by Lee et al.[41] used the encapsulation technique and fortified milk with ascorbic acid and iron complex, and ferric ammonium sulfate, as direct addition causes certain undesirable reactions and leads to off-flavor and instability problems.[58] Polyglycerol monostearate (PGMS) was used as a coating material for enhanced stability for microencapsulation. The study also involved the evaluation of lipid oxidation due to addition of iron complex and also the inhibitory effect of Vitamin C. In results, a decrease in the oxidation of the milk was observed as compared to the one where unencapsulated iron was used. The sensory properties of the milk were also found to be consistent when evaluated under storage.

A similar study on yogurt was performed by Kim et al.,[39] where fortification of yogurt was done by using microcapsules of iron complex and ascorbic acid with spray-dried PGMS as a coating material. Further studies reported a slight decrease in the stability of Vitamin C when the milk is pasteurized and more decrease when sterilized at 121°C for 15 min. A degradation during storage has also been observed. The solution to this problem was achieved by adding compensatory concentration before the sterilizing process.[27,73] Antioxidants possess several health benefits and improve the nutritive quality of the food. But they have poor membrane permeability and get cleared out of the cell. Incorporation in the nanoliposome is a solution.[8]

## 1.8 SUMMARY

Considering the proven ability of liposomes in the pharmaceutical and medical industry, food technologists are also utilizing their potential in the food industry. The use of the technology for the controlled release and delivery of various functions and bioactive components such as vitamins, carotenoids, enzymes as well as flavor components have been explored for various food applications. Bioactive components of the food that do not retain for long in the circulatory system or get degraded by the action of acid in the stomach are now preserved in the liposomal coat, which not only increases their retention but also promotes targeted delivery of the components. Though there are certain limitations to the technology, such as a high current cost, poor manufacturing, the possibility of leakage, and instability; however, the advantages of the techniques outstand the limitations involved. Having said that, the future recommendation is to conduct more in-depth research and trials for the optimization of the process so that the technology can be utilized at its full potential.

## KEYWORDS

- **bioavailability**
- **delivery system**
- **nanoliposomes**
- **nutraceuticals**

## REFERENCES


1. Akbarzadeh, A.; Rezaei-Sadabady, R.; Davaran, S.; Joo, S. W.; Zarghami, N.; Hanifehpour, Y.; Samiei, M.; Kouhi, M.; Nejati-Koshki, K. Liposome: Classification, Preparation, and Applications. *Nanosc. Res. Lett.* **2013**, *8*, 102–106.
2. Albrecht, T. R.; Grütter, P.; Horne, D.; Rugar, D. Frequency Modulation Detection Using High-Q Cantilevers for Enhanced Force Microscope Sensitivity. *J. Appl. Physics,* **1991**, *69*, 668–673.
3. Allen, T. M.; Cullis, P. R. Liposomal Drug Delivery Systems: From Concept to Clinical Applications. *Adv. Drug Deliv. Rev.* **2013**, *65*, 36–48.
4. Allen, T. M.; Hansen, C.; Martin, F.; Redemann, C.; Yau-Young, A. Liposomes Containing Synthetic Lipid Derivatives of Polyethylene Glycol Show Prolonged Circulation Half-Lives In Vivo. *Biochimica et Biophysica Acta—Biomembranes* **1991**, *1066*, 29–36.
5. Augustin, M. A.; Bhail, S.; Cheng, L. J.; Shen, Z.; Øiseth, S.; Sanguansri, L. Use of Whole Buttermilk for Microencapsulation of Omega-3 Oils. *J. Funct. Foods* **2015**, *19*, 859–867.
6. Bangham, A. D.; Standish, M. M.; Watkins, J. C. Diffusion of Univalent Ions across the Lamellae of Swollen Phospholipids. *J. Mol. Biol.* **1965**, *13*, 238–252.
7. Baruch, L.; Machluf, M. Alginate–Chitosan Complex Coacervation for Cell Encapsulation: Effect on Mechanical Properties and on Long-Term Viability. *Biopolym. Original Res. Biomol.* **2006**, *82*, 570–579.
8. Bhosale, S.; Fulpagare, Y. G.; Desale, R. J. Nanoliposomes: Applications in Food and Dairy Industry. *Int. J. Adv. Res. Biol. Sci.* **2019**, *6*, 79–84.
9. Brandelli, A.; Taylor, T. M. Nanostructured and Nanoencapsulated Natural Antimicrobials for Use in Food Products. Chapter 11; In *Handbook of Natural Antimicrobials for Food Safety and Quality*; Taylor, M., Ed., Vol. 1; Woodhead Publishing: London, 2015; pp 229–257.
10. Brar, S. K.; Verma, M. Measurement of Nanoparticles by Light-Scattering Techniques. *TrAC Trends Analyt. Chem.* **2011**, *30*, 4–17.
11. Camire, M. E. Nutraceuticals for Health Promotion and Disease Prevention. *Council Agric. Sci. Technol.* **2003**, *24*, 1–16.
12. Chaudhry, Q.; Scotter, M.; Blackburn, J.; Ross, B.; Boxall, A.; Castle, L.; Aitken, R.; Watkins, R. Applications and Implications of Nanotechnologies for the Food Sector. *Food Additives Contaminants* **2008**, *25*, 241–258.
13. Chaudhry, Q.; Watkins, R.; Castle, L. Nanotechnologies in Food: What, Why and How? Chapter 1; In *Nanotechnologies in Food*; Chaudhry, Q.; Castle, L.; Watkins, R., Eds., Vol. 2; Royal Society of Chemistry: London, 2017; pp 1–19.
14. Chauhan, B.; Kumar, G.; Kalam, N.; Ansari, S. H. Current Concepts and Prospects of Herbal Nutraceutical: A Review. *J. Adv. Pharm. Technol. Res.* **2013**, *4* (1), 4–8.
15. Chen, C.; Zhu, S.; Huang, T.; Wang, S.; Yan, X. Analytical Techniques for Single-Liposome Characterization. *Analyt. Methods* **2013**, *5*, 2150–2157.
16. Chen, C.; Zhu, S.; Wang, S.; Zhang, W.; Cheng, Y.; Yan, X. Multiparameter Quantification of Liposomal Nanomedicines at the Single-Particle Level by High-Sensitivity Flow Cytometry. *ACS Appl. Mater. Interfaces* **2017**, *9*, 13913–13919.
17. Cheng, Z.; Zaki, A. A.; Hui, J. Z.; Tsourkas, A. Simultaneous Quantification of Tumor Uptake for Targeted and Nontargeted Liposomes and their Encapsulated Contents by ICPMS. *Analytical Chemistry,* **2012**, *84*, 7578–7582.

18. Cochemé, H. M.; Kelso, G. F.; James, A. M.; Ross, M. F.; Trnka, J.; Mahendiran, T.; Asin-Cayuela, J.; Blaikie, F. H.; Manas, A. R. B.; Porteous, C. M. Mitochondrial Targeting of Quinones: Therapeutic Implications. *Mitochondrion* **2007**, *7*, S94–S102.
19. Cortesi, R. Preparation of Liposomes by Reverse-Phase Evaporation Using Alternative Organic Solvents. *J. Microencapsulation* **1999**, *16*, 251–256.
20. Danaei, M.; Dehghankhold, M.; Ataei, S.; Hasanzadeh Davarani, F.; Javanmard, R.; Dokhani, A.; Khorasani, S.; Mozafari, M. R. Impact of Particle Size and Polydispersity Index on the Clinical Applications of Lipidic Nanocarrier Systems. *Pharmaceutics* **2018**, *10*, 1–17.
21. Danaei, M.; Kalantari, M.; Raji, M.; Fekri, H. S.; Saber, R.; Asnani, G. P.; Mortazavi, S. M.; Mozafari, M. R.; Rasti, B.; Taheriazam, A. Probing Nanoliposomes Using Single Particle Analytical Techniques: Effect of Excipients, Solvents, Phase Transition and Zeta Potential. *Heliyon* **2018**, *4* (e01088), 1–32.
22. Deshpande, N. M.; Gangrade, M. G.; Kekare, M. B.; Vaidya, V. V. Determination of Free and Liposomal Amphotericin B in Human Plasma by Liquid Chromatography–Mass Spectroscopy with Solid Phase Extraction and Protein Precipitation Techniques. *J. Chromatogr. B* **2010**, *878*, 315–326.
23. Espín, J. C.; García-Conesa, M. T.; Tomás-Barberán, F. A. Nutraceuticals: Facts and Fiction. *Phytochemistry* **2007**, *68*, 2986–3008.
24. Filipe, V.; Hawe, A.; Jiskoot, W. Critical Evaluation of Nanoparticle Tracking Analysis (NTA) by Nanosight for the Measurement of Nanoparticles and Protein Aggregates. *Pharma Res.* **2010**, *27*, 796–810.
25. Gentine, P.; Bubel, A.; Crucifix, C.; Bourel-Bonnet, L.; Frisch, B. Manufacture of Liposomes by Isopropanol Injection: Characterization of the Method. *J. Liposome Res.* **2012**, *22*, 18–30.
26. Gibbs, F.; Kermasha, S.; Alli, I.; Mulligan, C. N.; Bernard. Encapsulation in the Food Industry: A Review. *Int. J. Food Sci. Nutr.* **1999**, *50*, 213–224.
27. Gliguem, H.; Birlouez-Aragon, I. Effects of Sterilization, Packaging, and Storage on Vitamin C Degradation, Protein Denaturation, and Glycation in Fortified Milks. *J. Dairy Sci.* **2005**, *88*, 891–899.
28. Gokce, E. H.; Ozyazici, M.; Souto, E. B. Nanoparticulate Strategies for Effective Delivery of Poorly Soluble Therapeutics. *Therap. Deliv.,* **2010**, *1*, 149–167.
29. Hood, R. R.; Andar, A.; Omiatek, D. M.; Vreeland, W. N.; Swaan, P. W.; DeVoe, D. L. Pharmacy-On-A-Chip: Microfluidic Synthesis of Pegylated and Folate Receptor-Targeted Liposomes for Drug Delivery. In: *Proceedings of 16th International Conference on Miniaturized Systems for Chemistry and Life Sciences*; 2012; pp 1870–1872.
30. Jaafar-Maalej, C.; Diab, R.; Andrieu, V.; Elaissari, A.; Fessi, H. Ethanol Injection Method for Hydrophilic and Lipophilic Drug-Loaded Liposome Preparation. *J. Liposome Res.* **2010**, *20*, 228–243.
31. Jesorka, A.; Orwar, O. Liposomes: Technologies and Analytical Applications. *Annu. Rev. Analyt. Chem.* **2008**, *1*, 801–832.
32. Jiang, Y.; Genin, G. M.; Pryse, K. M.; Elson, E. L. Atomic Force Microscopy of Phase Separation on Ruptured, Giant Unilamellar Vesicles. *Biorxiv (Cold Spring Harbour Laboratory)*, **2018**, *250944*, 1–32.
33. Jincheng, W.; Xiaoyu, Z.; Sihao, C. Preparation and Properties of Nanocapsulated Capsaicin by Complex Coacervation Method. *Chem. Eng. Commun.,* **2010**, *197*, 919–933.

34. Jones, M. N. The Surface Properties of Phospholipid Liposome Systems and Their Characterisation. *Adv. Colloid Interface Sci.* **1995**, *54*, 93–128.
35. Kanásová, M.; Nesměrák, K. Systematic Review of Liposomes' Characterization Methods. *Monatshefte für Chemie-Chemical Monthly,* **2017**, *148*, 1581–1593.
36. Kendall, K.; Dhir, A.; Du, S. A New Measure of Molecular Attractions Between Nanoparticles Near kT Adhesion Energy. *Nanotechnology* **2009**, *20*, E-article 275701.
37. Kheadr, E. E.; Vuillemard, J. C.; El-Deeb, S. A. Acceleration of Cheddar Cheese Lipolysis by Using Liposome-Entrapped Lipases. *J. Food Sci.* **2002**, *67*, 485–492.
38. Khorasani, S.; Danaei, M.; Mozafari, M. R. Nanoliposome Technology for the Food and Nutraceutical Industries. *Trends Food Sci. Technol.* **2018**, *79*, 106–115.
39. Kim, S. J.; Ahn, J.; Seok, J. S.; Kwak, H. S. Microencapsulated Iron for Drink Yogurt Fortification. *Asian-Australasian J. Anim. Sci.* **2003**, *16*, 581–587.
40. Koshani, R.; Jafari, S. M. Ultrasound-Assisted Preparation of Different Nanocarriers Loaded with Food Bioactive Ingredients. *Adv. Colloid Interface Sci.* **2019**, *270*, 123–146.
41. Lee, J. B.; Ahn, J.; Lee, J.; Kwak, H. S. L-Ascorbic Acid Microencapsulated with Polyacylglycerol Monostearate for Milk Fortification. *Biosci. Biotechnol. Biochem.* **2004**, *68*, 495–500.
42. Lin, J.; Shigdar, S.; Fang, D. Z.; Xiang, D.; Wei, M. Q.; Danks, A.; Kong, L.; Li, L.; Qiao, L.; Duan, W. Improved Efficacy and Reduced Toxicity of Doxorubicin Encapsulated in Sulfatide-Containing Nanoliposome in a Glioma Model. *PLOS One,* **2014**, *9* (e103736), 1–13.
43. Mahdi-Jafari, S.; He, Y.; Bhandari, B. Nano-Emulsion Production by Sonication and Microfluidization: A Comparison. *Int. J. Food Prop.* **2006**, *9*, 475–485.
44. Maherani, B.; Arab-Tehrany, E.; R Mozafari, M.; Gaiani, C.; Linder, M. Liposomes: A Review of Manufacturing Techniques and Targeting Strategies. *Curr. Nanosci.* **2011**, *7*, 436–452.
45. Mahmud, A.; Xiong, X.-B.; Aliabadi, H. M.; Lavasanifar, A. Polymeric Micelles for Drug Targeting. *J. Drug Target.* **2007**, *15*, 553–584.
46. Mamvura, T. A.; Iyuke, S. E.; Paterson, A. E. Energy Changes During Use of High-Power Ultrasound on Food Grade Surfaces. *SA J. Chem. Eng.* **2018**, *25*, 62–73.
47. Mason, T. G.; Wilking, J. N.; Meleson, K.; Chang, C. B.; Graves, S. M. Nanoemulsions: Formation, Structure, and Physical Properties. *J. Phys. Condensed Matter* **2006**, *19* (7), R635–R640.
48. Mohammadi, M.; Ghanbarzadeh, B.; Hamishehkar, H. Formulation of Nanoliposomal Vitamin D3 for Potential Application in Beverage Fortification. *Adv. Pharma. Bull.* **2014**, *4*, 569–575.
49. Mozafari, M. R. Nanoliposomes: Preparation and Analysis. *Methods Mol. Biol.* **2010**, *605*, 29–50.
50. Mozafari, M. R.; Danaei, M.; Javanmard, R.; Raji, M.; Maherani, B. Nanoscale Lipidic Carrier Systems: Importance of Preparation Method and Solvents. *Global J. Nanomed.* **2017**, *2*, 555–593.
51. Mozafari, M. R., Ed. *Nanomaterials and Nanosystems for Biomedical Applications*; Springer Nature: Switzerland, 2007; p 166.
52. Mozafari, M. R. Liposomes: An Overview of Manufacturing Techniques. *Cell. Mol. Biol. Lett.* **2005**, *10* (4), 711–719.
53. Mozafari, M. R.; Johnson, C.; Hatziantoniou, S.; Demetzos, C. Nanoliposomes and Their Applications in Food Nanotechnology. *J. Liposome Res.* **2008**, *18*, 309–327.

54. Mozafari, M. R.; Pardakhty, A.; Azarmi, S.; Jazayeri, J. A.; Nokhodchi, A.; Omri, A. Role of Nanocarrier Systems in Cancer Nanotherapy. *J. Liposome Res.* **2009**, *19*, 310–321.
55. Mozafari, M. R., Ed. *Nanoliposomes: From Fundamentals to Recent Developments*; Oxford Publishing Co. Ltd.: Oxford, 2005; p 238.
56. Mundargi, R. C.; Babu, V. R.; Rangaswamy, V.; Patel, P.; Aminabhavi, T. M. Nano/micro Technologies for Delivering Macromolecular Therapeutics Using Poly (D, L-lactide-co-glycolide) and Its Derivatives. *J. Controlled Release* **2008**, *125*, 193–209.
57. Noomhorm, A.; Ahmad, I.; Anal, A. K., Eds. *Functional Foods and Dietary Supplements: Processing Effects and Health Benefits*, 1st ed.; John Wiley & Sons: New York, 2014; p 516.
58. Olivares, M. Bioavailability of Microencapsulated Ferrous Sulfate in Milk. *Nutrition* **2002**, *18* (3), 285–296.
59. Ozer, A. Y. Applications of Light and Electron Microscopic Techniques in Liposome Research. Chapter 10; In *Nanomaterials and Nanosystems for Biomedical Applications*; Mozafari, M. R., Ed.; Springer Nature: Dordrecht, 2007; pp 145–153.
60. Patist, A.; Bates, D. Ultrasonic Innovations in the Food Industry: From the Laboratory to Commercial Production. *Innov. Food Sci. Emerg. Technol.* **2008**, *9*, 147–154.
61. Prathyusha, K.; Muthukumaran, M.; Krishnamoorthy, B. Liposomes as Targeted Drug Delivery Systems Present and Future Prospective: A Review. *J. Drug Deliv. Therap.* **2013**, *3*, 195–201.
62. Rangan, A.; Manjula, M. V.; Satyanarayana, K. G. Trends and Methods for Nanobased Delivery for Nutraceuticals. In *Emulsions*; Grumezescu, A. M., Ed., Vol. 3; Elsevier: San Diego, CA, 2016; Chapter 17, pp 573–609.
63. Reshetov, V.; Zorin, V.; Siupa, A.; D'Hallewin, M. A.; Guillemin, F.; Bezdetnaya, L. Interaction of Liposomal Formulations of Meta-Tetra (Hydroxyphenyl) Chlorin (Temoporfin) with Serum Proteins: Protein Binding and Liposome Destruction. *Photochem. Photobiol.* **2012**, *88*, 1256–1264.
64. Riaz, M. Liposomes Preparation Methods. *Pak. J. Pharma. Sci.* **1996**, *9*, 65–77.
65. Ruozi, B.; Belletti, D.; Tombesi, A.; Tosi, G.; Bondioli, L.; Forni, F.; Vandelli, M. A. AFM, ESEM, TEM, and CLSM in Liposomal Characterization: A Comparative Study. *Int. J. Nanomed.* **2011**, *6*, 557–562.
66. Rupert, D. B. L. M.; Mapar, M.; Shelke, G. V.; Norling, K.; Elmeskog, M.; Lötvall, J. O.; Block, S.; Bally, M.; Agnarsson, B. R.; Höök, F. Effective Refractive Index and Lipid Content of Extracellular Vesicles Revealed Using Optical Waveguide Scattering and Fluorescence Microscopy. *Langmuir* **2018**, *34*, 8522–8531.
67. Sahu, A.; Bora, U.; Kasoju, N.; Goswami, P. Synthesis of Novel Biodegradable and Self-Assembling Methoxy Poly (Ethylene Glycol)—Palmitate Nanocarrier for Curcumin Delivery to Cancer Cells. *Acta Biomaterialia* **2008**, *4*, 1752–1761.
68. Salapaka, S. M.; Salapaka, M. V. Scanning Probe Microscopy. *IEEE Control Syst. Magaz.* **2008**, *28*, 65–83.
69. Salzman, G. C.; Wilder, M. E.; Jett, J. H. Light Scattering with Stream-In-Air Flow Systems. *J. Histochem. Cytochem.* **1979**, *27*, 264–267.
70. Schiffelers, R. M.; Metselaar, J. M.; Fens, M. H. A. M.; Janssen, A. P. C. A.; Molema, G.; Storm, G. Liposome-Encapsulated Prednisolone Phosphate Inhibits Growth of Established Tumors in Mice. *Neoplasia* **2005**, *7*, 118–127.
71. Schrooyen, P. M. M.; van der Meer, R.; De Kruif, C. G. Microencapsulation: Its Application in Nutrition. *Proc. Nutr. Soc.* **2001**, *60*, 475–479.

72. Shade, C. W. Liposomes as Advanced Delivery Systems for Nutraceuticals. *Integr. Med.: Clin. J.* **2016**, *15* (1), 33–36.
73. Sharma, R.; Lal, D. Fortification of Milk with Microencapsulated Vitamin C and Its Thermal Stability. *J. Food Sci. Technol.* **2005**, *42*, 191.
74. Siegel, D. P.; Tenchov, B. G. Influence of the Lamellar Phase Unbinding Energy on the Relative Stability of Lamellar and Inverted Cubic Phases. *Biophys. J.* **2008**, *94*, 3987–3995.
75. Singh, P.; Bodycomb, J.; Travers, B.; Tatarkiewicz, K.; Travers, S.; Matyas, G. R.; Beck, Z. Particle Size Analyses of Polydisperse Liposome Formulations with a Novel Multispectral Advanced Nanoparticle Tracking Technology. *Int. J. Pharma.* **2019**, *566*, 680–686.
76. Solans, C.; Izquierdo, P.; Nolla, J.; Azemar, N.; Garcia-Celma, M. J. Nano-Emulsions. *Curr. Opin. Colloid Interface Sci.* **2005**, *10* (3–4), 102–110.
77. Su, C.; Liu, Y.; He, Y.; Gu, J. Analytical Methods for Investigating In Vivo Fate of Nanoliposomes: A Review. *J. Pharma. Analy.* **2018**, *8*, 219–225.
78. Tadros, T.; Izquierdo, P.; Esquena, J.; Solans, C. Formation and Stability of Nano-Emulsions. *Adv. Colloid Interface Sci.* **2004**, *108*, 303–318.
79. Tang, S.; Gao, D.; Zhao, T.; Zhou, J.; Zhao, X. An Evaluation of the Anti-Tumor Efficacy of Oleanolic Acid-Loaded Pegylated Liposomes. *Nanotechnology* **2013**, *24*, 235102.
80. Vélez, M. A.; Perotti, M. C.; Santiago, L.; Gennaro, A. M.; Hynes, E. Bioactive Compounds Delivery Using Nanotechnology: Design and Applications in Dairy Food. Chapter 6; In *Nutrient Delivery*; Grumezescu, A. M., Ed., Vol. 5; Elsevier: San Diego, CA, 2017; pp 221–250.
81. Vemuri, S.; Yu, C. D.; Wangsatorntanakun, V.; Roosdorp, N. Large-Scale Production of Liposomes by a Microfluidizer. *Drug Dev. Ind. Pharm.* **1990**, *16*, 2243–2256.
82. Vorauer-Uhl, K.; Wagner, A.; Borth, N.; Katinger, H. Determination of Liposome Size Distribution by Flow Cytometry. *J. Int. Soc. Analy. Cytol.* **2000**, *39*, 166–171.
83. Xie, Y.; Shao, N.; Jin, Y.; Zhang, L.; Jiang, H.; Xiong, N.; Su, F.; Xu, H. Determination of Non-Liposomal and Liposomal Doxorubicin in Plasma by LC–MS/MS Coupled with an Effective Solid Phase Extraction: In Comparison with Ultrafiltration Technique and Application to a Pharmacokinetic Study. *J. Chromatogr. B* **2018**, *1072*, 149–60.
84. Xing, F.; Cheng, G.; Yi, K.; Ma, L. Nanoencapsulation of Capsaicin by Complex Coacervation of Gelatin, Acacia, and Tannins. *J. Appl. Polym. Sci.* **2005**, *96*, 2225–2229.
85. Yu, H.; Huang, Q. Enhanced In Vitro Anti-Cancer Activity of Curcumin Encapsulated in Hydrophobically Modified Starch. *Food Chem.* **2010**, *119*, 669–674.
86. Yurdugul, S.; Mozafari, M. R. Recent Advances in Micro- and Nanoencapsulation of Food Ingredients. *Cell. Mol. Biol. Lett.* **2004**, *9*, 64–65.

# CHAPTER 2

# ENCAPSULATION METHODS FOR BIOACTIVE COMPOUNDS FROM SPENT COFFEE GROUNDS

CINTIA DA SILVA ARAÚJO, LEANDRO LEVATE MACEDO, and WALLAF COSTA VIMERCATI

## ABSTRACT

The coffee beverage is recognized for its stimulating effect, mainly attributed to caffeine. However, several other compounds have already been identified and many have beneficial effects on our body. Coffee grounds consist of by-products produced at the time of beverage preparation. These by-products have several extractable bioactive compounds with high antioxidant action, which can be recovered and added to food and beverage formulations. The extraction of phenolic compounds and antioxidants from SCGs (spent coffee grounds) has been studied as a promising way to obtain these important compounds for human health from a low-cost raw material, while adding value to this by-product. Microencapsulation is a technique that can be applied to SCG extracts as a way to stabilize them and ensure greater preservation.

## 2.1 INTRODUCTION

World coffee production has increased each year, with more than 175 million bags being produced in 2020, an increase of 6.3% compared to 2019.[24] In the world ranking, Brazil stands out as the largest producer and the second largest consumer market for this product. In Brazil, domestic consumption

Plant-Based Bioactive Compounds and Food Ingredients: Encapsulation, Functional, and Safety Aspects.
Junaid Ahmad Malik, Megh R. Goyal, Preeti Birwal, & Ritesh B. Watharkar (Eds.)

grew 1.34% during November 2019 to October 2020, when compared to the same period of the previous year, reaching 21.2 million bags. Per capita consumption is estimated to be 5.99 kg/year of raw coffee and 4.79 kg/year of roasted and ground coffee. In addition to consuming traditional coffees, Brazilian consumers have shown greater interest in purchasing high-quality coffees. It is documented that the consumption of specialty and gourmet coffees will continue to grow and gain more and more importance in the kitchen.[1]

The high popularity of the beverage means that it is produced and consumed in different regions of the world, contributing to the product playing an important role in the economy of many countries. However, together with the large production and consumption volumes, large amounts of by-products are also generated from the various processing stages. Among such by-products, spent coffee grounds (SCG) deserve mention, due to the volume produced and their richness in bioactive compounds.[15]

The ingestion of bioactive compounds can have several beneficial effects on health. However, there are many difficulties in their maintenance under food processing and storage conditions, given their instability to agents such as light, oxygen, and temperature, for example. Thus, it is important to seek methods that allow to increase the stability of these compounds, allowing their incorporation and controlled release during the shelf-life of the product.[4,43]

One of the most efficient and suitable methods to increase the stability of bioactive compounds is microencapsulation. In this process, bioactive compounds are involved by a polymeric matrix forming a protective layer, preventing their contact with factors capable of degrading them. In addition, the microencapsulated material is obtained in powdered form, facilitating its handling and application.[29]

Encapsulation can be performed using different drying methods, promoting stabilization, protection, solubility, and controlled release of the encapsulated compound. The drying method and the wall material used are decisive for the characteristics of the dry particles. Thus, the choice of these parameters must be done carefully for each compound, whether they are lipids, vitamins, fatty acids, minerals, antioxidants, or probiotics.[36]

Considering the importance of drying for the microencapsulation process and that some disadvantages related to commonly used drying processes exist, this study on drying technologies can be of great interest, opening up new application possibilities in management of coffee waste and SCG.

## 2.2 SPENT COFFEE GROUNDS

Throughout the entire coffee production chain, several by-products are generated, whether they are related to the production of beans, such as husk and mucilage, or to the preparation of the beverage. Thus, as a result of the high demand for this product, which provides one of the most consumed and appreciated beverages in the world, it is essential from an environmental and economic point of view that its by-products are used.[33]

One of the main by-products generated is the SCG, which is the insoluble material obtained after contact between hot water and roasted and ground coffee beans for the production of the beverage.[6,15,32] It is estimated that around 6 million tons of SCG are produced per year in the world.[21] The soluble coffee industry is the main producer of this by-product, concentrating about 50% of world production, the rest is produced domestically and in commercial establishments.[15] For each kilo of soluble coffee produced, approximately two kilos of wet SCG are generated.[33]

SCG has received greater attention in recent years, as it is produced in large quantities and contains several interesting compounds. These compounds can be extracted and applied by various industrial segments, such as the food, pharmaceutical, cosmetic, and chemical industries.[5,39] In addition, its use is desirable, as when discarded into the environment, the SCG becomes a source of environmental contamination. Some compounds, such as caffeine, phenolics, and tannins, are responsible for the toxicity of this material.[6,33]

The main compounds found in SCG are polysaccharides, mainly cellulose and hemicellulose, which together account for about 50% of the dry matter. In addition, it has considerable amounts of lignin, insoluble dietary fiber, proteins, lipids, caffeine, low ash contents, and considerable amounts of bioactive compounds.[6,39]

Studies carried out in recent years have shown some potential uses for this by-product. Some proposed uses for SCG include application as a fertilizer,[12] biofuel production,[31] ion adsorbent,[27] and activated carbon production.[34] In addition, recent studies have demonstrated the potential of this material as a source of bioactive compounds. Bravo et al.[8] evaluated SCG from different beverage preparation methods and found that although the quantity of bioactives was different among different processing methods, it was possible to extract considerable amount of antioxidant compounds.

López-Barrera et al.[28] investigated the anti-inflammatory benefits of SCG. They found that SCG has a high dietary fiber content, allowing its fermentation by the intestinal microbiota. In this process, the release of inflammatory mediators was reduced, demonstrating its ability to prevent inflammation and indicating the potential use of SCG in the food industry as a source of dietary fiber. Martinez-Saez et al.[30] found fiber, antioxidant compounds, essential amino acids, and low glycemic index sugars in the SCG. These authors evaluated the use of coffee grounds as an ingredient in bakery products and concluded that the final quality of the product was not affected by the incorporation of this ingredient.

### *2.2.1 EXTRACTION PROCESS ENGINEERING*

Solvent extraction method is a commonly used process for extracting bioactive compounds. In this process, the choice of the extracting solvent is a crucial step, as it has a significant influence on several variables, such as the extraction yield and type of compounds extracted.[40]

The extraction processes can be used to provide extracts from various by-products. In these processes can be validated, for example, the effects of pressure, temperature, extraction time, and type of solvent in the final quality two extracts obtained.[39] The optimization of these variables results in the best conditions for exploring the raw material, generating reliability, saving time, and labor, thus making the process attractive for large-scale applications.[5]

Recent studies on the extraction of compounds from SCG have shown promising results. Xu et al.[43] used subcritical water to extract phenolic compounds from SCG. Among the extraction conditions tested, the optimum level was determined for the following conditions: temperature of 179 C, extraction time of 36 min, and 14.08 g of SCG per liter of water, where it was possible to extract 88.34 mg of GAE/g of raw material. Mussatto[14] used hydrothermal pretreatment to recover phenolic compounds and flavonoids from SCG, which resulted in a phenolic extraction yield of 32.92 mg GAE/g and flavonoids yield of 8.29 mg QE/g. Bravo et al.[9] evaluated different engineering methods (soxhlet, solid–liquid, and filter coffeemaker) and solvents (water, methanol, and ethanol) for the extraction of phenolic compounds and antioxidants from SCG and it was concluded that the use of water together with the filter coffeemaker extraction method resulted in highest recovery of these compounds (13.94 mg GAE/g and 82.40 μmol of Trolox/g).

On the other hand, Al-Dhabi et al.[3] evaluated the extraction yield of phenolic compounds using ultrasound-assisted extraction process. Under optimal conditions (244 W of ultrasonic power, temperature of 40°C, time of 34 min, and ratio of 1:17 g of SCG/mL of water), it was possible to obtain

36.25 mg of GAE/g. These results are quite expressive when compared to other agro-industrial residues, such as grape pomace and carrot and apple peels, where levels were lower than 15 mg GAE/g of dry sample.[41]

Shang et al.[39] optimized the extraction of caffeine from SCG using high pressure and a mixture of water and ethanol. According to these authors, it was possible to recover from 3 to 9 mg of caffeine/g of SCG, for the 10 different coffee cultivars under study.

## 2.3 PHENOLIC COMPOUNDS IN SCG

An antioxidant compound is defined as one capable of inhibiting or minimizing oxidative degradation. In this context, phenolic compounds are main antioxidants present in coffee waste. In products (such as oils and fats), they are found mainly in the form of monophenols, such as tocopherols. On the other hand, in fruits, vegetables, teas, coffee, and wine, they are more commonly present as water-soluble polyphenols.[37]

Polyphenols consist of secondary metabolites naturally present in plants and that play role in energy production, electron transfer, and metal absorption from the soil. Another function attributed to these compounds is the protection of cells against oxidative stress.[26] In this way, antioxidant compounds from the diet help in cellular defense, preventing oxidative damage to cellular components.[38]

Coffee is a food considered a source of phenolic compounds in the diet, whose main representatives are chlorogenic acids. These acids consist of soluble polyols, which are formed by the esterification of caffeic and quinic acids, totaling 71 natural compounds.[20,26] Chlorogenic acids (according to their chemical identity, number, and position of acyl residues) are named as[25]

- caffeoylquinic acids (3-caffeoylquinic; 4-caffeoylquinic; and 5-caffeoylquinic);
- dicaffeoylquinic acids (3,4-dicaffeoylquinic; 3,5-dicaffeoylquinic; and 4,5-dicaffeoylquinic);
- feruloylquinic acids (3-feruloylquinic; 4-feruloylquinic; and 5-feruloylquinic);
- *p*-coumaroylquinic acids (3-*p*-coumaroylquinic; 4-*p*-coumaroylquinic; and 5-*p*-coumaroylquinic); and
- the mixed di-esters of caffeoyl-feruloylquinic acids.

As reported by Farah et al.,[17] the main chlorogenic acid in green coffee beans is caffeoylquinic acid, followed by dicaffeoylquine and feruloylquinic acids, which account for 80% and 76%, 15% and 18%, and 5.2% and 6.2% of the total chlorogenic acids found in arabica and conilon coffees, respectively. In general, conilon coffee has 28% more chlorogenic acids compared to arabica coffee.

During the roasting process, the high temperature causes the breakdown of the carbon–carbon bonds of the chlorogenic acids, resulting in isomerization and degradation.[17] Although these compounds undergo major changes during roasting, beverage antioxidant activity is ensured by the production of new compounds, such as melanoidins, with such activity.[42]

As reported by Borsato et al.,[7] high notes regarding coffee aroma and flavor were obtained in samples that contained high concentration of chlorogenic acids and sugars, so that the roasting conditions were sufficient for the sugar degradation products to mask the acid and bitter flavors from the degradation of chlorogenic acids. However, these authors emphasize that samples with low chlorogenic acid content and high sugar concentrations were better evaluated in the global evaluation of the beverage.

Carlsen et al.[10] evaluated the antioxidant activity of hundreds of food products consumed in different parts of the world. According to these authors, the antioxidant activity of the coffee beverage showed the greatest variation (0.89 mmol/100 g to 16.33 mmol/100 g) among all the analyzed beverages. The latter value was obtained for double espresso coffee, which presented the highest value among all drinks. In addition, they found that filtered coffee had the same antioxidant capacity as red wine (2.5 mmol/100 g).

## 2.4 MICROENCAPSULATION ENGINEERING

The food now has the function not only of nourishing and providing energy but also of improving health and promoting well-being. In this context, functional foods have received increasing attention and gained market importance.[44] In addition, the increased interest on the part of consumers in acquiring healthier, nutritionally balanced foods that are free of artificial additives has led the food industry to seek alternative sources of bioactive compounds and antioxidants for food formulations.[8]

Antioxidant compounds contribute to the prevention of chronic diseases related to oxidative stress, such as diabetes, cancer, cardiovascular, and neurodegenerative diseases. Oxidative stress appears when there is an imbalance between free radical production and antioxidant defense. Thus, free radicals can damage macromolecules, such as proteins, lipids and DNA, generating potential risks for the occurrence of various pathologies.[13,35]

The environmental pollution and risk behaviors (such as smoking and alcohol) can contribute to a tendency to increase the production of free radicals in our body, and although there is an endogenous antioxidant production system, this may not be enough to eliminate these radicals. Thus, the consumption of exogenous antioxidants, such as polyphenols, is necessary.[11] However, antioxidants that are naturally present in foods alone may not be sufficient to prevent oxidative stress, and thus the popularity of antioxidant-added engineered foods has grown.[2]

Despite the beneficial effects of natural antioxidants, they are generally very prone to degradative processes, limiting their use in the beverage industry.[4] Thus, the microencapsulation process can be an alternative to protect these compounds from degradation, allowing their incorporation into the food matrices. Microencapsulation is a widespread technique used to protect bioactive compounds, consisting of involving these compounds in a polymeric matrix and creating a microparticle to avoid or delay their contact with harmful factors, such as light, heat, and oxygen.[19,29] It also has the advantages of improving the retention time of bioactive compound in the food, allowing its controlled release, providing stability during processing and storage, and masking undesirable flavors.[4] Thus, microencapsulation enables the elaboration of food products with greater added-value and highly stable and innovative functional design products.[16]

Microencapsulation can be performed for solid particles, liquid droplets, or even compounds in a gaseous state. The ingredient encapsulated within the microparticles is called the core, internal phase or filler material. The core can be formed from one or more ingredients and can consist of a crystalline material, an emulsion, a suspension of solids, or a suspension of microparticles.

On the other hand, the material that coats these microparticles can be called a coating, wall material, or membrane and can have one or more layers. The composition of the core and wall material as well as the microencapsulation method will determine the final shape of the microparticle.[22] The size of the microcapsule is variable, ranging from submicrons to millimeters, depending on the technique used for the microencapsulation process.[44]

Considering that microcapsules carry active compounds of interest, it is important that they possess desirable functional properties, such as encapsulation efficiency, storage stability, and good release properties in vitro and in vivo, since one of the major challenges of the technique is to ensure that the active compounds remain stable and are released into the human body at the proper time and location.[44]

Microencapsulation usually occurs through drying process, given the liquid nature of extracts containing the bioactive compounds. Some common drying processes for microencapsulation are spray-drying, fluid bed, and freeze-drying.[2,23] The type of drying and the coating material used are decisive for the retention capacity of the compound within the matrix. In this way, the proper choice of these variables ensures maximum incorporation and retention of the compound of interest.[18,23]

## 2.5 GENERAL CONSIDERATIONS

The optimization of extraction processes of bioactive compounds has become an important strategy to make industrial processes increasingly sustainable and economical. The need for the proper use of raw materials and by-products that are still little explored encourages the development of research related to these materials. In this context, the by-products of coffee production chain (especially SCG) have shown great potential for exploration and application in various industrial segments.

To provide better stabilization of active compounds, microencapsulation engineering has also shown promise. Thus, the use of different drying techniques and wall materials can bring new perspectives for the application of these extracts.

## 2.6 SUMMARY

SCG is a by-product of beverage preparation and these contain several biocompounds that can be extracted and microencapsulated using novel engineering methods. These bioactive compounds possess health benefits due to antioxidant action and can be added to food and beverage formulations. The SCGs are low-cost raw materials and can be used for several purposes to reduce biowaste.

## KEYWORDS

- **bioactive extraction**
- **drying**
- **encapsulation**
- **spent coffee grounds**

## REFERENCES

1. ABIC. *Indicadores da Indústria de Café* 2020 (2020 Indicators of Coffee Industry). https://www.abic.com.br/estatisticas/indicadores-da-industria/indicadores-da-industria-de-cafe-2020/ (accessed Sept 30, 2021).
2. Aguiar, J.; Estevinho, B. N.; Santos, L. Microencapsulation of Natural Antioxidants for Food Application (Specific Case of Coffee Antioxidants): A Review. *Trends Food Sci. Technol.* **2016**, *58*, 21–39.
3. Al-Dhabi, N. A.; Ponmurugan, K.; Jeganathan, P. M. Development and Validation of Ultrasound-Assisted Solid-Liquid Extraction of Phenolic Compounds from Waste Spent Coffee Grounds. *Ultrasonics Sonochem.* **2017**, *34*, 206–213.
4. Ballesteros, L. F.; Ramirez, M. J.; Orrego, C. E.; Teixeira, J. A.; Mussatto, S. I. Encapsulation of Antioxidant Phenolic Compounds Extracted from Spent Coffee Grounds by Freeze-Drying and Spray-Drying Using Different Coating Materials. *Food Chem.* **2017**, *237*, 623–631.
5. Ballesteros, L. F.; Ramirez, M. J.; Orrego, C. E.; Teixeira, J. A.; Mussatto, S. I. Optimization of Autohydrolysis Conditions to Extract Antioxidant Phenolic Compounds from Spent Coffee Grounds. *J. Food Eng.* **2017**, *199*, 1–8.
6. Ballesteros, L. F.; Teixeira, J. A.; Mussatto, S. I. Chemical, Functional, and Structural Properties of Spent Coffee Grounds and Coffee Silverskin. *Food Bioprocess Technol.* **2014**, *7* (12), 3493–3503.
7. Borsato, D.; Pina, M. V. R.; Spacino, K. R.; Scholz, M. B.; Androcioli-Filho, A. Application of Artificial Neural Networks in the Geographical Identification of Coffee Samples. *Eur. Food Res. Technol.* **2011**, *233* (3), 533–543.
8. Bravo, J.; Juániz, I.; Monente, C.; Caemmerer, B.; Kroh, L. W.; De Peña, M. P.; Cid, C. Evaluation of Spent Coffee Obtained from the Most Common Coffeemakers as a Source of Hydrophilic Bioactive Compounds. *J. Agric. Food Chem.* **2012**, *60* (51), 12565–12573.
9. Bravo, J.; Monente, C.; Juániz, I.; Peña, M. P.; Cid, C. Influence of Extraction Process on Antioxidant Capacity of Spent Coffee. *Food Res. Int.* **2013**, *50* (2), 610–616.
10. Carlsen, M. H.; Halvorsen, B. L.; Holte, K.; Bøhn, S. K.; Dragland, S.; Sampson, L.; Blomhoff, R. The Total Antioxidant Content of More Than 3100 Foods, Beverages, Spices, Herbs and Supplements Used Worldwide. *Nutr. J.* **2010**, *9* (3), 1–11.
11. Carocho, M.; Morales, P.; Ferreira, I. C. F. R. Antioxidants: Reviewing the Chemistry, Food Applications, Legislation and Role as Preservatives. *Trends Food Sci. Technol.* **2018**, *71*, 107–120.
12. Cervera-Mata, A.; Navarro-Alarcón, M.; Delgado, G.; Pastoriza, S.; Montilla-Gómez, J.; Llopis, J.; Rufián-Henares, J. Á. Spent Coffee Grounds Improve the Nutritional Value in Elements of Lettuce (*Lactuca sativa* L.) and Are an Ecological Alternative to Inorganic Fertilizers. *Food Chem.* **2019**, *282*, 1–8.
13. Chobotova, K. Aging and Cancer: Converging Routes to Disease Prevention. *Integr. Cancer Therap.* **2009**, *8*(2), 115–122.
14. Conde, T.; Mussatto, S. I. Isolation of Polyphenols from Spent Coffee Grounds and Silverskin by Mild Hydrothermal Pretreatment. *Preparative Biochem. Biotechnol.* **2016**, *46* (4), 406–409.
15. Cruz, R.; Cardoso, M. M.; Fernandes, L.; Oliveira, M.; Mendes, E.; Baptista, P.; Casal, S. Espresso Coffee Rsidues: A Valuable Source of Unextracted Compounds. *J. Agric. Food Chem.* **2012**, 7777–7784.

16. Eratte, D.; Dowling, K.; Barrow, C. J.; Adhikari, B. Recent Advances in the Microencapsulation of Omega-3 Oil and Probiotic Bacteria Through Complex Coacervation: A Review. *Trends Food Sci. Technol.* **2018**, *71*, 121–131.
17. Farah, A.; Paulis, T.; Trugo, L. C.; Martin, P. R. Effect of Roasting on the Formation of chlorogenic acid lactones in coffee. *J. Agric. Food Chem.* **2005**, *53*, 1505–1513.
18. Fernandes, R. V. D. B.; Borges, S. V.; Botrel, D. A. Gum Arabic/Starch/maltodextrin/inulin as Wall Materials on the Microencapsulation of Rosemary Essential Oil. *Carbohyd. Polym.* **2014**, *101* (1), 524–532.
19. Fuchs, M.; Turchiuli, C.; Bohin, M.; Cuvelier, M. E.; Ordonnaud, C.; Peyrat-Maillard, M. N.; Dumoulin, E. Encapsulation of Oil in Powder Using Spray Drying and Fluidised Bed Agglomeration. *J. Food Eng.* **2006**, *75* (1), 27–35.
20. George, S. E.; Ramalakshmi, K.; Rao, L. J. M. A Perception on Health Benefits of Coffee. *Crit. Rev. Food Sci. Nutr.* **2008**, *48*, 464–486.
21. Getachew, A. T.; Chun, B. S. Influence of Pretreatment and Modifiers on Subcritical Water Liquefaction of Spent Coffee Grounds: A Green Waste Valorization Approach. *J. Clean. Prod.* **2017**, *142*, 3719–3727.
22. Gharsallaoui, A.; Roudaut, G.; Chambin, O.; Voilley, A.; Saurel, R. Applications of Spray-Drying in Microencapsulation of Food Ingredients: An Overview. *Food Res. Int.* **2007**, *40* (9), 1107–1121.
23. Gonçalves, A.; Estevinho, B. N.; Rocha, F. Microencapsulation of Vitamin A: A Review. *Trends Food Sci. Technol.* **2016**, *51*, 76–87.
24. ICO. Coffee Market Report. Retrieved June 4, 2021, from https://www.ico.org/trade_statistics.asp?section=Statistics (accessed Sept 30, 2021).
25. Jeszka-Skowron, M.; Zgoła-Grzeskowiak, A.; Grzeskowiak, T. Analytical Methods Applied for the Characterization and the Determination of Bioactive Compounds in Coffee. *Eur. Food Res. Technol.* **2015**, *240*, 19–31.
26. Karpinska, J.; Swisłocka, R.; Lewandowski, W. Mystery of a Cup of Coffee: An Insight Look by Chemist. *Int. Union Biochem. Mol. Biol.* **2017**, *43* (5), 621–632.
27. Le, V. T.; Pham, T. M.; Doan, V. D.; Lebedeva, O. E.; Nguyen, H. T. Removal of Pb(II) Ions from Aqueous Solution Using a Novel Composite Adsorbent of Fe(3) o Fe(4) / PVA/Spent Coffee Grounds. *Sep. Sci. Technol.* **2019**, *54* (18), 3070–3081.
28. López-Barrera, D. M.; Vázquez-Sánchez, K.; Loarca-Piña, M. G. F.; Campos-Vega, R. Spent Coffee Grounds, an Innovative Source of Colonic Fermentable Compounds, Inhibit Inflammatory Mediators In Vitro. *Food Chem.* **2016**, *212*, 282–290.
29. Lozano-Vazquez, G.; Lobato-Calleros, C.; Escalona-Buendia, H.; Chavez, G.; Alvarez-Ramirez, J.; Vernon-Carter, E. J. Effect of the Weight Ratio of Alginate-Modified Tapioca Starch on the Physicochemical Properties and Release Kinetics of Chlorogenic Acid Containing Beads. *Food Hydrocoll.* **2015**, *48*, 301–311.
30. Martinez-Saez, N.; García, A. T.; Pérez, I. D.; Rebollo-Hernanz, M.; Mesías, M.; Morales, F. J.; del Castillo, M. D. Use of Spent Coffee Grounds as Food Ingredient in Bakery Products. *Food Chem.* **2017**, *216*, 114–122.
31. Marx, S.; Venter, R.; Karmee, S. K.; Louw, J.; Truter, C. Biofuels from Spent Coffee Grounds: Comparison of Processing Routes. *Biofuels* **2020**, *2020*, 1–7.
32. Mussatto, S. I.; Ballesteros, L. F.; Martins, S.; Teixeira, J. A. Extraction of Antioxidant Phenolic Compounds from Spent Coffee Grounds. *Sep Purif. Technol.* **2011**, *83*, 173–179.

33. Mussatto, S. I.; Machado, E. M. S.; Martins, S.; Teixeira, J. A. Production, Composition, and Application of Coffee and Its Industrial Residues. *Food Bioprocess Technol.* **2011**, *4* (5), 661–672.
34. Pagalan Jr, E.; Sebron, M.; Gomez, S.; Salva, S. J.; Ampusta, R.; Macarayo, A. J.; Arazo, R. Activated Carbon from Spent Coffee Grounds as an Adsorbent for Treatment of Water Contaminated by Aniline Yellow Dye. *Ind. Crops Products* **2020**, *145*, 111953.
35. Porrini, M.; Riso, P. Lymphocyte Lycopene Concentration and DNA Protection from Oxidative Damage Is Increased in Women After a Short Period of Tomato Consumption. *J. Nutr.* **2000**, *130*, 189–192.
36. Ray, S.; Raychaudhuri, U.; Chakraborty, R. An Overview of Encapsulation of Active Compounds Used in Food Products by Drying Technology. *Food Biosci.* **2016**, *13*, 76–83.
37. Roginsky, V.; Lissi, E. A. Review of Methods to Determine Chain-Breaking Antioxidant Activity in Food. *Food Chem.* **2005**, *92* (2), 235–254.
38. Sato, Y.; Itagaki, S.; Kurokawa, T.; Ogura, J.; Kobayashi, M.; Hirano, T.; Iseki, K. In Vitro and In Vivo Antioxidant Properties of Chlorogenic Acid and Caffeic Acid. *Int. J. Pharma.*, **2011**, *403*, 136–138.
39. Shang, Y. F.; Xu, J. L.; Lee, W. J.; Um, B. H. Antioxidative Polyphenolics Obtained from Spent Coffee Grounds by Pressurized Liquid Extraction. *SA J. Bot.* **2017**, *109*, 75–80.
40. Singh, M.; Jha, A.; Kumar, A.; Hettiarachchy, N.; Rai, A. K.; Sharma, D. Influence of the Solvents on the Extraction of Major Phenolic Compounds (Punicalagin, Ellagic Acid and Gallic Acid) and Their Antioxidant Activities in Pomegranate Aril. *J. Food Sci. Technol.* **2014**, *51*(9), 2070–2077.
41. Stylianou, M.; Agapiou, A.; Omirou, M.; Vyrides, I.; Ioannides, I. M.; Maratheftis, G.; Fasoula, D. Converting Environmental Risks to Benefits by Using Spent Coffee Grounds (SCG) as a Valuable Resource. *Environ. Sci. Pollut. Res.* **2018**, *25* (36), 35776–35790.
42. Vignoli, J. A.; Viegas, M. C.; Bassoli, D. G.; Benassi, M. de T. Roasting Process Affects Differently the Bioactive Compounds and the Antioxidant Activity of Arabica and Robusta Coffees. *Food Res. Int.* **2014**, *61*, 279–285.
43. Xu, H.; Wang, W.; Liu, X.; Yuan, F.; Gao, Y. Antioxidative Phenolics Obtained from Spent Coffee Grounds (*Coffea Arabica* L.) by Subcritical Water Extraction. *Ind. Crops Products* **2015**, *76*, 946–954.
44. Ye, Q.; Georges, N.; Selomulya, C. Microencapsulation of Active Ingredients in Functional Foods: From Research Stage to Commercial Food Products. *Trends Food Sci. Technol.* **2018**, *78*, 167–179.

# PART II

# FUNCTIONAL ASPECTS OF PLANT-BASED FOODS

## CHAPTER 3

# PRINCIPLES OF EXTRUSION TECHNOLOGY: DEVELOPMENT OF PLANT-BASED FUNCTIONAL AND ENGINEERED FOODS

ARYASREE SUKUMAR and K. A. ATHMASELVI


## ABSTRACT

This chapter focuses on the principles of extrusion technology for the development of plant-based functional/engineered food products. It also includes types of extruded products (co-extruded, directly expanded, unexpanded, modified, half-products, candy, and texturized products), characterization of extruded food products (nutritional properties, physicochemical and functional properties, microbiological properties, sensory properties, and morphological properties). It also includes discussions on: factors affecting the quality of extruded food products; digestibility of extruded food products; packaging/storage/stability of extruded functional food products.


## 3.1 EXTRUSION TECHNOLOGY

The extrusion technique is a multivariate process in which dampened powdery or particulate raw material modifies under different process parameter combinations inside an extruder. The extrusion process gelatinizes starch and form complexes with starch, protein, and lipids in turn changes the physical, chemical, functional, and structural characteristics of the raw material resulting in the development of a novel food product. Advancement

Plant-Based Bioactive Compounds and Food Ingredients: Encapsulation, Functional, and Safety Aspects.
Junaid Ahmad Malik, Megh R. Goyal, Preeti Birwal, & Ritesh B. Watharkar (Eds.)

in this technology provides the opportunity to control the process as per product requirements. Deep understanding of the process helps in the formulation of the recipe based on nutrient content in them.[14] The extruder consists of a cylinder envelops called a barrel, inside which Archimedes screws (1 or 2) are fitted and at the end, a small orifice called a die these are the three major components.[23] A wide range of raw materials can be used for extrusion food processing resulting in a large variety of food products for human consumption with a different texture, flavor, color, and shape. The area of application includes both ready to eat foods and ready to cook foods hence it becomes a vital processing technique in food industries.[43]

The extrusion process can be done in two ways, one with high temperature and pressure called cooking extrusion, and the other one is under pressure called forming extrusion or cold extrusion. Cooking extrusion is a high temperature short time process which eliminates unwanted denaturation of protein, carbohydrate, vitamins, amino acids, and enzymes.[30] The advantages of this technique are product quality, variability, versatility, high productivity, low cost, and fewer effluents.[30,31,38]

This chapter focuses on the extrusion processing and its potential in the development of vegan food products.

## 3.2 CLASSIFICATION OF EXTRUSION SYSTEMS

### 3.2.1 COLD EXTRUSION

Extrusion processing under a temperature of 100°C is called forming extrusion or cold extrusion. It is mainly prepared on single screw extruders whose dimension ratio, length of screw to its diameter (L/D) is about 6 to 7.[30] In cold extrusion, the temperature remains constant and is used for mixing and forming of the ingredients. The main use of cold extrusion is for the manufacturing of pasta products like macaroni, spaghetti, etc., low-pressure conditions are also used in cold extrusion.[43] Temperature about 5–10°C is used for the development of chocolates and meat analogs. To get sufficient texture to these products, long dies are used for these processes.[7]

### 3.2.2 HOT EXTRUSION

Extrusion processing carried out above temperature of 100°C is called cooking extrusion or hot extrusion. Commonly used cooking extruders are co-rotating or counter-rotating twin-screw extruders. To achieve a quick rise in temperature frictional heating or other external heating methods are used. Small flight inside the barrel induces high pressure and shear on the material when it comes out of the die sudden expansion happens due to steam release. Products can be made in variety of shapes as per die configuration. Extrusion cooked products include puffed cereals, expanded snacks, breakfast cereals, etc.[43].

### *3.2.3 SINGLE SCREW EXTRUSION*

In single screw extrusion, extruder configuration consists of one piece of a screw or a screw made of multiple pieces. Parameters used for classifying single-screw extruders are moisture content (MC) of feed, screw type (single solid, segments), level of shear, and source of heat. A common classification is based on the source of heat and level of shear. Generally, these type of extruders are used for forming extrusion.[40]

### *3.2.4 TWIN SCREW EXTRUSION*

Twin-screw extruder composed of two screws in two axes enclosed in a barrel with a smooth interior. Depending on the configuration of the screws, these extruders can be classified into four types on the basis of position of the screw and its direction of rotation: (1) co-rotating intermeshing screws; (2) co-rotating non-intermeshing screws; (3) counter-rotating intermeshing screws; and (4) counter-rotating nonintermeshing screws. Screws with intermeshing result in better residence time inside the extruder and screws with nonintermeshing result in better degree of shear.[45]

## 3.3 WORKING PRINCIPLE OF AN EXTRUSION SYSTEM

The extruder is a continuous reactor in which it converts plasticized biopolymer into a uniformly processed viscoelastic mass which forms or shapes into the product by the die. The extruder is a versatile and well-organized machine that controls the reactions happening inside it. Fundamentally extruder can be considered as a screw pump in which screw(s) rotate inside a compactly fitted barrel. Moistened raw material (powdery/particulate) is introduced into the extruder through the feeding section, material moves forward inside the barrel by rotation of flighted screw(s). Ways of heat generation are by external supply or mechanical energy dissipation due to viscous flow thus mass temperature will be always higher than barrel temperature. The temperature controller is given at various sections of the extruder and also at die end. Unit operations involved in extrusion processing are mixing, compression, shearing, cooking, and shaping. Major components of a typical extruder are feeder, extruder barrel, screws, driving system, die, temperature controller system, and safety control facilities.[7]

By the equipment design and type of feed material need to be handled processing conditions have to be changed to get desired product characteristics. For extrusion formed products, cooking is not the target but sizing and shaping of the moistened flour is important. Consumers are supposed to cook the product before eating (pasta). But in the case of extrusion-cooked products like expanded products and ready to eat product cooking is the target for setting processing conditions.

Heating inside the extruder is attained in several ways through external heating, steam injection, or dissipation of mechanical energy into thermal energy by rotation of the screws. Die at the extruder exit shape the food product and rotating knives will cut the product into small pieces. The speed of the cutting system can be adjusted according to the product's length requirement. When the product comes out of an extrusion cooker it expands because of the sudden release of steam but extrusion formed products will not expand much. So expanded products have a porous structure with low density. The texture and overall sensory acceptability of the final products are thus affected by process parameters. Different postextrusion treatments like frying, drying, toasting, roasting, baking, and favoring/coating are done to increase product stability during storage and also for increasing acceptability.

## 3.4 CLASSIFICATION OF EXTRUDED PRODUCTS

### *3.4.1 COEXTRUDED PRODUCTS*

Coextruded food products are the one which is having an extruded outer shell with a filling inside. Different varieties of coextruded products can be made with different dies and also postextrusion equipment like rollers, belt-type cutters, and stampers. Varieties include sandwich shape, multiple shells, and fillings, pinched ends for filling, exotic shapes, etc. The shelf life of coextruded products is comparatively low because of the movement of MC of filling to the outer shell. Advanced technologies are available to tackle these problems to get products with different MC fillings.[21]

### 3.4.2 DIRECTLY EXPANDED PRODUCTS

Direct expanded snacks are the most popular extruded snacks. Twin-screw extruders are mainly used for manufacturing directly expanded snacks. Flour is used as the starting material for this process. Additives like sugar, salt, flavors, and color can also be added with the formulation dough is created inside the extruder. As it exposed to shear, pressure, and temperature inside the extruder it becomes a fluidized mass. Fluidized mass shapes into the expanded product as it exits die due to the sudden release of steam and pressure drop. Different shapes of die can be used to get various shapes.[17]

### 3.4.3 UNEXPANDED PRODUCTS

Unexpanded products are cold extruded products or formed products. It is a noncooking method in which ingredients are mixed and shaped into the final product. For manufacturing unexpanded products, the temperature of the extruder (less than 100°C) remains constant and is used for mixing and forming operations. Major commercially available unexpanded extruder products are pasta products (macaroni, spaghetti, etc.).[43]

### 3.4.4 MODIFIED PRODUCTS

The best technological application of extrusion processing is for the production of modified food products. The conventional process of starch modification possesses difficulties and disadvantages. Through extrusion, processing modification can be done most easily. Twin-screw extruders are used for this purpose because of their consistent product quality. It can be operated as a continuous process with excellent mixing capacity for handling highly viscous fluids (gelatinized starch). Starch gelatinizes at a temperature above 60°C effects in the disintegration of starch molecules to form a consistent extremely viscous paste that is tough to handle. An advanced method for developing modified products is called reactive extrusion. Reactive extruders have decent heat transfer characteristics, plug flow characteristics, and starch modification that can be done in a consistent medium.[29]

### 3.4.5 HALF PRODUCTS

Half products are also called third-generation snacks or pellets. Complete cooking is not achieved during extrusion processing and drying is required after extrusion. Postextrusion treatments like oil frying, puffing, and microwave heating are needed before consumption. Salting and flavoring can be given externally. In some products, the flavor is given inside the pellets itself. Half products have extended stability during storage than expanded products due to low MC and also they have high bulk density. Third-generation snacks have interesting market potential due to its characteristics.[21]

### 3.4.6 CANDY/CONFECTIONERY PRODUCTS

Extruded confectionery products are made with cooker extruders. Using a single-step extrusion process, physicochemical characteristics of confectionery products can be achieved by adjusting ingredient formulation, which is not possible with conventional confectionery production methods. Secondary reactions in confectionery processing like Maillard reaction, sugar inversion, protein breakdown in whipping and gelling, sugar crystallization, fat polymorphism, enzymatic changes, and starch crystallization become controllable in extrusion processing. Commercially available food products in this category are licorice, toffees, caramels, fudges, boiled sweets, sugar crust liqueurs, candy sticks, gums, jellies, and chocolates.[6]

### 3.4.7 TEXTURIZED PRODUCTS

Texturized vegetable proteins are of attaining importance in the food market. Vegetable proteins are texturized by restructuring molecules of protein into a cross-linked layered mass which become heat resistant on later processing. Stabilization of this structure can be possible due to hydrophobic nature, cation-mediated electrostatic interactions, and covalent bonds formation during the extrusion process. Protein extrusion does not possess a melt transition phase and the dependence of process parameters and product characteristics are also different from other extruded products. This shows varied and complex macromolecular interaction in plant protein extrusion.[4]

## 3.5 PROCESSING PARAMETERS USED IN EXTRUSION PROCESSING

### 3.5.1 TEMPERATURE AND SCREW SPEED OF EXTRUDER

Extruder variables such as screw speed, feed rate, and geometry of die used are those which regulates the residence time of the feed material inside the extruder and also its mechanical parameters. The product's physical and functional characteristics are also dependent on these processing parameters. Physical and functional characteristics determine the success of a product. The shear stress of the final product is highly sensitive to extrusion variables. Torque values are high at high extrusion temperatures. Specific mechanical energy is also high at high screw speed. Temperature and screw speed have a curvilinear effect going on extrusion variables and extrudates characteristics.[1] Barrel temperature of the process and screw speed are significant in optimizing a process for product development.

In a study for the development of minerals and fiber-rich breakfast cereal, screw speed and temperature are taken as optimization parameters. Results show that a screw speed of 350 RPM, the temperature of 120°C are ideal parameters for getting all the desired physicochemical properties of the product.[10] In another study for the development of gluten-free breakfast cereal showed a barrel temperature of 159°C for black rice and 150°C for red-rice-based extruded product as an optimized parameter.[27] But in the case of breakfast cereal with naranjitha pomace showed temperature between 121°C and 131°C as optimal for getting a product with the desired physicochemical and nutritional properties.[15] Depending on the feed formulation extrusion parameter values vary.

### 3.5.2 SPECIFIC MECHANICAL ENERGY (SME)

Specific mechanical energy is the total power supplied for extrusion processing; it is given by eq 3.1.[46]

$$\mathrm{SME(kWh/kg)} = \frac{\text{Screw speed} \times \text{Power}(\text{kW}) \times \text{Torque}(\%)}{\text{Max. screw speed} \times \text{Through put}(\text{kg/h}) \times 100} \tag{3.1}$$

At high specific mechanical energy intermolecular bond between starch molecules ruptures resulting in gelatinization. Due to starch gelatinization, melt become viscoelastic under high temperature. Starch gelatinization is directly proportional to product expansion; thus SME is an indicator of expansion. As the screw speed increases shear rate increases inside the barrel resulting in higher SME.[14] In a study for the development of black and red rice extruded products showed SME values of around 42–70 kW/kg for the products. According to the study with low feed moisture under low temperature resulted in higher SME values, because under these conditions shear rate becomes high resulting in high energy requirement.[29]

### 3.5.3 DIE PRESSURE AND TORQUE

Die pressure is a sign of resistance to material flow over die opening and is considerably affected by the die geometry. Torque is the work needed to force the product through the die; hence, it is an indicator of safety in extrusion operations. Die pressure and torque reduce with a rise in temperature of system, screw speed, and MC of feed. High temperature of processing and screw speed decrease the viscosity of the melt due to an increase in shear rate lead to torque drop.[14] A study on high moisture extrusion of peanut protein showed that energy input for an extruder system has a significant effect on die pressure and SME of the process. The study concluded that parameters that play a decisive role on pressure and temperature of die are shear way, extrusion process temperature, and feed MC especially for textured vegetable proteins.[51]

### 3.5.4 MOISTURE CONTENT (MC) OF FEED

Depending on the MC of the feed and type of product, a single screw or twin-screw extruder can be selected.[8] Moisture is directly added to the feed formulation or is injected as steam/water into the extruder barrel separately. Moisture acts as a lubricant, prerequisite for gelatinization of starch and protein denaturation, reduces specific mechanical energy, increases extruder capacity, and provides energy for cooking. When moisture is added to the feed it softens and reduces the wear of equipment parts resulting in less cost of operation. Depending on the dough viscosity, glass transition temperature and melt transition temperature of dough changes.[14]

For the development of high quality, texturized products MC should be greater than 40%. In a study for the development of extruded peanut protein moisture given was between 50 to 60% and found out that MC control SME of the system and plays important role in color and hardness of the product.[52] In another study for the development of fiber-rich breakfast cereal snacks, the optimum MC is found to be 12%.[10] In another study for the development of extruded cereal which has to be reconstituted before consumption, results showed that MC is the main factor that affects the bowl-life hardness of the product and optimum MC is found to be 16%.[33]

## 3.6 CHARACTERIZATION OF EXTRUDED PRODUCTS

### 3.6.1 NUTRITIONAL PROPERTIES

#### 3.6.1.1 CARBOHYDRATES

Starch is the major carbohydrate in many of the cereal-based products. Since starch is the major structural constituent any change in starch results in structural modification. Amylose and amylopectin are the starch molecule that is responsible for viscosity and gelatinization of melt. Major changes happening to starch during extrusion processing are gelatinization, dextrinization, decrystallization, and depolymerization. Starch gelatinization results in expanded products because of increased gas holding capacity. Intermolecular hydrogen bonds are broken with starch gelatinization resulting in increased water absorption capacity. In high-temperature extrusion, there will be a significant increase in starch digestibility due to the breakdown of hydrogen bonds in starch molecules. But in the case of starch-protein, starch-lipid, and resistant starch formation digestibility of starch decrease.

A study on rice-based extrudates showed a reduction in starch molecular weight of extruded wheat flour in low moisture and low temperature of die.[16] In reactive extrusion granular structure of starch is replaced with a rougher irregular structure having aggregates and agglomerates. Crystalline complexes were formed with the existing lipid granules and esterification reagents. Structural changes are beneficial for decreasing the digestibility of starch.[12] Addition of protein to starch extrusion results in some interaction effect. In a study on millet, starch extrusion showed that retrogradation of starch molecules is retarded by protein content. Results propose that protein addition can extend shelf life of starch rich foods. Starch and protein interaction decreases rapidly digesting starch and increases resistant starch content.[53]

#### 3.6.1.2 PROTEINS

Proteins are affected extrusion cooking, in a study for the development of brown rice pasta showed that at a process temperature of 135°C as well as 150°C degradation of alkali-soluble proteins was higher than at 120°C. But in the case of MC, protein aggregation increases with an increase in MC by forming cross-linking disulfide bonds thus changes the structure of the starch protein complex. Protein content directly correlates to the textural properties of extrudates. High temperature and low MC resulting in loss of amino acids, but with high MC amino acids retain due to low shearing. Mechanical shear breakdown protein molecule thereby increasing its digestibility.[13] Peanut protein extrusion showed the effect of exogenous polysaccharides in textural changes of extrudates. It enhances the protein aggregation to get a better fibrous structure.[52]

#### 3.6.1.3 LIPIDS

Lipids act as plasticizers or emulsifiers in extrusion cooking. Proper texture and stickiness of extrudates are provided by lipids. Lipids help with the extrusion process it may be present in ingredients or can be added externally. Lipid decreases slip in the barrel resulting in torque reduction. Lipid affects product expansion due to inadequate pressure caused by reduces torque. High temperature and mechanical disruption cause lipid release during extrusion which affects final product quality.[16] A study on extruded cereal with apple pomace fortification showed that temperature has a significant effect on lipid oxidation of extrudates.[24]

#### 3.6.1.4 ANTIOXIDANTS AND TOTAL PHENOLIC CONTENT

Total phenolic content is accountable for the antioxidant activity of a product. Total phenolic content found to be affected by the process parameters of extrusion. A study showed that soluble free and conjugated phenolic acid converted to insoluble bound phenolic acids resulting in a significant increase in total phenolic acids with a rise in temperature and a drop in die pressure of the process.[20]

A study for the extrusion of rice incorporated with *Agriophyllum squarrosum* flour showed temperature of extrusion had a significant positive influence on total phenolic content and total flavonoid content.[25] A study on the antioxidant and phenolic content of apple pomace in an extruded cereal showed that *p*-coumaric acid increased with increasing temperature and different moisture levels reach a maximum of about 0.48 mg/100 g at 144°C and decreased thereafter with a rise in temperature. Caffeic acid increased with temperature rise and decreased after 154°C. Ferulic acid increases with a rise in MC irrespective of temperature variation. Temperature and MC significantly affected total antioxidant activity.[24]

#### *3.6.1.5 VITAMINS AND MINERALS*

Vitamin stability differs due to their chemical composition and structure at different extrusion process parameters. A study on the production of texturized rice protein reported higher retention of vitamins riboflavin and thiamine in high moisture extrusion. In high moisture extrusion, low shear results in less mechanical energy dissipation result in the retention of micronutrients. Nutrient bioavailability is increased during extrusion processing due to the destruction of antinutritional factors.[26] In another study on sorghum extrusion, extrudates showed greater content of γ-tocopherols (59.5%) and α-tocopherols (28.9%) of total vitamin E than that of unprocessed sorghum.[3]

A study of cold extrusion of fortified corn extrudates showed that thiamine and riboflavin retention was higher than conventional extrusion process and an increase in MC did not affect much on vitamin retention on extrudates.[11] Minerals are heat stable during high-temperature processing. A study on the extrusion of cornflour and brown rice pasta showed a significant increase in mineral content.[13] Another advantage of the extrusion process is it will enhance mineral absorption by eliminating antinutritional factors like phytates and condensed tannins.[2] In another study on fortified pasta showed more than 94% retention of iron in pasta.[42]

#### *3.6.1.6 FIBER CONTENT*

Demand for fiber-rich products is increasing because researches show the intake of dietary fiber reduces the risk of diseases like diabetes, constipation, and colon cancer. Incorporation of fiber in formulation reduces its calorific value thus controls weight gain. Studies showed greater content of dietary fiber in formulation negatively affected the rapidly digestible starch content in rice extrudates. The interaction effect of nutrients is also there in extrusion processing if there is no protein in the formulation fiber directly attach to starch otherwise expansion of the product got affected.[5] In another study showed that the incorporation of whole grain wheat flour has a significant positive effect on total dietary fiber, insoluble dietary fiber, and soluble dietary fiber. Insoluble dietary fiber is found to be higher than soluble dietary fiber.[32] In another study for the development of fiber-rich breakfast cereal with naranjita pomace showed an increase in insoluble dietary fiber with an increase in extrusion temperature.[15]

## 3.6.2 PHYSICOCHEMICAL AND FUNCTIONAL PROPERTIES

### 3.6.2.1 BULK DENSITY (BD)

BD can be found out by dividing the mass by volume of the extrudates (kg/cm$^3$). BD shows how porous the product is and it is inversely proportional to the extrudates expansion. When the food product exit the extruder water evaporates immediately resulting in the incorporation of air bubbles in the product and it expands. A study for extrusion of whole-grain millets also showed similar results.[22] An increase in temperature and screw speed of the extruder reduces the BD of the product. In a study for the development of fiber-rich breakfast, cereal snacks showed a similar result.[10] As the screw speed rises, shear rate increases and melts viscosity decreases affecting expansion ratio. With a rise in feed rate melt, rheological characteristics vary due to starch degradation and melt become elastic resulting in density changes. In a study for the development breakfast cereal based on whole-grain sorghum showed that as the concentration of sorghum rises the BD of the final product increases, this may be due to greater fiber content in them and also expansion reduced.[28] The BD of a product influences the types of packaging required for that product and its design.

### 3.6.2.2 EXPANSION RATIO (ER)

The expansion ratio is the diameter of extrudates to the die diameter and is also called the sectional expansion of extrudates.[16] As the melt exit die end it expands due to water evaporation and the difference in vapor pressure inside and outside. Dynamic steps involved in the expansion are bubble nucleation, bubble growth, coalescence, and shrinkage, and melt setting. Expansion is characterized by a cellular and airy structure with low density. Extrudates are a solid form that can be characterized by a vertical expansion index which is equal to the product of longitudinal and radial expansion index. Expansion in flow direction of die is a longitudinal expansion index and a cross-sectional increase in the orthogonal plane of flow direction is the radial expansion index.[35] The expansion ratio is a critical factor for an expanded snack and is related to its eating characteristics. In a study for the development of millet-based extrudates showed maximum expansion for foxtail millet with ER 4.41concluded the effect of fiber content in expansion.[22] An increase in feed moisture decreases the expansion of extrudates. A study on whole grain wheat flour extrudates development showed a negative linear effect of moisture on the sectional expansion index.[32]

#### *3.6.2.3 WATER ABSORPTION INDEX (WAI) AND WATER SOLUBILITY INDEX (WSI)*

The WAI is the volume occupied by swelled starch granule in water. It is a factor that indicates product interaction with water and starch conversion after processing. WAI decreases with a decrease in water holding capacity because of molecular size reduction due to starch degradation. The WSI is the number of soluble polysaccharides remains in the product after processing.[50] The WSI is a sign of starch degradation during processing. A rise in temperature and a drop in feed moisture results in high WSI, and WAI also increase with an increase in temperature and MC.[27] Another study for millet extrusion also revealed a significant positive impact of temperature and screw speed on WAI; but feed moisture was not significant. With an increase in feed, moisture WSI decreases; and with an increase in temperature, WSI increases.[22]

#### *3.6.2.4 TEXTURAL PROPERTIES*

Textural properties of extruded food products are vital for the sensory approval of the product. Textural properties mainly depend on the structural characteristics of the product. Important textural properties of extruded expanded snack products are crispiness, hardness, chewiness, adhesiveness, and springiness. Hardness is the force necessary to break the extrudates and is directly related to the expansion of the product. In a study for the development of extruded food products, there was a direct correlation between density and hardness of extrudates. Processing temperature increases extrudates expansion, decreases the density, and results in decreased hardness values.[10] An antagonic interaction effect is shown by whole grain wheat flour and MC of feed on the hardness of extrudates because of the fiber content in the flour. Wheat flour and moisture have a negative effect on the crispiness of the extrudates.[32] In another study for the production of rice-based breakfast cereal also shows the effect of MC on the textural properties of extrudates. Moisture decreases expansion properties resulting in increased hardness and reduced crispiness values.[27]

#### *3.6.2.5 COLOR*

The color of the extrudates is an important parameter in getting consumer acceptance. Color is mainly dependent on ingredients used in the formulation of the product and its processing conditions. The main reactions happening inside the extruder that affect product color are nonenzymatic brownings like caramelization, Maillard reaction, and pigment degradation. Process conditions like temperature and MC of the formulation also affect the color of the extruded product. As the temperature increases browning reaction increases. Instrumental color value L* represents lightness, a* represent redness, and b* represent yellowness of the product. Whole grain wheat extrudates showed a significant positive effect with L* and a* value of the extrudates. But b* value was significantly affected by the concentration of whole grain wheat flour.[32] Another study on the extrusion of rice flour incorporated with *Agriophyllum squarrosum* flour showed a significant effect of L* and b* values by temperature and MC but a* value was affected significantly by MC only.[25]

#### *3.6.2.6 PASTING AND VISCOELASTIC PROPERTIES*

The pasting profile gives the macro structural arrangement in a product. The pasting profile includes details about pasting temperature, peak viscosity, breakdown viscosity, set back viscosity, hot paste viscosity, and cold paste viscosity. Pasting temperature is the lowest temperature at which starch gelatinization begins. Peak viscosity, hot paste viscosity, and cold paste viscosity decrease with a rise in temperature and screw speed because it boosts starch gelatinization. Break-down viscosity refers to stability of paste or granular starch disintegration during the holding stage. Set-back viscosity refers to recovered viscosity value for the period of the cooling of starch that is heated. Set-back viscosity specifies the recrystallization of starch molecules that gelatinized during heating. So high set back value means high retrogradation.[9]

A study on the extrusion of wholegrain wheat flour showed the effect of different formulation on the viscosity profile of extrudates. Pasting parameters were lower for the extrudates than that of the flours, which narrates the susceptibility of native starches on viscosity during heating and cooling than that of pregelatinized flours. Wheat flour concentration, moisture, and temperature of extrusion and their interaction showed a significant effect on pasting properties. Whole grain wheat flour showed a significant positive linear effect on peak viscosity and trough viscosity, negative effect on setback viscosity. Except peak viscosity feed moisture have a significant positive effect on other pasting properties.[33] In another study on blended flours, extrusion showed that extrudates of sorghum rich blend have a low tendency for retrogradation because of high dietary fiber content. Extrudates of maize rich blend showed low cold viscosity and break down viscosity due to the high lipid content in them.[47]

#### *3.6.2.7 THERMAL PROPERTIES*

In high-temperature processing, cereal-based products undergo structural changes causing in swelling of starch granules called gelatinization. In high temperatures, high shear and low MC extrusion processing disrupt the starch granules resulting in gelatinization or dextrinization.[34] Melting enthalpy ($\Delta H$) of the product indicates its degree of gelatinization. In a study on the extrusion of chickpea flour showed that extrusion improved the thermal stability of chickpea flour than that of raw flour. The particle size of feed also showed a significant effect on the thermal stability of extrudates.[39] In another study on the effect of enzymatic treatment on rice and soya, bean extrusion showed that in low MC degree of gelatinization increases up to 105°C and decreases with further increase in temperature.[49]

### *3.6.3 MICROSTRUCTURE*

The molecular structural properties of a product are disclosed by micrographs. Structural changes during extrusion like the presence of fracture, size of pores, structural orientation, and space between cells can be visualized through micrographs. Studies on the extrusion of rice starch and bran reported a porous and rough structure of extrudates.[48] Micrographs of extruded products with very few voids indicate poor expansion. But in high temperature, extrusion results in expansion and larger voids in structure. A study on low moisture extrusion of soy protein with gelatin and sugar shows more compact structures due to the Maillard reaction between sugar and protein.[19] A study on high moisture extrusion of wheat gluten showed anisotropic structures due to the formation of dispersed phases at high temperature 155°C and high protein content as a result of wheat gluten polymerization.[36]

### 3.6.4 MICROBIOLOGICAL PROPERTIES

One of the important safety aspects of developing a food product is its microbiological and is a consumer requirement. Conventional food products like snack ready to eat snack foods and breakfast cereals are exposed to high temperature (greater than 130°C) other extruded formed products their MC is less than 5%, so these are safe to eat. Most of the vegetative organisms, yeast, and mold got damaged below usual extrusion conditions of 55–145°C.[18]

### 3.6.5 SENSORY PROPERTIES

Sensory analysis corresponds to sensations on physiological reactions by responses of people. Interpretation of peoples such responses on certain stimuli generates sensory properties of a food product. The stimulus can be measured by physical and chemical processes and sensations by the physiological effect. Sensations aggravated by-product differs in duration, pleasantness, quality, and intensity. The sensory analysis depends on the sensory properties of a particular product by the targeted population. Consumer testing is categorized based on acceptability and preference. Acceptability depends on like and dislike of the product by consumer and preference based on a comparison of two or more products by a consumer. Consumer testing is an important parameter in analyzing the quality of a product, optimizing a process/product, and new product development. 9-point hedonic scale be used to measure consumer preference.[41] In a 9-point hedonic scale analysis, sensory acceptance of fiber-rich breakfast cereal with seeded banana and carambola pomace is found to be 7.8 for optimized product.[10]

## 3.7 DIGESTIBILITY OF EXTRUDED FOOD PRODUCTS

Cereal-based extruded food products can be classified according to their starch digestibility. Starch digestibility is described by the rate and duration of glycemic response, so is an important factor for the development of diabetic foods. Predicting the glucose absorption and its control is an important aspect of health concern. Researches support the relationship between in vitro digestion of food and its glycemic response. Enzymatic digestibility of starch is affected by textural and rheological properties of food, amount of proteins, lipids, carbohydrates, and their interactions during processing. Studies showed a significant increase in starch digestion upon extrusion cooking. Extrusion cooking changes the structural integrity of starch molecules which increases its susceptibility to enzyme hydrolysis. Sometimes interaction effect of starch lipid complex formation, starch protein interaction, and low MC extends starch digestibility during enzymatic hydrolysis.[44] A study for the extrusion of sorghum showed an increase in sorghum protein digestibility upon extrusion. Resistant starch content in white sorghum cereal extrudates is found to be higher than that of any other cereals, i.e., 208 g/100 g starch. Results also showed that tannins do not show any effect on starch digestibility of extruded sorghum cereals.[28]

## 3.8 PACKAGING, STORAGE, AND STABILITY OF EXTRUDED FOOD PRODUCTS

Multilayer flexible pouches are used as packaging material for extruded products. Depending on the type of extruded product, packaging material changes. Lipid containing products have to be packed in a material that should mask light to limit lipid oxidation otherwise product becomes rancid. Type of packaging materials is polyethylene, polypropylene, and polyethylene terephthalate pouches. For packing ready to eat products multilayer trays or polyethylene terephthalate thermoform trays having heat-sealed lid stock are used. To determine the stability of extruded products during storage accelerated storage stability tests are used. In a study conducted to determine the shelf life of extruded weaning food is as follows. The 50 g of products were sealed in polyethylene bags for stability check. Sealed samples were stored at 30, 45, and 50°C incubators. Samples were analyzed for loss of quality at different intervals of time. Nonenzymatic browning and free fatty acid content were used as an index for calculating the shelf life of the product.[37]

## 3.9 SUMMARY

Extrusion technology has a wide range of applications to develop plant-based functional/engineered foods in the processing sector. In recent times, demand for vegan products is increasing; therefore, extrusion processing can be effectively used for developing wide range of plant-based food products. Apart from commonly available extruded products centered on corn and rice, further research is required in the field of pseudo-cereal-based extruded products.

## KEYWORDS

- **engineered food**
- **expanded food**
- **extrusion technology**
- **functional food**
- **texturized food**
- **twin screw extruder**

## REFERENCES


1. Alam, M. S.; Kaur, J.; Khaira, H.; Gupta, K. Extrusion and Extruded Products: Changes in Quality Attributes as Affected by Extrusion Process Parameters: A Review. *Crit. Rev. Food Sci. Nutr.* **2016**, *56* (3), 445–473.
2. Alonso, R.; Rubio, L. A.; Muzquiz, M.; Marzo, F. The Effect of Extrusion Cooking on Mineral Bioavailability in Pea and Kidney Bean Seed Meals. *Anim. Feed Sci. Technol.* **2001**, *94* (1–2), 1–13.
3. Anunciacao, P. C.; de Morais Cardoso, L.; Gomes, J. V. P.; Della Lucia, C. M.; Carvalho, C. W. P.; Galdeano, M. C.; Pinheiro-Santana, H. M. Comparing Sorghum and Wheat Whole Grain Breakfast Cereals: Sensorial Acceptance and Bioactive Compound Content. *Food Chem.* **2017**, *221*, 984–989.
4. Areas, J. A. Extrusion of Food Proteins. *Crit. Rev. Food Sci. Nutr.*, **1992**, *32* (4), 365–392.
5. Arribas, C.; Cabellos, B.; Sanchez, C.; Cuadrado, C.; Guillamon, E.; Pedrosa, M. M. The Impact of Extrusion on the Nutritional Composition, Dietary Fiber and In Vitro Digestibility of Gluten-Free Snacks Based on Rice, Pea and Carob Flour Blends. *Food Funct.* **2017**, *8* (10), 3654–3663.
6. Best, E. T. Confectionery Extrusion. In *The Technology of Extrusion Cooking*; Springer: Boston, MA, 1994; pp 190–236.

7. Bhattacharya, S. Extrusion Technology and Glass Transition. In *Non-Equilibrium States and Glass Transitions in Foods*; Woodhead Publishing, 2017; pp 137–152.
8. Bhattacharya, S. Raw Materials. In: *Advances in Extrusion Technology*; Maskan, M., Altan, A., Eds., CRC Press: Boca Raton, FL, 2012; pp 69–85.
9. Bhattacharya, S; Das, H; Bose, A. N. Effect of Extrusion Process Variables on Microstructure of Blends of Minced Fish and Wheat Flour. *J. Food Sci. Technol. (Mysore)* **1990**, *27* (1), 22–28.
10. Borah, A.; Mahanta, C. L.; Kalita, D. Optimization of Process Parameters for Extrusion Cooking of Low Amylose Rice Flour Blended with Seeded Banana and Carambola Pomace for Development of Minerals and Fiber-Rich Breakfast Cereal. *J. Food Sci. Technol.* **2016**, *53* (1), 221–232.
11. Boyaci, B. B.; Han, J. Y.; Masatcioglu, M. T.; Yalcin, E.; Celik, S.; Ryu, G. H.; Koksel, H. Effects of Cold Extrusion Process on Thiamine and Riboflavin Contents of Fortified Corn Extrudates. *Food Chem.* **2012**, *132* (4), 2165–2170.
12. Cai, C.; Wei, B.; Tian, Y.; Ma, R.; Chen, L.; Qiu, L.; Jin, Z. Structural Changes of Chemically Modified Rice Starch by One-Step Reactive Extrusion. *Food Chem.* **2019**, *288*, 354–360.
13. da Silva, E. M. M.; Ascheri, J. L. R.; Ascheri, D. P. R. Quality Assessment of Gluten-Free Pasta Prepared with a Brown Rice and Corn Meal Blend via Thermoplastic Extrusion. *LWT—Food Sci. Technol.*, **2016**, *68*, 698–706.
14. Dalbhagat, C. G.; Mahato, D. K.; Mishra, H. N. Effect of Extrusion Processing on Physicochemical, Functional and Nutritional Characteristics of Rice and Rice-Based Products: A Review. *Trends Food Sci. Technol.* **2019**, *85*, 226–240.
15. Delgado-Nieblas, C.; Ruiz-Beltran, K.; Sanchez-Lizarraga, J.; Zazueta-Morales, J. D. J.; Aguilar-Palazuelos, E.; Carrillo-López, A.; Quintero-Ramos, A. Effect of Extrusion on Physicochemical, Nutritional and Antioxidant Properties of Breakfast Cereals Produced from Bran and Dehydrated Naranjita Pomace. *CyTA—J. Food* **2019**, *17* (1), 240–250.
16. Ding, Q. B.; Ainsworth, P.; Tucker, G.; Marson, H. The Effect of Extrusion Conditions on the Physicochemical Properties and Sensory Characteristics of Rice-Based Expanded Snacks. *J. Food Eng.* **2005**, *66* (3), 283–289.
17. Fast, R. B. Manufacturing Technology of Ready-to-Eat Breakfast Cereals. In: *Breakfast Cereals and How They are Made*; Fast, R. B., Caldwell, E. F., Eds., Vol. 1, 2nd ed.; American Association of Cereal Chemists: St. Paul, Minnesota; **2000**; pp 17–54.
18. Fraiha, M.; Ferraz, A. C. D. O.; Biagi, J. D. Pre-treatment of Thermoduric Spores in $CO_2$ Modified Atmosphere and Their Survivability During Food Extrusion. *Food Sci. Technol.* **2011**, *31* (1), 167–171.
19. Guerrero, P.; Beatty, E.; Kerry, J. P.; De la Caba, K. Extrusion of Soy Protein with Gelatin and Sugars at Low Moisture Content. *J. Food Eng.*, **2012**, *110* (1), 53–59.
20. Hu, Z.; Tang, X.; Zhang, M.; Hu, X.; Yu, C.; Zhu, Z.; Shao, Y. Effects of Different Extrusion Temperatures on Extrusion Behavior, Phenolic Acids, Antioxidant Activity, Anthocyanins and Phytosterols of Black Rice. *RSC Adv.* **2018**, *8* (13), 7123–7132.
21. Huber, G. R.; Rokey, G. J; Extruded Snacks. In: *Snack Foods*; Springer, Boston, MA, 1990; pp 107–138.
22. Kharat, S.; Medina-Meza, I. G.; Kowalski, R. J.; Hosamani, A.; Ramachandra, C. T.; Hiregoudar, S.; Ganjyal, G. M. Extrusion Processing Characteristics of Whole Grain Flours of Select Major Millets (Foxtail, Finger, and Pearl). *Food Bioproducts Process.* **2019**, *114*, 60–71.

23. Kristiawan, M.; Chaunier, L.; Sandoval, A. J.; Della Valle, G. Extrusion: Cooking and Expansion. In *Breakfast Cereals and How They Are Made*; AACC International Press: St Paul, MN, 2020; pp 141–167.
24. Leyva-Corral, J.; Quintero-Ramos, A.; Camacho-Davila, A.; de Jesus Zazueta-Morales, J.; Aguilar-Palazuelos, E.; Ruiz-Gutierrez, M. G.; de Jesus Ruiz-Anchondo, T. Polyphenolic Compound Stability and Antioxidant Capacity of Apple Pomace in an Extruded Cereal. *LWT—Food Sci. Technol.* **2016**, *65*, 228–236.
25. Li, W.; Chen, S.; Peng, J.; Pan, L.; Tu, K. Effects of Twin-Screw Extrusion Processing Variables on Physicochemical Properties and Antioxidant Activity of Rice Incorporated with *Agriophyllum squarrosum* Flour. *J. Food Process. Preserv.* **2020**, *2020*, E-article 14524.
26. Liu, C.; Zhang, Y.; Liu, W.; Wan, J.; Wang, W.; Wu, L.; Yin, Z. Preparation, Physicochemical and Texture Properties of Texturized Rice Produce by Improved Extrusion Cooking Technology. *J. Cereal Sci.* **2011**, *54* (3), 473–480.
27. Mcza, S. L. R.; Sinnecker, P.; Schmiele, M.; Massaretto, I. L.; Chang, Y. K.; Marquez, U. M. L. Production of Innovative Gluten-Free Breakfast Cereals Based on Red and Black Rice by Extrusion Processing Technology. *J. Food Sci. Technol.* **2019**, *56* (11), 4855–4866.
28. Mkandawire, N. L.; Weier, S. A.; Weller, C. L.; Jackson, D. S.; Rose, D. J. Composition, In Vitro Digestibility, and Sensory Evaluation of Extruded Whole Grain Sorghum Breakfast Cereals. *LWT—Food Sci. Technol.* **2015**, *62* (1), 662–667.
29. Moad, G. Chemical Modification of Starch by Reactive Extrusion. *Progress Polym. Sci.* **2011**, *36* (2), 218–237.
30. Moscicki, L.; van Zuilichem, D. J. Extrusion Cooking and Related Technique. In *Extrusion Cooking Techniques: Applications, Theory and Sustainability*; Wiley, Weinheim: Germany, 2011; pp 1–24.
31. Navale, S. A.; Swami, S. B.; Thakor, N. J. Extrusion Cooking Technology for Foods: A Review. *J. Ready Eat Food* **2015**, *2* (3), 66–80.
32. Oliveira, L. C.; Rosell, C. M.; Steel, C. J. Effect of the Addition of Whole-Grain Wheat Flour and of Extrusion Process Parameters on Dietary Fiber Content, Starch Transformation and Mechanical Properties of a Ready-To-Eat Breakfast Cereal. *Int. J. Food Sci. Technol.* **2015**, *50* (6), 1504–1514.
33. Oliveira, L. C.; Schmiele, M.; Steel, C. J. Development of Whole Grain Wheat Flour Extruded Cereal and Process Impacts on Color, Expansion, and Dry and Bowl-Life Texture. *LWT*, **2017**, *75*, 261–270.
34. Pardhi, S. D.; Singh, B.; Nayik, G. A.; Dar, B. N. Evaluation of Functional Properties of Extruded Snacks Developed from Brown Rice Grits by Using Response Surface Methodology. *J. Saudi Soc. Agric. Sci.* **2019**, *18* (1), 7–16.
35. Perdon, A. A.; Schonauer, S. L.; Poutanen, K., Eds. *Breakfast Cereals and How They Are Made: Raw Materials, Processing, and Production*; AACC International Press: St Paul, MN, 2020; p 262.
36. Pietsch, V. L.; Werner, R.; Karbstein, H. P.; Emin, M. A. High Moisture Extrusion of Wheat Gluten: Relationship Between Process Parameters, Protein Polymerization, and Final Product Characteristics. *J. Food Eng.* **2019**, *259*, 3–11.
37. Plahar, W. A.; Okezie, B. O.; Annan, N. T. Nutritional Quality and Storage Stability of Extruded Weaning Foods Based on Peanut, Maize and Soybean. *Plant Foods Human Nutr.* **2003**, *58* (3), 1–16.

38. Ramachandra, H. G.; Thejaswini, M. L. Extrusion Technology: A Novel Method of Food Processing. *Int. J. Innov. Sci. Eng. Technol.* **2015**, *2* (4), 358–369.
39. Raza, H.; Ameer, K.; Zaaboul, F.; Sharif, H. R.; Ali, B.; Shoaib, M.; Zhang, L. Effects of Ball-Milling on Physicochemical, Thermal and Functional Properties of Extruded Chickpea (*Cicer arietinum* L.) Powder. *CyTA—J. Food* **2019**, *17* (1), 563–573.
40. Riaz, M. N. Introduction to Extruders and Their Principles. *Extruders Food App.* **2000**, *2020*, 1–23.
41. Santos, I. L.; Steel, C. J.; Aguiar, J. P. L.; Schmiele, M.; de Sales Ferreira, J. E. C.; de Amaral Souza, F. D. C. Sensory Analysis of Extruded Corn-Based Breakfast Cereals with Whole Peach Palm Fruit (*Bactris gasipaes*) Powder. *Afr. J. Food Sci.* **2017**, *11* (9), 310–317.
42. Sharma, N.; Sharma, S.; Singh, B. Stability Evaluation of Iron and Vitamin-A During Processing and Storage of Fortified Pasta. *Quality Assurance Safety Crops Foods* **2020**, *12* (2), 50–60.
43. Shelar, G. A.; Gaikwad, S. T. Extrusion in Food Processing: An Overview. *Pharma Innov. J.*, **2019**, *8* (2), 562–568.
44. Singh, J.; Dartois, A.; Kaur, L. Starch Digestibility in Food Matrix: A Review. *Trends Food Sci. Technol.* **2010**, *21* (4), 168–180.
45. Steel, C. J.; Leoro, M. G. V.; Schmiele, M.; Ferreira, R. E.; Chang, Y. K. Thermoplastic Extrusion in Food Processing. *Thermoplast. Elastomers* **2012**, *2012*, 265–271.
46. Stojceska, V.; Ainsworth, P.; Plunkett, A.; Ibanoglu, S. The Recycling of Brewer's Processing By-Product Into Ready-To-Eat Snacks Using Extrusion Technology. *J. Cereal Sci.* **2008**, *47* (3), 469–479.
47. Toledo, V. C.; Carvalho, C. W.; Vargas-Solorzano, J. W.; Ascheri, J. L.; Comettant-Rabanal, R. Extrusion Cooking of Gluten-Free Whole Grain Flour Blends. *J. Food Process Eng.* **2020**, *43* (2), E-article 13303.
48. Wang, P.; Fu, Y.; Wang, L.; Saleh, A. S.; Cao, H.; Xiao, Z. Effect of Enrichment with Stabilized Rice Bran and Extrusion Process on Gelatinization and Retrogradation Properties of Rice Starch. *Starch-Stärke* **2017**, *69* (7–8), E-article 1600201.
49. Xu, E.; Wu, Z.; Pan, X.; Long, J.; Wang, F.; Xu, X.; Jiao, A. Effect of Enzymatic (Thermostable α-Amylase) Treatment on the Physicochemical and Antioxidant Properties of Extruded Rice Incorporated with Soybean Flour. *Food Chem.* **2016**, *197*, 114–123.
50. Yousf, N.; Nazir, F.; Salim, R.; Ahsan, H.; Sirwal, A. Water Solubility Index and Water Absorption Index of Extruded Product from Rice and Carrot Blend. *J. Pharmacogn. Phytochem.*, **2017**, *6* (6), 2165–2168.
51. Zhang, J.; Liu, L.; Jiang, Y.; Faisal, S.; Wang, Q. A New Insight Into the High-Moisture Extrusion Process of Peanut Protein: From the Aspect of the Orders and Amount of Energy Input. *J. Food Eng.*, **2020**, *264*, E-article 109668.
52. Zhang, J.; Liu, L.; Jiang, Y.; Shah, F.; Xu, Y.; Wang, Q. High-Moisture Extrusion of Peanut Protein-/Carrageenan/Sodium Alginate/Wheat Starch Mixtures: Effect of Different Exogenous Polysaccharides on the Process Forming a Fibrous Structure. *Food Hydrocoll.* **2020**, *99*, E-article 105311.
53. Zheng, M.; Xiao, Y.; Yang, S.; Liu, M.; Feng, L.; Ren, Y.; Liu, J. Effect of Adding Zein, Soy Protein Isolate and Whey Protein Isolate on the Physicochemical and In Vitro Digestion of Millet Starch. *Int. J. Food Sci. Technol.*, **2020**, *55* (2), 776–784.

## CHAPTER 4

# FUNCTIONAL ASPECTS OF PLANT-BASED FOOD PRODUCTS

SANIKA BHOKARIKAR and K. A. ATHMASELVI


## ABSTRACT

The plant-based diet (PBD) and vegan foods are becoming popular for their health and environmental benefits. Plant-based food industry targets vegan consumers, but it is not limited to them. Plant-based dairy products are mainly developed for lactose-intolerant persons. The plant-based meat alternatives are preferred by health-conscious people and patients with heart ailments to reduce cholesterol levels. The development of different technologies for the processing of these products is required to increase the growth of the plant-based food industry. This chapter presents the scenario of plant-based food products in the global market, the importance of PBD, and future trends.


## 4.1 INTRODUCTION

Vegetarians are persons who do not eat any fish, meat, or poultry. Lacto–ovo vegetarians and vegans are two types of vegetarians. Vegans do not consume any animal products, whereas Lacto–ovo vegetarians take dairy and/or eggs. Geographic location, cultural and religious beliefs, climatic circumstances, and availability of vegetarian food products all influence the amounts of lacto vegetarians/vegans. India has the largest population of vegetarians (more than 30% of the total population) in the world.[15] In other regions of the world, plant-based diets (PBD) and vegan diets are less widespread. It gained popularity in the last 10 years.

Plant-Based Bioactive Compounds and Food Ingredients: Encapsulation, Functional, and Safety Aspects.
Junaid Ahmad Malik, Megh R. Goyal, Preeti Birwal, & Ritesh B. Watharkar (Eds.)

Whole grains, pulses and legumes, fruits, nuts and oilseeds, and vegetable seeds are the mainstays of a PBD. Dairy, meat, fish, and any other animal-based food products are avoided in the vegan or PBD. The PBD is rich in fiber, vitamin C, folic acids, and other micronutrients. People prefer a PBD for not only health benefits but also for environmental issues, and cultural and religious beliefs. For people with lactose intolerance, PBD is a very convenient option. There are three categories of people who can eat plant-based food products—omnivorous, vegetarians, and vegans. There are a few types of vegetarians as follows: Lacto–ovo vegetarians who consume milk, milk products, and eggs, ovo–vegetarians who consume eggs but not dairy; Lacto vegetarians who consume milk and milk products but not eggs; Semi-vegetarians generally eat vegetarian food products but occasionally consume meat and poultry. Excluding vegans, others eat plant-based food products by choice. Nutritionists recommend these plant-based food products in their daily diet to reduce calorie intake and other health issues.

Worldwide, 2 billion people live primarily on a meat-based diet, and 4 billion live primarily on a PBD. Also, animal-based food products need more energy and water for processing than plant-based food products.[4] The percentage of cognitive vegetarians hold similar beliefs about meat and vegetarian diets as do vegetarians and consume less meat than the general population in Australia, the UK, and the USA.[21]

The purpose of this chapter is to study trends of plant-based food products in the market, types of vegan alternatives, and their scope in the future.

## 4.2 PLANT-BASED DIET

Veganism is a lifestyle that attempts to reduce the exploitation of animals by excluding all animal products. Grains, fruits, vegetables, seeds, pulses, spices, and other plant-based foods are included in a PBD. This diet excludes all animal products, such as milk and milk products, eggs, meat, fish, and so on. Vegans are persons who exclusively eat foods that come from plants. People eat a PBD for a variety of reasons. By avoiding dairy, lactose-intolerant people switch to plant-based foods. Few individuals consume plant-based foods because of their potential health and environmental benefits. Plant-based foods are usually higher in dietary fiber, minerals, vitamins, and phytochemicals but PBD contains less amount of saturated fats, calories, cholesterol, omega-3 fatty acids, vitamin D, and vitamin B-12.

Fruits, vegetables, legumes, spices, cereals, and oil seeds are an essential part of everyone's daily meal. There are some food substitutes for vegans which help them to maintain a vegan lifestyle as well as provide the required nutrients. The plant-based substitutes are similar in taste, appearance, texture, and color to animal-based food products. These substitutes can be used in most food preparations where dairy, eggs, or meat is used to satisfy the cravings of consumers. There are many plant-based food products available in the market, but the main products are meat substitutes, dairy substitutes, and animal protein substitutes.

### 4.2.1 ALTERNATIVES TO MEAT

Traditionally, only vegetarians liked meat alternatives or meat substitute products; however, in today's market, these products are popular and enjoyed by a wide spectrum of consumers. Pulses, nuts, vegetables, and grains are the main components of plant-based meat substitutes. Soy protein isolate, tofu, soy protein isolate, tempeh, and textured tofu are all popular meat substitutes. The dry form of protein-rich, fat-free, and carbohydrate-free soy protein isolate is widely available. It's used to make vegan sausages, chicken-style nuggets, and chicken-style breasts, among other things. Defatted soy flour is used to make textured soy protein, then extruded and dehydrated, to give it a sponge-like structure. The flavored textured soy protein is used as chunks or in meat-free sausages. Tofu prepared by the coagulation method is used in the preparation of burger patties or cutlets. The wheat protein made from gluten is similar to the texture and consistency of meat, and hence it is used in the preparation of vegan or plant-based sausages and burgers. Pea protein combined with gluten is also a significant ingredient to develop chunks, lasagna, and beef-style chunks. Peanuts and other nuts are used in nut cutlets, but these products do not resemble meat. These oil seeds or nut-based food products are healthy options as they provide less saturated fat than meat-based food products.[23]

Currently, existing meat substitutes are low in fats, as defatted flours are used in the preparation of the products. The fat or oil is added during the processing to form a fibrous structure. Additional fats also provide juiciness, tenderness, and flavor development but sometimes it turns out to be slippery also. The fats used in plant-based meat are canola oil, corn oil, sunflower oil, and rapeseed oil according to the desired characteristics and flavor. Binding agents are often used to hold the main ingredients together. The concentration is optimized according to the characteristics of binding agents. Gluten, whey protein, and albumin are protein binders majorly used in meat alternatives, especially for chicken equivalents. Carrageenan, soy protein concentrate, casein, 5% fat, and xanthan gum are used to make mushroom-based sausage or meat analogs. Gelling and thickening functions, rheological properties, and water binding capacity are the parameters that affect the texture, taste, and other properties and are considered while using polysaccharides as binding agents such as pectin, guar gum, and carrageenan.

Different flavoring agents and colors are used to improve the appearance and taste of the meat products. Extrusion is majorly used in meat alternative preparation. Impossible Foods (USA), Beyond Meat (USA), Morning Star Farms (USA), Gardein (Canada), and The Vegetarian Bucher (UK) are a few established enterprises in meat analogs or meat alternatives production using plants.[14]

### *4.2.2 MILK ALTERNATIVES OR MILK SUBSTITUTES*

Plant-based milk substitutes are water extracts of oil seeds, legumes, and cereals, which resemble animal milk. The most popular and widely consumed milk alternative is soy milk than other milk alternatives. The general process of preparation of plant-based milk includes the following steps, such as soaking, wet milling, filtering, stable suspension, enzyme treatment (if required), homogenization, pasteurization, and packaging. Sometimes pre-treatments are needed, such as dehulling, soaking, and blanching. Blanching is done to inactivate the trypsin inhibitors. Excess fat is removed if the milk is nut-based. Phase separation in milk is avoided by homogenization or using binding agents like xanthan gum, guar gum, carrageenan, etc. If the milk is prepared from cereals like rice milk or oat milk, enzymes are used to reduce starch because heating of milk causes gelatinization of starch present in cereals which is problematic for new food product development.[28] Coconut milk and soy milk are rich in antioxidants and proteins.

Different kinds of products are then developed from these types of milk by processing methods and using additives. Beverages and drinks are prepared by adding additives, such as stabilizers, sweetening agents, flavors, oil or fat, and salt fermentation of milk, and the use of diverse flavors is done for yogurt preparation. Ice creams are prepared by using different types of plant-based fats like palm oil, or cold-pressed coconut oil to give a creamy and rich texture to the product.

Many non-thermal processing technologies and innovative techniques for preservation like the pulsed electric field, and ultrasound need to be explored for the preservation of plant-based milk alternatives because thermal processing methods reduce the nutritional value of the product.[24]

### 4.2.3 OTHER FOOD PRODUCTS

A whole-grain oat beverage was prepared by fermentation of oat substrate with lactic acid bacteria (LAB). The beverage contained $7.5 \times 10^{10}$ CFU $ml^{-1}$ of LAB. Sweeteners such as aspartame, sodium cyclamate, saccharine, and Huxol (12% cyclamate and 1.2% saccharine) were added to the beverage, and they did not show any effect on the dynamics of the fermentation process and also no alteration occurred in any other sensory characteristics during storage. The amount of beta-glucan was stable in the developed drink during 21 days of storage under refrigeration.[3]

The yogurt powder (mango flavored) was obtained from different processing methods, such as convective drying, conditioning, and grinding. The dried powder mango yogurt was packed in HDPP pouches, and the predicted shelf-life of the powder was 45 days. The quality parameters were tested for free fatty acids, thiobarbituric acid, and also for color fluctuations. The kinetics of quality parameter change was zero order.[13]

## 4.3 RESEARCH STUDIES ON PLANT-BASED DIET

The study looked into the benefits and side effects of a PBD on humans. Veganism causes a few changes in the human body. People, who use PBD, have a lower risk of developing ailments, such as cardiovascular disease, type 2 diabetes, obesity, and a few forms of cancer. This diet's phytochemicals and nutrients protect customers from a variety of chronic diseases.[7]

The United Nations has recommended PBD for preventing obesity and chronic diseases. A study was conducted on 41,387 individuals for 2 years to determine the effects of a vegan diet on diabetic patients which concluded that vegans had a lower risk of developing type 2 diabetes compared to non-vegetarians. Nurses' Health Study, Nurses' Health Study II, and health professionals monitored the 4.1 million people for years. After the study, they found that individuals who consume a high amount of animal protein had a 13% increased risk of type 2 diabetes. This study also witnessed that switching 5% of energy intake to plant-based protein from animal protein resulted in 23% reduced risk of type 2 diabetes. Furthermore, PBD reduces insulin resistance, improves glycemic control, and reduces cholesterol levels also. As these plant-based food products are high in fiber, they show beneficial metabolic effects by promoting gut microbial flora.[16]

A study conducted by European Prospective Investigation into Cancer and Nutrition stated that vegans gain less weight than omnivores as they age. The Swedish Mammography Cohort found that omnivores are more susceptible to obesity than individuals following a more PBD. A study was conducted in Taiwan among 49,098 individuals, of which 10.9% were on a vegetarian diet and 15.4% individuals were on a non-vegetarian or animal-based diet. Vegetarians had a body mass index (BMI) of 27 $kg/m^2$, which was comparatively lower than non-vegetarians. This study also found that the risk of obesity was decreased by 7% per year with a vegan diet.[29]

A German vegan study found that vegans generally avoid smoking and consume less alcohol. They also engage in higher levels of physical activity than the general population. In this study, they also found the presence of unique gut microbial flora in vegans than omnivorous. Vegetarians and vegans have similarities in gut microbial flora with diverse characteristics including, a reduced number of pathobionts as well as plenty of protective microbial strains or species, such as *Faecalibacterium prausnitzii*.[6]

A systematic review was conducted, which included the comparison of vegan, vegetarian and omnivorous diets and their impact on the environment by studying 16 different studies and 18 reviews. This study was conducted on the basis of research done in the US and Europe. This review concluded that the vegan diet, or PBD, which is solely based on plants, is the finest and safe diet for the environment. The making of plant-based food products, not only food but cosmetics, fabric, accessories, etc., causes the emission of the lowest level of greenhouse gases. The water used for diets other than vegan is more, hence more the consumption of animal proteins, the higher the amount of water used. Thus, locally developed vegan products should be preferred as those are more environmentally friendly than the remaining options.[8]

Vegan products are important not only for vegans and vegetarians but also for omnivores, according to these studies. In comparison to western countries, India is heavily reliant on plant-based foods. As a result, plant-based food product research, awareness, marketing, and production are critical.

## 4.4 MARKET SURVEY FOR PLANT-BASED FOODS

The major goal of plant-based food products is to develop items for vegan consumers, who are a possible target market. Lactose-intolerant people are the ones who buy plant-based milk products. According to a poll of experts and consumers performed in Germany in 2018, purchasing plant-based cuisine is mostly motivated by ethical and environmental concerns. Many people purchased plant-based items to support entrepreneurs' ideas, such as local and biocyclic-vegan agriculture. High prices, a lack of understanding, and a rejection of the vegan lifestyle are all obstacles.[11]

The Vegan Society is the world's oldest association for veganism, founded in 1944. The founder of this association, Donald Watson, and the other five members coined the modern word vegan for vegetarians, who do not consume dairy. In 2019, the Global meat alternative market was priced at $4532.6 million, and it is expected to touch $7106.6 million by 2025. A large share is contributed by Europe and America, which is 70%. There are many meat alternatives (mainly made up of soybean), that is, Soy meat, textured vegetable protein, Tofu, Mycoprotein, Tempeh, and Seitan.[26]

Plant-based milk is quite popular, and there is a huge demand for all other plant-based alternatives. Vegan drinks, milk (almond, soy, peanut, and oat), curds, yogurts, ice cream, creamers, and several types of butter are all included. The increasing population of lactose-intolerant persons, together with innovative product development in milk replacements and their goods, resulted in massive market expansion in a short period of time. The global revenue was 416,130.9 in 2019. According to reports (2019), soy milk and soy milk products each contributed approximately 40% of global market revenue while other plant milk such as rice, oats, hazelnut, and coconut took the remaining 20%. Almond milk is becoming more popular than soy milk.[28]

According to the Vegan Food Market Size and Industry Analysis Report of 2020, a total of 2026 plant-based food has a lot of scope in the future, and the vegan market is going to witness steady growth in recent years. In Canada, more than 8,000,000 people were classified as vegans in the year 2018. The demand for plant-based food has increased in the USA by around $3.3 billion during 2017–2018. Asia Pacific held the largest share of vegan consumers in the world, that is, more than 8.5% of people were following only PBD in 2016.[25] There are many international brands for plant-based food products some have been compiled in Table 4.1.

## 4.5 NEW PRODUCT DEVELOPMENT, CHALLENGES, AND FUTURE TRENDS

Plant-based food is a challenging notion for many people to accept, thus launching and introducing it successfully is a difficult endeavor. Alternatives are more difficult to market and accept than traditional plant-based food products. Customers have high sensory expectations of these alternatives, especially when it comes to meat replacements. Meeting sensory attributes while preserving nutrients and reducing process costs is a critical element

**TABLE 4.1** Some Popular Brands for Plant-Based Food Products.[1,2,5,10,12,17,18,19,22,25]

| Brand name | Country | Type of products | Products |
|---|---|---|---|
| Aduna | Africa | Chocolates; powders; tea | Energy bars; plant-based powder |
| Alpro | UK | Alternative dairy | Drinks; yogurt; ice creams; desserts; cream from oats; almonds; soybean |
| Beyond Meat | USA | Meat alternatives | Sausages, sausages for breakfast; burgers prepared from beans (Mung and Faba), brown rice, coconut oil, etc. |
| Good Mylk | India | Dairy alternatives | Curd, beverages, milk, butter from oats, peanuts, etc. |
| Just | US | Egg alternatives | Plant-based egg alternative in liquid and folded form |
| Mozzarella | UK | Alternative dairy products; ready meals. | Plant-based cheese and yogurt; ready meals; condiments |
| New Wave Foods | USA | Alternative seafood. | Plant-based shrimp alternative |
| Oatly | Sweden | Dairy alternatives | Ice creams, yogurt, milk, coffee whitener prepared from oats, coconut oil, etc. |
| Quorn | USA | Frozen meat alternative | Meatless buffalo dippers, patties, fillets and sticks prepared from oats, wheat, pea protein, etc. |
| Silk | USA | Alternative dairy | Beverages; yogurts alternatives; creamers; latte from coconut; oats; almonds |

of product development. According to studies, it is necessary to design unique replacements, and these items must be consumed by customers on a regular and consistent basis to gain acceptance. Few studies recommend the development of entirely diverse products than existing or conservative food products to attract customers.[10]

According to a survey, experts and consumers said that vegan food products are pricy and hard to access than animal-based food products. Few consumers reject vegan products to avoid PBDs as they think it is absurd.[9]

More research is being done on product development, but not on inadequacies caused by consuming exclusively PBDs or vegan products. Plant-based food products could be fortified to help fix the problem without affecting the product's other sensory qualities.

Plant-based food knowledge is limited in society, and specific efforts are needed to increase awareness. Different preservation and processing technologies should be used in product development to lower the product's price, which is still a worry for many people. Vegan items that are affordable will undoubtedly promote plant-based food product usage. In order to acquire the trust of consumers, every country must have strong labeling and other quality regulations. Many industries are devoted solely to the production of plant-based goods. Because the trend of plant-based or vegan products is only a few decades old, these manufacturing enterprises are relatively new. Vegans can benefit from mainstream and established industries that produce animal or dairy products investing in plant-based food products and promoting a sustainable lifestyle. The involvement of widespread and well-established food brands or industries can gain the trust of customers and promote healthy, vegan products.

The accessibility of plant-based food products in supermarkets is a crucial issue. Many vegan food products are available only on their brand's websites and E-commerce websites which sometimes changes consumers' minds to other easily accessible options. So, to promote plant-based food products, a separate section should be dedicated to these vegan products. Local vegan brands should promote their food products in supermarkets and restaurants. Vegan-friendly restaurants and cafes should also be one of the targets for the industry. Vegan alternatives to cheese, premixes, and other raw ingredients should be available to design plant-based menus for consumers. The development of the vegan cosmetic industry is also increasing awareness about PBD These products are far better for the skin than chemically loaded products. Though this industry is not related to the food industry directly, the overall impact of marketing can boost PBD.

Social media is also important for promoting the PBD and the vegan lifestyle, and it may be used to promote new vegan companies. People are encouraged to live a vegan lifestyle via a variety of websites, blogs, pages, and organizations on social media.

According to People for the Ethical Treatment of Animals (PETA), new trends in plant-based food in 2020 will be Innovative Ice creams or frozen desserts, eco-friendly packaging, fast food, jerky, unusual oilseed; and nut butter, vegan fish from bananas, and so on.[20]

In the future, there is excellent scope for research in vegan fortified food, functional foods, egg powder replacers, perfection in sensory qualities, and protein powders from new plant-based sources.

## 4.6 SUMMARY

Vegan food is becoming more popular around the world every day. The growth of the vegan food sector can be attributed to a variety of factors, including lactose intolerance, nut allergies, environmental concerns, and lifestyle changes. These industries are developing plant-based alternatives with the goal of achieving the same sensory and nutritional attributes as non-vegan products. However, new product development faces numerous technical and formulation-related problems. Soybean, oats, peanut, almond, hemp (milk substitutes), pulses and wheat (meat substitutes), and pea protein (animal protein substitutes) are just a few of the components commonly utilized in vegan products. Vegan diets have been shown in studies to help people lose weight. The vegan food business should develop new technology to extend the shelf-life or refrigeration life of products while also improving the nutritional content of vegan foods.

## KEYWORDS

- **dairy alternatives**
- **lactose intolerance**
- **meat alternatives**
- **oats**
- **plant protein**
- **soybean**
- **vegan**

## REFERENCES


1. Aduna. https://aduna.com/ (accessed on July 2, 2021).
2. Alpro. https://www.alpro.com/uk/ (accessed on July 3, 2021).
3. Angelov, A.; Gotcheva, V.; Kuncheva, R.; Hristozova, T. Development of a New Oat-Based Probiotic Drink. *Int. J. Food Microbiol.* **2006**, *112* (1), 75–80.
4. Appleby, P. N.; Key, T. J. The Long-Term Health of Vegetarians and Vegans. *Proc. Nutr. Soc.* **2016**, *75* (3), 287–293.
5. Beyondmeat. https://www.beyondmeat.com/. (accessed on July 6, 2021).
6. Chai, B. C.; van der Voort, J. R.; Grofelnik, K.; Eliasdottir, H. G.; Klöss, I.; Perez-Cueto, F. J. Which Diet Has the Least Environmental Impact on Our Planet? A Systematic Review of Vegan, Vegetarian and Omnivorous Diets. *Sustainability* **2019**, *11* (15), 1–18.
7. Craig, W. J. Health Effects of Vegan Diets. *Am. J. Clin. Nutr.* **2009**, *89* (5), 1627S–1633S.
8. Glick-Bauer, M.; Yeh, M. C. The Health Advantage of a Vegan Diet: Exploring the Gut Microbiota Connection. *Nutrients* **2014**, *6* (11), 4822–4838.
9. Global Market Insights. https://www.gminsights.com/industry-analysis/vegan-food-market (accessed on July 10, 2021).
10. Good Mylk. https://goodmylk.in/ (accessed on July 7, 2021).
11. Jürkenbeck, K.; Schleicher, L.; Meyerding, S. G. Marketing Potential for Biocyclic-Vegan Products? A Qualitative, Explorative Study with Experts and Consumers. *Ger. J. Agric. Econ.* **2019**, *68*, 289–298.
12. Just. https://www.ju.st/en-us (accessed on July 2, 2021).
13. Kumar, P.; Mishra, H. N. Storage Stability of Mango Soy Fortified Yoghurt Powder in Two Different Packaging Materials: HDPP and A.L.P. *J. Food Eng.* **2004**, *65* (4), 569–576.
14. Kyriakopoulou, K.; Dekkers, B.; van der Goot, A. J. Plant-Based Meat Analogues. Chapter 6; In *Sustainable Meat Production and Processing*; Galanakis, C. M., Ed.; Academic Press: New York, 2019; pp 103–126.
15. Lea, E. J., Crawford, D.; Worsley, A. Public Views of the Benefits and Barriers to the Consumption of a Plant-Based Diet. *Eur. J. Clin. Nutr.* **2006**, *60* (7), 828–837.
16. McMacken, M; Shah, S. A Plant-Based Diet for the Prevention and Treatment of Type 2 Diabetes. *J. Geriatric Cardiol.* (*J.G.C.*), **2017**, *14* (5), 342–354.
17. Mozzarisella; http://www.mozzarisella.co.uk/ (accessed on July 3, 2021).
18. New Wave Foods. https://www.newwavefoods.com/ (accessed on July 5, 2021).
19. Oatly. https://www.oatly.com/int/ (accessed on July 4, 2021).
20. PETA. https://www.peta.org/living/food/plant-based-vegan-food-trends-2020/ (accessed on July 15, 2021).
21. Pimentel, D.; Pimentel, M. Sustainability of Meat-Based and Plant-Based Diets and the Environment. *Am. J. Clin. Nutr.* **2003**, *78* (3), 660S–665S.
22. Quorns. https://www.quorn.us/products/vegan-food (accessed on July 5, 2020)
23. Sadler, M. J. Meat Alternatives: Market Developments and Health Benefits. *Trends Food Sci. Technol.* **2004**, *15* (5), 250–260.
24. Sethi, S.; Tyagi, S. K.; Anurag, R. K. Plant-Based Milk Alternatives an Emerging Segment of Functional Beverages: A Review. *J. Food Sci. Technol.* **2016**, *53* (9), 3408–3423.
25. Silk. https://silk.com/ (accessed on July 1, 2021).

26. The Vegan Society. https://www.vegansociety.com/news/market-insights/meat-alternative-market (accessed on July 2, 2021).
27. Turner-McGrievy, G.; Mandes, T.; Crimarco, A. Plant-Based Diet for Overweight and Obesity Prevention and Treatment. *J. Geriatric Cardiol.: J.G.C.*, **2017**, *14* (5), 369–374.
28. Yadav, D. N.; Bansal, S.; Jaiswal, A. K.; Singh, R. Plant-based Dairy Analogues: An Emerging Food. *Agric. Res. Technol.* **2017**, *2017*, 1–4.

## CHAPTER 5

# FUNCTIONAL ASPECTS OF FOODS FORTIFIED WITH OMEGA-3 FATTY ACIDS

PANDULE VISHAL SHRIRANG and MONIKA SHARMA

### ABSTRACT

A fortification is an approach to fulfilling our needs for essential nutrients. Omega-3 fatty acids (ω-3 fatty acids) are one of such essential nutrients, which offer several health benefits to humans. To complete the deficit of these essential fatty acids, various products have been used for fortification and studied their effect on various attributes of the food products, such as sensory quality, textural/thermal/physicochemical properties, etc. Bakery items, eggs, meat, baby meals, cereal bars, chocolates, milk, and milk products (curd, cheese, yogurt, ice cream, and butter) have all been enriched with ω-3 fatty acids to benefit people of all ages. Fortification provides various health benefits such as anti-diabetic properties, reduces atherosclerosis, brain development, anti-tumor properties, prevention of obesity and kidney disease, reduces cardiovascular diseases, and exhibits anti-inflammatory, antipyretic, and analgesic activities. Also, ω-3 fatty acids are unsaturated fatty acids, which are more prone to oxidation. To prevent oxidation and adverse effects on sensory properties, it is added in either direct form, emulsion, or microencapsulated powder form for fortification. This chapter discusses the structure, nature, sources, health benefits, and food products fortified with ω-3 fatty acids.

Plant-Based Bioactive Compounds and Food Ingredients: Encapsulation, Functional, and Safety Aspects.
Junaid Ahmad Malik, Megh R. Goyal, Preeti Birwal, & Ritesh B. Watharkar (Eds.)

## 5.1 INTRODUCTION

Consumption of food provides energy and fulfills the requirements of organisms. Food is the only source to get some nutrients, which are, not synthesized in our body. Our body is dependent on such sources of essential nutrients. Furthermore, ω-3 fatty acids are essential nutrients for carrying out several body functions and provide many health benefits. The ω-3 fatty acids play a significant role in the prevention of cardiovascular diseases (CVDs), and eye health, reduces the risk of hypertension, inflammatory bowel disease (IBD), coronary heart diseases, neurogenerative disorder, cancers, diabetes,[17] anti-inflammatory,[86] and hypocholesterolemic effect.[25] Due to the changing eating habits, the ratio of ω-6 to ω-3 fatty acid is very high in the range of 38–50:1[104,105] among the Indian urban population compared to the level of 5:1.[32]

Food items can be fortified with ω-3 fatty acids to attain the correct ratio. Fortification could be achieved through various sources of ω-3 fatty acids such as oily fishes which majorly contain eicosapentaenoic acid (EPA) and docosahexaenoic acid (DHA) while α-linolenic acid (ALA) is found in vegetarian sources such as flaxseeds, soybean oil, wheat germ oil, canola oil, and nuts.[46,82] Adipose tissues, skin, and carcass contain ALA in small quantities as most of it is degraded by β-oxidation.[103]

There are certain ethical issues related to the consumption of fish oil among vegetarian groups of people. This has paved the way for vegetarian alternatives. Through a procedure of elongation and desaturation, ALA is transformed into DHA and EPA in the human body. Many food items have been enriched with ω-3 fatty acids to meet daily need. The instability of the PUFA toward oxidation makes its fortification into food products difficult as it affects the shelf life. Thus, to avoid this, food products were fortified using emulsion as well as microencapsulated ω-3-fatty-acids-rich oils.

This chapter reviews the current status of polyunsaturated fatty acids (PUFAs), especially ALA, one of the ω-3 fatty acids, its health benefits, sources, and major challenges for fortification in dairy products. This chapter also includes a review of various food products fortified with ω-3 fatty acid and the effect of ALA fortification on various physicochemical and functional properties of butter.

## 5.2 POLYUNSATURATED FATTY ACIDS

Dietary fatty acids show a significant role in supplying energy to the body, as they are the main source of energy. Along with providing energy, these are the precursor of some biomolecules and involve in the formation of the cell membrane.[55] Dietary fatty acids not only help in survival but also the well-being of humans.[56] The presence or lack of double bonds inside the alkyl chain divides fatty acids into two categories. Saturated fatty acids have no double bonds in the alkyl chain, whereas unsaturated fatty acids have one or more double bonds. Fatty acids with at least two unsaturated bonds in their structure are called as PUFAs.[20] The fatty acids can be further classified as long chain, medium chain or short chain depending upon the length of alkyl groups.

Structural arrangement further classified PUFAs into different groups; conjugated fatty acids, methylene-interrupted polyenes, and other PUFAs. Methylene-interrupted polyenes showed a divinyl-methane pattern having at least two *cis* double bonds partitioned from each other with single methylene ($-CH_2-$) bridge.[47] Another group contains more than two conjugated double bonds in their structure and is thus called as conjugated fatty acids, while the third group includes podocarpic acids and pinolenic acid. As this chapter focuses on ω-3 and ω-6 fatty acids, they are discussed in detail in the subsequent sections.

### 5.2.1 OMEGA-3 FATTY ACIDS

Omega-3 fatty acids are unsaturated fatty acid with more than one double bond. Between the two ends of fatty acid, one end with a carboxyl group (–COOH), and on the other methyl group ($-CH_3$) exists. The first double bond appears after the third carbon atom from the methyl end of the alkyl chain. The methyl end is called "omega-" (ω) end; therefore, the nomenclature ω-3 fatty acids are used for such PUFAs. The ω-3 fatty acids vary in the alkyl chain length and number of double bonds (Figure 5.1). When three conventionally double bonds are present in an 18-carbon fatty acid it is said to be ALA (C18:3); while five and six double bonds containing 20 and 22 carbon fatty acids are said to be EPA (C20:5) and DHA (C22:6), respectively.[83] The long chain ω-3 fatty acids can be either directly taken from diet or synthesized in the human body from ALA by elongation and desaturation pathways. However, this conversion may be limited to levels even below 10% in some individuals.[12]

### 5.2.2 OMEGA-6 FATTY ACIDS

Omega-6 fatty acids also termed as ω-6 or n-6 fatty acids contain first double bond in the alkyl chain, sixth position from the methyl terminus of the chain (Figure 5.1). Linoleic acid (LA) accounts 85–90% of ω-6 PUFA and is an essential fatty acid with 18 carbon and 2 double bonds in the alkyl chain structure.[50] Absence of ω-3 denaturase in the mammalian cells results in inability to convert ω-6 fatty acid into ω-3 fatty acid.[10] Moreover, LA undergoes desaturation and elongation to produce γ-linolenic and dihomo-γ-linolenic acids, which include another group of ω-6 PUFAs. These could be further converted into arachidonic acid (C20:4),various types of eicosanoids such as thromboxane $A_2$, leukotriene $B_4$, and prostaglandin $E_2$. They exhibit pro-inflammatory, pro-aggregatory, and/or vasoconstrictive properties responsible for inflammation and hypertension.[50]

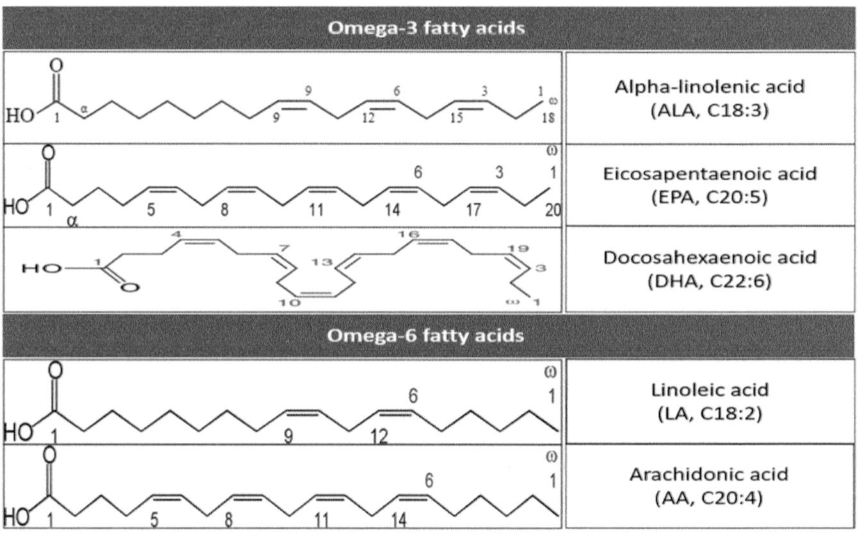


**FIGURE 5.1** Chemical structure of ω-3 and ω-6 fatty acids; modified from Refs. 120–124.

### *5.2.3 METABOLIC CONVERSION OF LINOLEIC ACID AND α-LINOLEIC ACID INTO EICOSAPENTAENOIC ACID AND DOCOSAHEXAENOIC ACID*

The ω-3 and ω-6 fatty acids are considered as essential owing to their involvement in various structural and physiological functions. To begin with the structural components of cell membranes help in prostaglandin formation.[101] Also, LA upon conversion into two- and four-series prostacyclin and prostaglandin compounds, that is, eicosanoids such as prostaglandin, thromboxanes, and leukotrienes encourage inflammation and platelet aggregation while opposite effect is observed for ω-3 fatty acids.[40]

Prostaglandin is the hormone-like substance, which involve in the relaxation and contraction of smooth muscle, blood vessels constriction, and causes inflammation. Thromboxane is a lipid that causes constriction of the blood vessels. Leukotrienes is also a lipid and responsible for inflammation.[39,50] Among all ω-6 fatty acids, arachidonic acid shows maximum inflammatory activity and stimulates the production of glutamate (neurotransmitter) causing neuron destruction by undue free radical production.

The omega-3 PUFAs produce Prostaglandin-1, which has numerous beneficial effects such as calcium movement, controls cell growth and hormone regulation (Figure 5.1). It also produces prostaglandin-3 and prostacyclin-3 having similar benefits such as prostaglandin-1, that is, anti-inflammatory activity and increases the level of trienoic prostaglandins, pentaenoic leukotrienes, and thromboxane A3.[109] Incorporation of ω-3 in the phospholipids affect the membrane properties of cell such as flexibility, fluidity, permeability, and most importantly, the activity of membrane bound enzymes.[107] Moreover, DHA elevates the extent of serotonin ("feel good" neurotransmitter) and acetylcholine (memory boosting chemical). It also carried out neutralization of the oxygen free radicals, and it decreases cellular and vascular inflammation in brain and keeps the cell membranes in the brain soft and pliable. Thus, ω-3 PUFAs function significantly in treatment in various diseases mainly (IBD), psoriasis, rheumatoid arthritis and eczema.[22]

The metabolic conversion of ω-3 fatty acids requires the involvement of several enzymes. In case of ω-3 fatty acids, α-linolenic undergoes desaturation and elongation by the enzymes Δ6-desaturase and elongase, respectively, to form eicosatetraenoic acid (20:4 ω-3). Moreover, it undergoes desaturation and elongation to form EPA (20:5 ω-3). Finally, elongation, desaturation and β-oxidation of EPA leads to the formation of DHA. Further, EPA produces anti-inflammatory compounds (3-series prostaglandins, 3-series thromboxane, 5-series leukotriene and E-resolvin) and DHA produces D-resolvin and D-protectin.[115]

In case of ω-6 fatty acid, LA is converted into the γ-linoleic acid by Δ6-desaturase enzyme and further into the dihomo-γ-linoleic acid by elongase. Dihomo-γ-linoleic acid is converted into the Arachidonic acid (20:4ω-6) by Δ5-desaturase. The arachidonic acid undergoes elongation, desaturation and β-oxidation to produce docosapentaenoic acid (22:5ω-6). Arachidonic acid is responsible for the production of inflammatory compounds such as 2-series prostaglandins, 2-series thromboxane, and 4-series leukotriene.[115]

## 5.3 HEALTH BENEFITS OF OMEGA-3 FATTY ACIDS

### *5.3.1 ANTIDIABETIC PROPERTIES OF Ω-3 FATTY ACIDS*

In diabetic patients, blood glucose level increases because of the body's inability to synthesis completely or partially insulin. In such case, insulin secretary glands called pancreas becomes non-functional because of damage to the insulin secreting cells. Diabetes categories into two types, that is, types 1 and 2 diabetes. The type 1 diabetes occurs due to the inability of producing insulin by the body while type 2 diabetes is caused by both the inability in insulin production and its utilization. Among the two, type 2 is very common among the population.

There are several reports where ω-3 fatty acids exhibited positive health benefits in diabetic patients. Han et al.[49] observed the influence of ω-3 fatty acids supplementation in patients having diabetes and hyper-triglyceridemia. In this study, 344 patients having type 2 diabetes were the test subjects. These patients were divided into subgroups based on their daily doses (1, 2, and 4 g per day) of ω-3 fatty acids. In this study, albumin to creatinine ratio and glomerular filtrate were examined immediately after administration and after 8 h of fasting. It was found that out of 344 patients 172 did not face the problem of renal non-functioning and 125 patients found positive slope in glomerular filtrate. Also, patients given with higher dose found better renal maintenance that those given with lower doses.

Vuksan et al.[117] studied effect of feeding Salba (*Salvia hispanica* L.), which is a rich source of ALA and fiber on improvement in the cardiovascular disease of patients having type 2 diabetes. This study was conducted on 20 individuals (11 males and 9 females), who were fed on Salba or wheat bran for 12 weeks. Control group provided with the wheat bran while the other with Salba (37 ± 4 g per day). Observations revealed that reduction in, high-sensitivity C-reactive protein, von Willebrand factor and systolic blood pressure. Similarly, significant decrease in the fibrinogen and $A_1C$ was observed in the *Salba* supplemented group not in control. Thus, supplementation of *Salba* for longer period was helpful in systolic blood pressure reduction, lipid control and maintaining proper glycemic index. Similar kind of relationship existed, when flaxseed or its products were added as a source of ω-3 fatty acids.

Kelley et al.[67] observed 20% glycemia reduction by supplementation of 0.5% conjugated linoleic acid (CLA) and 0.5% flaxseed in the rats susceptible to diabetic tumors. Kapoor et al.[65] supplemented flaxseed powder between

15 and 20 g of per day over a period of 2 months and observed the effect on blood sugar level. The blood sugar decreased by 7.9 and 19.1% in rats fed with 15 and 20 g, respectively. Nazni et al.[90] conducted similar study on 25 diabetic patients and there was significant reduction in the blood glucose level of diabetic patients for 90 days due to feeding of bread supplemented with flaxseed powder.

### 5.3.2 REDUCES ATHEROSCLEROSIS

Atherosclerosis is the disease in which blood flow is obstructed due to formation of plaque. Plaque is made of cholesterol, fat, other compounds present in the blood, which grows gradually and causes hardening of the arteries that supplies oxygenated blood to the heart and other parts of the body. This ultimately leads to stroke, heart attack and peripheral vascular disease. This is how, ω-3 fatty acids exert the beneficial effects in preventing atherosclerosis. Dyerberg et al.[29] studied the same behavior in case of healthy males. They conducted an intervention study for 8 weeks by supplementing three different diets (trans fat enriched diet using partially hydrogenated soybean oil, ω-3 fatty acids enriched diet and control diets) and observed the effect on high-density lipoprotein (HDL) cholesterol, blood pressure and other parameters. The HDL cholesterol level of the group supplemented with trans fatty acid was decreased and ω-3 fatty acid supplementation showed lower blood pressure, thereby preventing atherosclerosis.

### 5.3.3 BRAIN AND NEURAL DEVELOPMENT

The ω-3 fatty acids also play a very significant role in the brain, neural and cognitive development. Supplementation of new-born animals with ALA deficient milk formulas resulted in decrease in DHA as well as increase in level of ω-6 fatty acids (adrenic acid; osbond acid) in the retina and brain both.[5,119] The decreased level of DHA in the brain of the animals supplemented with deficient in ALA leads to alteration in the neurotransmitter metabolism along with serotonin and dopamine and membrane linked enzymes.[58] Supplementary studies indicate that deficiency of ω-3 fatty acid decreases the size of neurons in the various organs such as parietal cortex, hypothalamus and reduces the trouble of cortical dendritic arborization.[3,118]

### 5.3.4 ANTITUMOR AND ANTI-CANCER ACTIVITIES

Omega-3 fatty acids have a role to play in antitumor benefits. Being good source ω-3 fatty acids, flaxseed oil can be used in treatment of cancer. Truan et al.[110] established the effect of feeding flaxseed oil to rats for 8 weeks, with 33% reduction in tumor area, 38% reduce in proliferation of tumor cells and thereby causing elevation in the apoptosis (110%). Similarly, Chen et al.[18] also observed 26 and 38% reduction on tumor cells growth by feeding 5 and 10% flaxseed, respectively.

### 5.3.5 PREVENTION OF KIDNEY DISEASE

Chronic kidney disease (CKD) is a general condition, which increases the risk of kidney failure and other kidney related complications.[77,89,98] Anti-inflammatory properties of ω-3 fatty acids exhibit protective action against these diseases. Gopinath et al.[42] studied the effect of ω-3 PUFA on fibrosis and renal inflammation. Observations revealed that increased intake reduces occurrence of CKD. Cicero et al.[21] observed the declined in systolic and diastolic blood pressure with long-term feeding of ω-3 PUFA and thus helpful in prevention of CKD by decreasing hypertension, one of the risk factors for kidney disease. An et al.[4] reported that long-term supplementation of ω-3 fatty acids can reduce the upregulation of pro-oxidant, pro-fibrotic, and pro-inflammatory pathways and helpful in attenuation of tubule–interstitial fibrosis in the remnant kidney.

### 5.3.6 REDUCTION IN CARDIOVASCULAR DISEASES

Cardiovascular diseases affect the functioning of blood vessels and heart.[94] It includes a group of diseases, namely, congenital heart disease, heart failure, stroke, peripheral artery disease, venous thrombosis, etc..[38,94] There is direct relationship between lipid profile of the blood serum and CVDs. Hypocholesterolmic activity of flaxseed has been studied on various animal including mice, rats and rabbits.[8,15,51,68,72,78,81]

Dawczynski et al.[25] had given about 40-g fat, partially replaced by fish oil, rapseed oil containing a daily dose of 3-g omega-3 fatty acids daily to patients and observed that the consumption of dairy products containing n-3 PUFA decreases cardiovascular risk. Similarly, Vijaimohan et al.[116] fed rats with high fat diet containing flaxseed oil (1 mL/kg b.w.) for 60 days and observed the hypocholesterolemia activity through dramatically lowered plasma cholesterol, liver weight, LDL, VLDL, and HDL. Jennifer et al.[62] reported improvement in the vascular relaxation responses by inhibiting ventricular fibrillation incidences due to the supplementation of dietary flaxseed and thus protect from ischemic heart disease (IHD).

### 5.3.7 PREVENTION OF OBESITY

Obesity is the metabolic disorder of the human being and its prevalence is increasing day by day. According to World Health Organization (WHO) report of 2018, 18% children and adolescents, 39% men, and 39% women were obese. Flaxseed oil has laxative effect and therefore helpful in the problem related with obesity by improving metabolic functions. Owing to this advantage, traditionally flaxseed was used for treatment of obesity as well as diabetics.[97,105]

### 5.3.8 ANTI-INFLAMMATORY, ANTIPYRETIC, AND ANALGESIC ACTIVITIES

Flaxseed oil provides several health benefits, that is, anti-inflammatory, antipyretic and analgesic in addition to the above discussed beneficial effects. Wistar and Swiss albino rats when fed with 3-mg/kg body weight exhibited analgesic activity comparable to aspirin.[64]

## 5.4 SOURCES OF ω-3 FATTY ACIDS

### 5.4.1 ANIMAL SOURCES

Marine sources seaweed and algae are rich in EPA and DHA[1] which includes oily fishes (Cod, Salmon, Mackerel, Seabass, Sardines, Trout),[13] shrimps and oysters. The ω-3 fatty acids content of some fish oils is shown in Figure 5.2.

### 5.4.2 PLANT SOURCES

Plant sources contain ω-3 and ω-6 fatty acids in the form of ALA and LA; ALA is obtained from plant origin sources. Some of the plant sources of ALA includes certain oils such as flaxseed oil, wheat germ oil, soybean oil, canola oil, etc., seeds (flaxseed, hemp, and pumpkin) and nuts (walnuts)[24,46] content of the selected vegetable oil, seeds, and nuts is given in Figure 5.3.

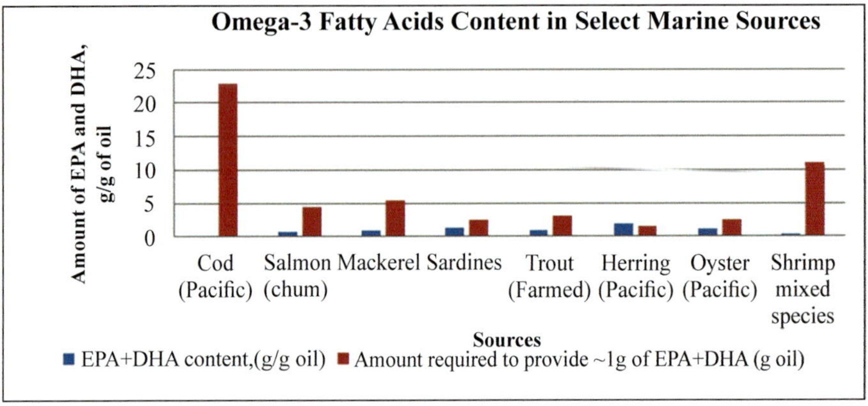


**FIGURE 5.2** Omega-3 fatty acids content in marine sources.
*Source:* Adapted from USDA Ref. 113. Nutrient Data Laboratory.

#### *5.4.2.1 FLAXSEED AND FLAXSEED OIL*

Flaxseed is an oilseed crop with maximum production by Canada, Ethiopia, China, United States and India. Canada ranks first in flaxseed production and contributing to about 80% of total production.[92] India produces 10.2% of total production and ranks third.[104] Depending upon the application, there two varieties of flaxseed, that is, seed flax and fiber flax. Oil flax is grown for oil production, which is used for direct consumption as well as utilized for production of soap, paints, oil cloths, varnishes and linoleum, etc.

Flaxseed is one of the rich plant sources of PUFA. Chemically, flaxseed is composed of fat 35–45%, total dietary fiber 28, protein 20–30, ash 3.4, and moisture 7.7%.[84] Flaxseed oil is extracted from the flaxseeds by cold pressing of dried seeds. Flaxseed oil also called as linseed oil; is used for human consumption because of its several health benefits. Flaxseed oil mostly contains unsaturated fatty acids, such as, ω-3 and ω-6 fatty acids. It contains nearly 16% LA, whereas approximately 57–60% ALA of flaxseed oil.[96]

## 5.5 DIETARY RECOMMENDATION OF Ω-3 FATTY ACIDS

A deficiency of essential ω-3 fatty acids causes various health problems, such as, dermatitis, rough skin and scaly skin. Decrease in the DHA concentration tissue and plasma is also cause by deficiency of ω-3 fatty acids. Also, ω-3 fatty acids prevent CVDs and improve the functioning and development of retina, brain, and testis.[7,23,102] Moreover, ω-6 to ω-3 ratio in Indian urban and Western diets is 38–50:1 and 20:1, respectively,[104,105] which is much than FAO/WHO[32] recommendation of ω-6 to ω-3 ratio (i.e., 5:1). In light of all these points various agencies have given different recommendation level.

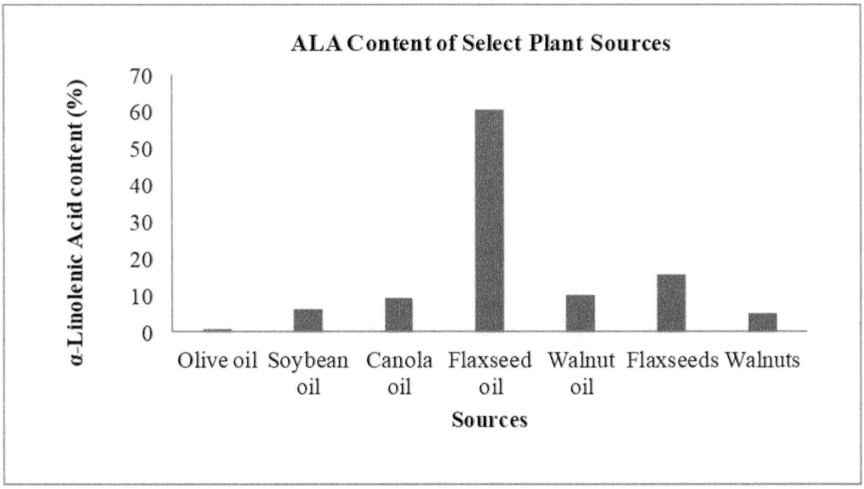


**FIGURE 5.3** ALA content of select vegetable oil, seeds, and nuts.

*Source:* Adapted from USDA Ref. 113. Nutrient Data Laboratory, 2002.

Indian Council of Medical Research has given Recommended Dietary Allowances (RDA) for ALA, 1.6 g/day and for EPA and DHA, above 250 mg/day.[57] Similarly, International Society for the Study of Fatty Acids and Lipids suggested RDA level for ALA, 2.2 g/day and for EPA/DHA, 500 mg/day.[60] The minimum intake levels given by ICMR,[57] is out of total energy, 2.5% from LA 0.5% from ALA for prevention of deficiency symptoms. Some of the countries such as Canada, United Kingdom, Sweden, Japan, and Australia and WHO have decided dietary recommendations for ω-3 fatty acids. This includes ALA about 0.8–1.1 g/day and EPA + DHA about 0.3–0.5 g/day.

The Dietary Reference Intakes (DRI) for energy and macronutrients was given by the Institute of Medicine and Food and Nutrition Board, in association with Health Canada. The Acceptable Macronutrient Distribution Range (AMDR) is given on the basis of energy utilized from ALA, that is,

0.6– 1.2% of energy, based on daily need, that is, 1.3–2.7 g/day and based on calorie, that is, 2000 calories of total diet.[71] This is much more than the current consumption of EPA + DHA.

## 5.6 FOOD FORTIFICATION WITH OMEGA-3 FATTY ACIDS

Food fortification process includes the addition of one or more essential nutrients to improve the functional properties of food. Food fortification plays a major role in dealing with problems related to deficiency diseases. The Codex general principles for the addition of essential nutrients to foods[61] defines "fortification or synonymously enrichment," as "*the addition of one or more essential nutrients to a food whether or not it is normally present in the food, for the purpose of preventing or correcting a demonstrated deficiency of one or more nutrients in the population or specific population groups*".[33]

### *5.6.1 APPROACHES FOR FORTIFICATION WITH ω-3 FATTY ACIDS*

Dietary lipids are prone to oxidation and form unsaturated carbonyls and other compounds during food processing.[31] Fatty acids present in the lipids are converted into lipid alkyl radicals with removal of a hydrogen atom from the structure. The factors responsible for conversion are metal catalysts present in the lipids, heat, and ultraviolet (UV) light exposure.[37] Atmospheric triplet oxygen reacts with lipid alkyl radicals and produces allylic hydroperoxide. This allylic hydroperoxide takes hydrogen atoms from another molecules and forms more reactive lipid peroxyl radicals.[37]

Transition metals such as iron and copper assist to increase the autooxidation rate by lowering the activation energy required for initiation. Additionally, a photosensitizer (type II) and light converted triplet oxygen into singlet oxygen which further reacts with lipids and form peroxyl radicals. The autoxidation process ends with the formation of stable end products from reaction of hydroperoxides and/or free radicals. Another reason for elevation of lipid oxidation is free fatty acids. Antioxidants prolongs the induction period of autooxidation and another mechanism of action is free radicals and/ or hydroperoxide scavenging effect to delay the process of autooxidation.[19] Preparation of emulsion makes the lipids unavailable for oxidant.[76]

Higher unsaturated fatty acids content in the flaxseed oil makes it more prone to the oxidation. Different approaches were found for fortification of food products, that is, emulsion, oil and microcapsules.

### 5.6.1.1 OMEGA-3 FATTY ACIDS RICH EMULSIONS

The oils either plant or marine origin (which are rich source of ω-3 fatty acids, that is, ALA and other unsaturated fatty acids) have higher susceptibility to oxidation. To overcome this problem, various efforts have been made for preparation of emulsion so that the oil is protected from external oxygen supply thereby reducing the extent of oxidation. Kartal et al.[66] prepared multilayer oil in water emulsion with 10% oil load (flaxseed oil) using pectin and Na-caseinate by microfluidization process and observed that multilayer membrane improves emulsion stability at pH values between (3 and 5) and also oxidative stability compared to a monolayer membrane.

Lee and Choo[75] prepared flaxseed oil emulsion with 40% oil load using WPI, gum Arabic, xanthan gum, potassium sorbate, citric acid and tween 80 and observed that surfactants have more emulsifying capacity compared to proteins in flaxseed oil. Lin et al.[80] fortified dairy products with soy lecithin-stabilized emulsions containing 50% algal oil, 6% soy lecithin, and 44% Milli-Q water. This resulted in increased stability of emulsion at gastric pH and improved bioaccessibility of ω-3 fatty acids. Horn et al.[54] compared the oxidative stability of phospholipids and protein-based fish oil emulsion (70% oil load) prepared using microfluidization. They observed that phospholipids function better at acidic pH while protein behaves better at neutral pH.

### 5.6.1.2 OIL

Divya et al.[27] prepared fortified milk using DHA and flaxseed oil and observed the rancid flavor after 3 days of refrigerated storage. Lim et al.[79] developed ice cream added with flaxseed oil and reported a decrease in the flavor score with increasing levels of addition. Thus, fortification of food products with ω-3 fatty acids using direct oil makes the product more prone to oxidation as well as affects the sensory quality of the product.

### 5.6.1.3 MICROCAPSULES

Goyal et al.[44] prepared ω-3 fatty acids fortified milk using flaxseed oil microcapsules and observed a similar sensory quality with the control up to 5 days of storage. Tambade et al.[108] developed spray-dried flaxseed oil microcapsules using soy protein isolate and modified starch as coat material. The average particle size of 37.917 m was found in microcapsules that were spherical in form. The oil recovered from microcapsule contained 61.67% ALA (18:3). In simulated gastrointestinal settings, the flaxseed oil microcapsules produced 38.24% oil, which rose to 60.86% when heated before digestion. The developed microcapsules were utilized for the fortification of milk. Estrada et al.[30] prepared strawberry-flavored yogurt by adding salmon oil microcapsules and observed the increase in the peroxide as well TBA (thiobarbituric acid) value without change in the color characteristics of the product. Thus, emulsion and microcapsules provide better oxidative stability over the direct oil without much affecting the sensory quality of the product.

## 5.7 FOOD AND DAIRY PRODUCTS FORTIFIED OMEGA-3 FATTY ACIDS

### *5.7.1 BAKERY PRODUCTS*

Wheat is one of the staple foods in the Western diet and comes in the third position after rice and corn in terms of production.[37] Bread making process includes additions of various oils during dough making. Replacement of these oils with ω-3 fatty acids enriched oils is difficult as it affects the sensory quality as well as durability of the product during storage.[99] Various bakery products such as biscuits, cookies, and breads have been prepared by replacing oils with ω-3 fatty acids enriched oils but only bread could grasp commercial production.

Gokmen et al.[41] prepared bread with enhanced ω-3 fatty acids content by adding flaxseed oil in the form of microparticle powder (made of flaxseed oil and corn starch rich in amylose nanosized complexes). Fortification with the nano-encapsulated powder of flaxseed oil improved bread quality and storage stability. Yep et al.[125] supplemented ω-3 fatty acids in bread using microencapsulated tuna oil and feed the same for a period of 3 weeks. They observed an increase in ω-3 fatty acid levels in the blood plasma of all the subjects.

Umesha et al.[112] carried out a comparative study for the fortification of biscuits (1-g ALA/100-g biscuits). Microencapsulated garden cress seed oil was used for fortification. Among the two, microencapsulated oil showed better sensory, physical, and oxidative quality during storage. Similarly, Rajiv et al.[95] prepared cookies by using dried and powdered flaxseed powder for part of the wheat flour. The fatty acid profile especially linolenic acid analyzed by gas chromatography was higher for cookies made with flaxseed powder (4.75–5.31%) than control values. Ramcharitar et al.[96] prepared muffins containing 11.6% milled flaxseed but partially acceptable. Further, Moraes et al.[85] found that up to 30% flaxseed four replacement of wheat flour for preparing cake results in acceptable sensory quality with functional benefits of linolenic acids and fiber.

### 5.7.2 INFANT FORMULA

Before conception, lipids from the mother are transferred to the placenta either in the form of triacylglycerol or free PUFAs with the help of membrane fatty acid binding protein.[53] After conception, infants are capable of synthesis and absorption of PUFAs from the mother's milk.[52] ω-3 PUFAs play an important role in visual and brain development in infants.[2,35,59] Human milk has a similar fat percentage that of cow milk but contains 7 times more DHA and is deficient in EPA.[48] So in the case of formula-based-diet fed infants devoid of others milk; ω-3 fatty acids need to be supplemented in the infant formula. PUFA content in the commercial formulae varies from 13% to 26% with fortification using ALA,[63] EPA as well as DHA.[28,106]

### 5.7.3 EGGS

Eggs contain high amount of protein (12.6%) and fat (10.6%). Chicken eggs are deficient in PUFA, that is, it contains about 10–15-mg/g fat, out of that above 95% of PUFA is LA.[14] The ALA content in the eggs is negligible and thus efforts have been made to fortify eggs with ω-3 PUFAs. Farrell[34] developed ω-3 PUFAs enriched eggs by feeding hens with fish oil or a combination of fish oil with vegetable oil. The ω-3 fatty acids level increased in the blood samples of people fed with ω-3 PUFAs enriched eggs. The malonaldehyde of eggs enriched in ω-3 PUFA increased significantly up to 28 days of storage but antioxidant makes it relatively stable during storage. Mridula et al.[87] also fed a flaxseed-rich diet to poultry and observed increased levels of ω-3 fatty acids in the eggs.

### 5.7.4 MEAT

Dietary lipids can be obtained from meat products as it contributes majorly to the adult diet of non-vegetarian people. However, ω-3 fatty acids cannot be synthesized in terrestrial animal and thus they are deficient in ω-3 PUFAs but contains a higher amount of linoleic and arachidonic acid in muscle and adipose tissue. Therefore, the level of ω-3 PUFAs in the meat and its products has been increased by feeding ω-3 PUFAs-enriched diets to the animals.

### 5.7.5 CEREAL BARS AND CHOCOLATES

Breakfast products including cereals and bars have also been fortified with ω-3 PUFAs. Mridula et al.[88] optimized a process for the preparation of energy bars with varying levels of flaxseed (0–20%) and sweeteners (45, 50, and 55%). The product made from 15% flaxseed, 45% sweeteners, and other components has acceptable sensory quality. Other products such as sauces and chocolates are commercially available in the market.

### 5.7.6 MILK

Milk is an ideal carrier for the fortification of essential nutrients and is almost consumed by all age groups peoples. Thus, various efforts were made to fortification of milk with ω-3 fatty acids. Milk fortification is carried out either by direct addition of oil or by feeding animals. Divya et al.[27] observed the occurrence of rancid flavor after 3 days of refrigerated storage when fortified milk with DHA and flaxseed oil.

Algal and fish oil are also good sources of ω-3 fatty acids and thus used for fortification. Baer et al.[6] attempted to fortify milk with fish oil and found a perceptible difference in the flavor of the control and fortified sample during storage. Considering the oxidation stability of algal oil and its impact on sensory quality, Gallaher et al.[36] studied the effect of fortification with algal oil emulsion on the storage stability of milk and no significant difference was observed between the fortified and control sample.

Goyal et al.[45] fortified market milk with ω-3 fatty acids by using flaxseed oil microencapsulated (made with milk protein) powder at 1% level. The fortified milk showed similar sensory qualities such as the control sample up to 5 days in storage. A similar kind of study has been conducted by Tambade et al.[108] for the fortification of milk with microencapsulated flaxseed oil powder prepared by soy protein isolate and modified starch. Pasteurized and sterilized flavored milk were stable for 6 and 28 days, respectively. Attempts have also been made to increase the natural level of ω-3 fatty acids in milk by feeding a diet rich in these fatty acids to the cattle. Nelson and Martini[91] supplied a diet containing fish oil and its calcium salts for 9 weeks and observed the effect on fatty acid content and flavor of the milk. Results showed that elevation in ω-3 fatty acids without affecting the flavor of milk during the experimental period.

### 5.7.7 CURD

Goyal et al.[44] optimized the process for fortification of *dahi* with ω-3 fatty acids using flaxseed oil in the form of the microcapsule. The addition of 2% microcapsule showed comparable sensory attributes to the control curd. A similar study was reported by Veena et al.[114] for the fortification of milk and curd using flaxseed oil and observed slightly lesser flavor in the case of fortified products when compared to the control.

### 5.7.8 CHEESE

Bermudez–Aguirre and Barbosa–Canovas[9] prepared cheddar (C), Queso fresco (QF), and mozzarella (M) cheeses fortified with ω-3 fatty acids by fish oil and ME fish oil and studied the effect of pasteurization, curdling and salting on ω-3 addition and concluded that the highest ω-3 fatty acid retention (8.69-mg/g MFO; 5.08-mg/g FO) was observed in cheddar cheese when added during salting.

### 5.7.9 YOGHURT

Chee et al.[16] manufactured strawberry-flavored yogurt containing algal oil emulsion and observed the perception of fishy flavor in the fortified sample after 15 days of storage. In another study, strawberry yogurt was prepared by Estrada et al.[30] using salmon oil microcapsules (2%). An increase in peroxide and TBA value was noticed without a change in the color characteristics during storage. Kumar et al.[74] developed ω-3 fatty acid-fortified fruit yoghurt using flaxseed flour and oil. Yogurt was prepared by adding 20% fruit and sugar mixture, 1% flaxseed flour, and 2% flaxseed oil optimized based on the sensory attributes of the product. Fortification elevated the ALA content in optimized yogurt, that is, 22.80%.

### 5.7.10 ICE CREAM

Lim et al.[79] manufactured ice cream containing flaxseed oil and reported a decrease in the flavor score with an increasing level of fortification. Furthermore, Ullah et al.[111] supplemented chia oil (olein fraction) into the ice cream and found improvement in the antioxidant properties with increasing levels of ω-3 fatty acids. Similarly, Gowda et al.[43] used microencapsulated flaxseed oil powder to make ω-3 fatty acid-supplemented ice cream, particularly ALA (MFOP). Fortification 4% MFOP provides approximately 45% of the RDA level of ALA in 100 g of fortified ice cream.

### 5.7.11 BUTTER

Kolanowski et al.[70] evaluated the sensory characteristics of butter fortified with 0.1–8% fish oil and found that butter fortified with 3% fish oil shows better sensory characteristics. Ivanov et al.[61] added flaxseed additive (0.8, 1.2, and 1.6%) in butter, and reported a reduction of structure destruction with improvement in convalescing ability of butter structure. Further, functional butter prepared with 4% CLA and 12% phytosterol was organoleptically acceptable.[26]

Bobe et al.[11] fed Holstein cows with flaxseed oil (0.9%) and roasted soybeans (5%) for 3 weeks and found that unsaturated fats were higher in the butter obtained from the milk of the cows with a soft texture. Silva–Kazama et al.[100] studied the changes in the fatty acid profile of butter obtained from cows fed with ground or intact flaxseed (0.02% on a dry matter basis) for 21 days storage period and observed that percentages of saturated as well as medium-chain fatty acids increased with the progression of the storage period, in the butter obtained from fed cows.

Kulkarni[73] prepared butter spread by blending butter with vegetable oil (20, 30, and 40%) and skim milk powder. The 30% addition of oil showed good spread based on sensory acceptability. In another study, Kolanowski et al.[69] used fish oil to make spreadable fats supplemented with ω-3 fatty acids at rates of 3, 4, and 5%, and found that there was no notable change in sensory acceptability of low-calorie spreadable oils loaded up to 1% EPA and DHA.

Flaxseed oil was used by Pandule[93] to make ω-3 fatty acid-enhanced butter. The influence of flaxseed oil and flaxseed oil emulsion into cream on texture, sensory, and color features was studied to enhance the procedure for ω-3 fatty acid-enriched butter. The sensory acceptability of fortified butter was comparable to control butter. Firmness decreased, while spread increased significantly ($p < 0.05$) with increasing levels of flaxseed oil. The ALA content was higher in the butter fortified with flaxseed oil (3.149%) as well as with emulsion (3.174%) than in control (0.86%) as determined by gas chromatography-mass spectrometry. During the 90-day storage period, no yeast, mold, or coliform count was found. In one serving, developed butter contains about 25% of the recommended dietary intake of ALA.

## 5.8 SUMMARY

Omega-3 fatty acids especially ALA is an essential nutrients for the body's growth and development. Various animal, as well as plant sources, contain these nutrients. These natural forms (oils and seeds) are difficult to use directly because of their susceptibility to oxidation as well as objectionable sensory attributes. However, knowing the importance of these nutrients, almost every food sector has worked out different ways for the fortification of food products with ω-3 fatty acids. Thus, ω-3 fatty acids fortified foods are available for all age groups peoples as discussed in this chapter. These ranges of food products play a significant role in fulfilling the recommended daily intake (RDA) level of ω-3 fatty acids, thereby providing numerous health benefits discussed in this chapter.

## KEYWORDS

- **alpha-linolenic acid**
- **dairy products**
- **fortification**
- **health benefits**
- **ω-3 fatty acid**

## REFERENCES

1. Adarme-Vega, T. C.; Thomas-Hall, S. R.; Schenk, P. M. Towards Sustainable Sources for Omega-3 Fatty Acids Production. *Curr. Opin. Biotechnol.* **2014**, *26*, 14–18.
2. Agostoni, C. Role of Long-Chain Polyunsaturated Fatty Acids in the First Year of Life. *J. Pediatric Gastroenterol. Nutr.* **2008**, *47*, S41–S44.

3. Ahmad, A.; Moriguchi, T.; Salem Jr, N. Decrease in Neuron Size in Docosahexaenoic Acid-Deficient Brain. *Pediatric Neurol.* **2002**, *26* (3), 210–218.
4. An, W. S.; Kim, H. J.; Cho, K.H.; Vaziri, N. D. Omega-3 Fatty Acid Supplementation Attenuates Oxidative Stress, Inflammation and Tubulointerstitial Fibrosis in the Remnant Kidney. *Am. J. Physiol.* **2009**, *297* (4), F895– F903.
5. Arbuckle, L. D.; Innis, S. M. Docosahexaenoic Acid in Developing Brain and Retina of Piglets Fed High or Low α-Linolenate Formula with and without Fish Oil. *Lipids* **1992**, *27* (2), 89–93.
6. Baer, R.; Ryali, J.; Schingoethe, D.; Kasperson, K.; Donovan, D.; Hippen, A.; Franklin, S. Composition and Properties of Milk and Butter from Cows Fed Fish Oil. *J. Dairy Sci.* **2001**, *84*, 345–353.
7. Banning, M. The Role of Omega-3-Fatty Acids in the Prevention of Cardiac Events. *Br. J. Nursing*, **2005**, *25*, 503–508.
8. Barakat, L. A. A.; Mehmoud, R. H. The Antiatherogenic, Renal Protective and Immunomodulatory Effects of Purslane, Pumpkin and Flax Seeds on Hypercholesterolemic Rats. *North Am. J. Med. Sci.* **2011**, *3* (9), 351–357.
9. Bermúdez-Aguirre, D.; and Barbosa-Cánovas, G. V. Quality of Selected Cheeses Fortified with Vegetable and Animal Sources of Omega-3. *LWT-Food Sci. Technol.* **2011**, *44* (7), 1577–1584.
10. Bibus, D.; Lands, B. Balancing Proportions of Competing Omega-3 and Omega-6 Highly Unsaturated Fatty Acids (HUFA) in Tissue Lipids. *Prostaglandins, Leukotrienes Essential Fatty Acids* **2015**, *99*, 19–23.
11. Bobe, G.; Zimmerman, S.; Hammond, E. G.; Freeman, A. E.; Porter, P. A.; Luhman, C. M.; Beitz, D. C. Butter Composition and Texture from Cows with Different Milk Fatty Acid Compositions Fed Fish Oil or Roasted Soybeans. *J. Dairy Sci.* **2007**, *90* (6), 2596–2603.
12. Brenna, J. T.; Salem Jr, N.; Sinclair, A. J.; Cunnane, S. C. α-Linolenic Acid Supplementation and Conversion to n-3 Long-Chain Polyunsaturated Fatty Acids in Humans. *Prostaglandins, Leukotrienes and Essential Fatty Acids*, **2009**, *80* (2–3), 85–91.
13. Calder, P. C. Marine Omega-3 Fatty Acids and Inflammatory Processes: Effects, Mechanisms and Clinical Relevance. *Biochimica et Biophysica Acta (BBA)—Mol. Cell Biol. Lipids* **2015**, *1851* (4), 469–484.
14. Cantor, A. H.; Decker, E. A.; Collins, V. P. Fatty Acids in Poultry and Egg Products. In *Fatty Acids in Foods and Their Health Implications*, Chow, C. K., Ed.; Marcel Drekker, Inc, 2000; pp 153–174.
15. Cardozo, L. F. M. D. F.; Chagas, M. A.; Soares, L. L.; Bonaventura, G. T. Exposure to Flaxseed During Lactation Does Not Alter Prostate Area or Epithelium Height but Changes Lipid Profile in Rats. *Nutrición Hospitalaria (Hospital Nutrition)*, **2010**, *25* (2), 250–255.
16. Chee, C. P.; Gallaher, J. J.; Djordjevic, D.; Faraji, H.; McClements, D. J.; Decker, E. A.; Coupland, J. N. Chemical and Sensory Analysis of Strawberry Flavored Yogurt Supplemented with an Algae Oil Emulsion. *J. Dairy Res.* **2005**, *72* (3), 311–316.
17. Chen, B.; Mc Clements, D. J.; Decker, E. A. Design of Foods with Bioactive Lipids for Improved Health. *Annu. Rev. Food Sci. Technol.* **2013**, *4*, 35–56.
18. Chen, J.; Power, K. A.; Mann, J.; Cheng, A.; Thompson, L. U. Flaxseed Alone or In Combination with Tamoxifen Inhibits MCF-7 Breast Tumor Growth in Ovariectomized Athymic Mice with High Circulating Levels of Estrogen. *Exp. Biol. Med.* **2007**, *232*, 1071–1080.

19. Choe, E.; Min, D. B. Chemistry of Deep-Fat Frying Oils. *J. Food Sci.*, **2007**, *72* (5), R77–R86.
20. Christie W. W. Composition and Structure of Milk Lipids. In: *Advances in Dairy Chemistry*; Lipids, Fox, P. F., Ed., Vol. 2; Chapman and Hall: London, 1995; pp 1–36.
21. Cicero, A. F.; Derosa, G.; Di, G.; Bove, M.; Gaddi, A.V.; Borghi, C. Omega-3 Polyunsaturated Fatty Acids Supplementation and Blood Pressure Levels in Hyper Triglyceridemic Patients with Untreated Normal-High Blood Pressure and with or without Metabolic Syndrome: A Retrospective Study. *Clin. Exp. Hypertension* **2010**, *32* (2), 137–144.
22. Cleland, L. G.; James, M. J.; Proudman, S.M. The Role of Fish Oils in the Treatment of Rheumatoid Arthritis. *Drugs* **2003**, *63*, 845–853.
23. Connor, W. E. Importance of n-3 Fatty Acids in Health and Disease. *Am. J. Clin. Nutr.* **2000**, *71* (S), 171–179.
24. Covington, M. B. Omega-3 Fatty Acids. *Am Fam Physician,* **2004**, *70* (1), 133–140.
25. Dawczynski, C.; Martin, L.; Wagner, A.; Jahreis, G. N-3 LC-PUFA-Enriched Dairy Products are Able to Reduce Cardiovascular Risk Factors: A Double-Blind, Cross-Over Study. *Clin. Nutr.* **2010**, *29*, 592–99.
26. Process Development for the Production of Functional Butter: National Dairy Research Institute (NDRI) Annual Report 2012–13; 2013; pp 44–45. http://www.ndri.res.in/ndri/Documents/Annul_Report_2012_2013.pdf (accessed on Aug 13, 2021).
27. Divya, K. B, Kumar, S.; Mittal, V.; Sabikhi, L. Sensory Attributes, Fatty Acid Profile and Oxidative Stability of Low Fat Buffalo Milk Supplemented with Omega 3 Fatty Acids and Dietary Fiber. *Ind. J. Dairy Sci.* **2013**, *66* (6), 469–476.
28. Douaud, C. First DHA/ARA-Fortified Organic Baby Formula Launched. 2006. www.nutraingredients-usa.com/Industry/First-DHA-ARAfortified-organic-babyformula-launched (accessed on Jan 30, 2021).
29. Dyerberg, J.; Eskesen, D. C.; Andersen, P. W.; Astrup, A.; Buemann, B.; Christensen, J. H. Effects of Trans- and n-3 Unsaturated Fatty Acids on Cardiovascular Risk Markers in Healthy Males. An 8 Weeks Dietary Intervention Study. *Eur. J. Clin. Nutr.* **2004**, *58*, 1062–1070.
30. Estrada, J. D.; Boeneke, C.; Bechtel, P.; Sathivel, S. Developing A Strawberry Yogurt Fortified with Marine Fish Oil. *Journal of Dairy Science*, **2011**, *94* (12), 5760–5769.
31. Eunok, C.; David, B. M. Mechanisms and Factors for Edible Oil Oxidation. *Compreh. Rev. Food Sci. Food Saf.* **2006**, *5* (4), 169–186.
32. FAO/WHO. Fats and Fatty Acids in Human Nutrition Rome: FAO Food and Nutrition Paper; 2010; pp 69–91. http://www.fao.org/3/i1953e/i1953e.pdf (accessed on May 12, 2021).
33. FAO/WHO. *Preparation and Use of Food-Based Dietary Guidelines Nutrition Programme*; WHO: Geneva, 1996. http://apps.who.int/iris/bitstream/handle/10665/42051/WHO_TRS_880.pdf;jsessionid=34221E628F570D536E0C6FB212EF24DA?sequence=1 (accessed on April 19, 2021).
34. Farrell, D. J. Enrichment of Hen Eggs with n-3 Long-Chain Fatty Acids and Evaluation of Enriched Eggs in Humans. *Am. J. Clin. Nutr.* **1998**, *68* (3), 538–544.
35. Fewtrell, M. S. Long-Chain Polyunsaturated Fatty Acids in Early Life: Effects on Multiple Health Outcomes. In *Primary Prevention by Nutrition Intervention in Infancy and Childhood*, Vol. 57; Karger Publishers: Basel, Switzerland, 2006; pp 203–221.
36. Gallaher, J. J.; Djordjevic, D.; Faraji, H.; McClements, D. J.; Decker, E. A.; Hollender, R.; Peterson, D. G.; Roberts, R. F.; Coupland, J. N. Chemical and Sensory Analysis of

Strawberry Flavored Yogurt Supplemented with An Algae Oil Emulsion. *J. Dairy Res.* **2005**, *72* (3), 311–316.
37. Ganesan, B.; Brothersen, C.; and McMahon, D. J. Fortification of Foods with Omega-3 Polyunsaturated Fatty Acids. *Crit. Rev. Food Sci. Nutr.* **2014**, *54* (1), 98–114.
38. GBD (Global Burden of Disease). Mortality and Causes of Death Collaborators: Global, Regional and National Age-Sex Specific All-Cause and Cause-Specific Mortality for 240 Causes of Death, 1990–2013: A Systematic Analysis for the GBD Study 2013. *Lancet* **2013**, *385* (9963), 117–171.
39. Ghafoorunissa. Requirements of Dietary Fats to Meet Nutritional Needs & Prevent the Risk of Atherosclerosis-An Indian Perspective. *Ind. J. Med. Res.* **1998**, *108*, 191–202.
40. Gogus, U.; Smith, C. n-3 Omega Fatty Acids: A Review of Current Knowledge. *Int. J. Food Sci. Technol.* **2010**, *45* (3), 417–436.
41. Gökmen, V.; Mogol, B. A.; Lumaga, R. B.; Fogliano, V.; Kaplun, Z.; Shimoni, E. Development of Functional Bread Containing Nanoencapsulated Omega-3 Fatty Acids. *J. Food Eng.* **2011**, *105* (4), 585–591.
42. Gopinath, B.; Harris, D. C.; Flood, V. M.; Burlutsky, G.; Mitchell, P. Consumption of Long-Chain n-3 PUFA, α-Linolenic Acid and Fish is Associated with the Prevalence of Chronic Kidney Disease. *Br. J. Nutr.* **2011**, *105* (9), 1361–1368.
43. Gowda, A.; Sharma, V.; Goyal, A.; Singh, A. K.; Arora, S. Process Optimization and Oxidative Stability of Omega-3 Ice Cream Fortified with Flaxseed Oil Microcapsules. *J. Food Sci. Technol.* **2018**, *55* (5), 1705–1715.
44. Goyal, A.; Sharma, V.; Sihag, M. K.; Singh, A. K.; Arora, S.; Sabikhi, L. Fortification of *Dahi* with Omega-3 Fatty Acids Using Microencapsulated Flaxseed Oil Microcapsules. *J. Food Sci. Technol.* **2016**, *53* (5), 2422–2433.
45. Goyal, A.; Sharma, V.; Sihag, M. K.; Tomar, S. K.; Arora, S.; Sabikhi, L.; Singh, A. K. Development and Physicochemical Characterization of Microencapsulated Flaxseed Oil Powder: A Functional Ingredient for Omega-3 Fortification. *Powder Technol.* **2015**, *286*, 527–537.
46. Gruenwald, J.; Petzold, E.; Busch, R.; Petzold, H. P.; Graubaum, H. J. Effect of Glucosamine Sulfate with or Without Omega-3 Fatty Acids in Patients with Osteoarthritis. *Adv. Ther.* **2009**, *26* (9), 858–862.
47. Gunstone, F. D. (Ed.). Vegetable Sources of Lipids. In *Modifying Lipids for Use in Food*; Woodhead Publishing: Swaston, Cambridge, 2006; pp 11–27.
48. Gunstone, F. D. Fatty Acids-Nomenclature, Structure, Isolation and Structure Determination, Biosynthesis and Chemical Synthesis. In *Fatty Acid and Lipid Chemistry*; Blackie Academic and Professional: Glasgow, 1996; pp 1–35.
49. Han, E.; Yun, Y.; Kim, G.; Lee, Y. H.; Wang, H. J.; Lee, B. W.; Kang, E. S. Effects of Omega-3 Fatty Acid Supplementation on Diabetic Nephropathy Progression in Patients with Diabetes and Hypertriglyceridemia. *PLoS One* **2016**, *11* (5), E-article 0154683.
50. Harris, W. S.; Mozaffarian, D.; Rimm, E.; Kris-Etherton, P.; Rudel, L. L.; Appel, L. J.; Sacks, F. Omega-6 Fatty Acids and Risk for Cardiovascular Disease: A Science Advisory from The American Heart Association Nutrition Subcommittee of The Council on Nutrition, Physical Activity, and Metabolism - Council on Cardiovascular Nursing, and Council on Epidemiology and Prevention. *Circulation* **2009**, *119* (6), 902–907.
51. Hassan, A. A.; Rasmy, N. M.; Foda, M. I.; Bahgaat, W. K. Production of Functional Biscuits for Lowering Blood Lipids. *World J. Dairy Food Sci.* **2012**, *7* (1), 01–20.

52. Herrera, E. Implications of Dietary Fatty Acids During Pregnancy on Placental, Fetal and Postnatal Development: A Review. *Placenta* **2002**, *23* (Suppl. A), S9–S19.
53. Herrera, E.; Amusquivar, E.; Lopez-Soldado, I.; Ortega, H. Maternal Lipid Metabolism and Placental Lipid Transfer. *Hormone Res. Pediatrics* **2006**, *65* (Suppl. 3), 59–64.
54. Horn, A. F.; Nielsen, N. S.; Andersen, U.; Søgaard, L. H.; Horsewell, A.; Jacobsen, C. Oxidative Stability of 70% Fish Oil in Water Emulsions: Impact of Emulsifiers and pH. *Eur. J. Lipid Sci. Technol.* **2011**, *113* (10), 1243–1257.
55. Hulbert, A. J.; Turner, N.; Storlien, L. H.; Else, P. L. Dietary Fats and Membrane Function: Implications for Metabolism and Disease. *Biol.Rev.* **2005**, *80* (1), 155–169.
56. Hwang, D.; Rhee, S. H. Receptor-Mediated Signaling Pathways: Potential Targets of Modulation by Dietary Fatty Acids. *Am. J. Clin. Nutr.* **1999**, *70* (4), 545–556.
57. ICMR. Nutrient Requirements and Recommended Dietary Allowances for Indians. New Delhi: 2009 Report of the Expert Group of the Indian Council of Medical Research (ICMR); 2010. http://icmr.nic.in/final/RDA-2010.pdf (accessed on August 05, 2021).
58. Innis, S. M.; Jacobson, K. Dietary Lipids in Early Development and Intestinal Inflammatory Disease. *Nutr. Rev.* **2007**, *65*, S188–S193.
59. Innis, S. M.; Sprecher, H.; Hachey, D.; Edmond, J.; Anderson, R. E. Neonatal Polyunsaturated Fatty Acid Metabolism. *Lipids* **1999**, *34* (2), 139–149.
60. ISSFAL. Recommendations for Intake of Polyunsaturated Fatty Acids in Healthy Adults. *Int. Soc. Study Fatty Acids Lipids (ISSFAL) News* **2004**, *11*, 12–18.
61. Ivanov, S.; Rashevskaya, T.; Makhonina, M. Flaxseed Additive Application in Dairy Products Production. *Procedia Food Sci.* **2011**, *1*, 275–280.
62. Jennifer, L. A.; Susan, J. W.; Bernhard, H. J.; Lilian, U. T.; Jane, A. Health Effects with Consumption of the Flax Lignan. *Br. J. Nutr.* **2010**, *103* (7), 929–939.
63. Jensen, R. G.; Ferris, A. M.; and Lammi-Keefe, C. J. Lipids in Human Milk and Infant Formulas. *Annu. Rev. Nutr.* **1992**, *12* (1), 417–441.
64. Kaithwas, G.; Mukherjee, A.; Chaurasia, A. K.; Majumdar, D. K. Anti-inflammatory, Analgesic and Antipyretic Activities of *Linum usitatissimum L.* (Flaxseed) Oil. *Ind. J. Exp. Biol.* **2011**, *49*, 932–938.
65. Kapoor, S.; Sachdeva, R.; Kochhar, A. Efficacy of Flaxseed Supplementation on Nutrient Intake and Other Lifestyle Pattern in Menopausal Diabetic Females. *Ethnomedicine* **2011**, *5* (3), 153–160.
66. Kartal, C.; Unal, M. K.; Otles, S. Flaxseed Oil-In-Water Emulsions Stabilized by Multilayer Membranes: Oxidative Stability and the Effects of pH. *J. Dispersion Sci. Technol.* **2016**, *37* (12), 1683–1691.
67. Kelley, D. S.; Vemuri, M.; Adkins, Y.; Gill, S. H. S.; Fedor, D.; Mackey, B. E. Flaxseed Oil Prevents *Trans*-10, *Cis*-12-Conjugated Linoleic Acid-Induced Insulin Resistance in Mice. *Br. J. Nutr.* **2009**, *101*, 701–708.
68. Khalesi, S.; Jamaluddina, R.; Ismailab, A. Effect of Raw and Heated Flaxseed (*Linum usitatissimum L.)* on Blood Lipid Profiles in Rats. *Int. J. Appl. Sci. Technol.* **2011**, *1* (4), 84–89.
69. Kolanowski, F. S.; Ewa, L.; Stanislaw Berger, W. Enrichment of Spreadable Fats with Polyunsaturated Fatty Acids Omega-3 Using Fish Oil. *Int. J. Food Sci. Nutr.* **2001**, *52* (6), 469–476.
70. Kolanowski, W.; Jaworska, D.; Wei Xbrodt, J. Importance of Instrumental and Sensory Analysis in The Assessment of Oxidative Deterioration of Omega-3 LC PUFA-Rich Foods. *J. Sci. Food Agric.* **2007**, *87*, 181–191.

71. Kris-Etherton, P. M.; Harris, W. S.; Appel, L. J. Fish Consumption, Fish Oil, Omega-3 Fatty Acids and Cardiovascular Disease. *Circulation* **2002**, *106* (21), 2747–2757.
72. Kristensen, M.; Jensen, M. G.; Aarestrup, J.; Petersen, K. E.; Søndergaard, L.; Mikkelsen, M. S.; Astrup, A. Flaxseed Dietary Fibers Lower Cholesterol and Increase Fecal Fat Excretion, but Magnitude of Effect Depends on Food Type. *Nutr. Metabol.* **2012**, *9* (1), 8–12.
73. Kulkarni, S. *Studies on Rheology of Butter as Influenced by Processing Conditions*; Doctoral Dissertation Thesis. ICAR-NDRI, Karnal, India, 1983; pp 218.
74. Kumar, S. S.; Balasubramanyam, B. V.; Rao, K. J.; Dhas, P. H. A.; Nath, B. S. Effect of Flaxseed Oil and Flour on Sensory, Physicochemical and Fatty Acid Profile of the Fruit Yoghurt. *J. Food Sci. Technol.* **2017**, *54* (2), 368–378.
75. Lee, P. E.; Choo, W. S. Characterization of Flaxseed Oil Emulsions. *J. Food Sci. Technol.* **2015**, *52* (7), 4378–4386.
76. Let, M. B.; Jacobsen, C.; Meyer, A. S. Lipid Oxidation in Milk, Yoghurt, and Salad Dressing Enriched with Neat fish Oil or Pre-Emulsified fish Oil. *J. Agric. Food Chem.* **2007**, *55* (19), 7802–7809.
77. Levey, A. S.; Eckardt, K.U.; Tsukamoto, Y. Definition and Classification of Chronic Kidney Disease: A Position Statement from Kidney Disease: Improving Global Outcomes (KDIGO). *Kidney Int.* **2005**, *67* (6), 2089–2100.
78. Leyvaa, D. R.; Zahradkab, P.; Ramjiawanc, B.; Guzmanc, R.; Alianid, M.; Piercea, G. N. The Effect of Dietary Flaxseed on Improving Symptoms of Cardiovascular Disease in Patients with Peripheral Artery Disease: Rationale and Design of the FLAX-PAD Randomized Controlled Trial. *Contemp. Clin. Trials* **2011**, *32* (5), 724–730.
79. Lim, C. W.; Norziah, M. H.; Lu, H. F. S. Effect of Flaxseed Oil Towards Physicochemical and Sensory Characteristics of Reduced Fat Ice-Cream and Its Stability in Ice Cream Upon Storage. *Int. Food Res. J.* **2010**, *17*, 393–403.
80. Lin, X.; Wang, Q.; Li, W.; Wright, A. J. Emulsification of Algal Oil with Soy Lecithin Improved DHA Bio Accessibility but Did Not Change Overall In Vitro Digestibility. *Food Funct.* **2014**, *5* (11), 2913–2921.
81. Mani, I.; Biswas, M.; Kumar, S. N. An Open-Label Study on The Effect of Flax Seed Oil Supplementation in The Management of Diabetes Mellitus. *J. Diet. Suppl.* **2011**, *8* (6), 77–95.
82. Mantzioris, E.; Cleland, L. G.; Gibson, R. A.; Neumann, M. A.; Demasi, M.; James, M. J. Biochemical Effects of a Diet Containing Foods Enriched With n–3 Fatty Acids. *Am. J. Clin. Nutr.* **2000**, *72* (1), 42–48.
83. Markiewicz-Kęszycka, M.; Czyżak-Runowska, G.; Lipińska, P.; Wójtowski, J. Fatty Acid Profile of Milk-A Review. *Bull. Vet. Inst. Pulawy* **2013**, *57* (2), 135–139.
84. Martinchik, A. N.; Baturin, A. K.; Zubtsov, V. V.; Molofeev, V. Nutritional Value and Functional Properties of Flaxseed. *Voprosy Pitaniia*, **2012**, *81* (3), 4–10.
85. Moraes, É. A.; Dantas, M. I. de, S.; Morais, D.; de, C.; Silva, C. O.; da Castro, F. A. F. de, Martino, H. S. D.; Ribeiro, S. M. R. Sensory Evaluation and Nutritional Value of Cakes Prepared with Whole Flaxseed Flour. *Ciência e Tecnologia de Alimentos* (Food Science & Technology) **2010**, *30* (4), 974–979.
86. Mori, T. A.; Burke V.; Puddey, I. B.; Watts, G. F.; O'Neal, D. N. Purified Eicosapentaenoic and Docosahexaenoic Acids Have Differential Effects on Serum Lipids and Lipoproteins LDL Particle Size Glucose, and Insulin in Mildly Hyperlipidemic Men. *Am. J. Clin. Nutr.* **2000**, *71*, 1085–1094.

87. Mridula, D.; Kaur, D.; Nagra, S.; Barnwal, P.; Gurumayum S.; Singh, K. Effect of Dietary Flaxseed Supplementation on Egg Production and Quality in Laying Hens. *Ind. J. Poultry Sci.* **2012**, *47*, 40–47.
88. Mridula, D.; Singh, K. K.; Barnwal, P. Development of Omega-3 Rich Energy Bar with Flaxseed. *J. Food Sci. Technol.* **2013**, *50* (5), 950–957.
89. National Kidney Foundation (NKF). K/DOQI Clinical Practice Guidelines for Chronic Kidney Disease Evaluation, Classification and Stratification. *Am. J. Kidney Dis.* **2002**, *39* (2), S212-S266.
90. Nazni, P.; Amrithaveni, M.; Poongodi, V. T. Impact of Flaxseed Based Therapeutic Food on Selected Type II Diabetic Patients. *Indian Journal of Nutrition and Dietetics*, **2006**, *43*, 141–145.
91. Nelson, K. A. S.; Martini, S. Increasing Omega Fatty Acid Content in Cow's Milk Through Diet Manipulation: Effect on Milk Flavor. *J. Dairy Sci.* **2009**, *92* (4), 1378–1386.
92. Oomah, B. D.; Mazza, G. Flaxseed products for disease prevention. In: *Functional Foods: Biochemical and Processing Aspects*; Mazza, G., Ed.; Technomic Publishing Co., Inc.: Lancaster, 1998; pp 91–138.
93. Pandule, V. S.; Sharma, M.; Devaraja, H. C.; Surendranath, B. Omega-3 Fatty Acid-Fortified Butter: Preparation and Characterisation of Textural, Sensory, Thermal and Physicochemical Properties. *Int. J. Dairy Technol.* **2020,** *74* (1), 181- 191.
94. Pekka, P.; Bo, N. *World Health Organization: Global Atlas on Cardiovascular Disease Prevention and Control (PDF)*; World Health Organization (WHO) in collaboration with the World Heart Federation and the World Stroke Organization: Rome, 2011; pp 3–18.
95. Rajiv, J.; Indrani, D.; Prabhasankar, P.; Rao, G. V. Rheology, Fatty Acid Profile and Storage Characteristics of Cookies as Influenced by Flax Seed (*Linum usitatissimum*). *J. Food Sci. Technol.* **2012**, *49* (5), 587–593.
96. Ramcharitar, A.; Badrie, N.; Mattfeldt Beman, M.; Matsuo, H.; Ridley, C. Consumer Acceptability of Muffins with Flaxseed (*Linum usitatissimum*). *J. Food Sci.* **2005**, *70* (7), s504-s507.
97. Santos, A. P.; Rogero, M. M.; Bastos, D. H. Edible Plants, Their Secondary Metabolites and Antiobesogenic Potential. *Recent Patents Food, Nutr. Agric.* **2010**, *2*, 195–212.
98. Sarnak, M. J.; Levey, A. S.; Schoolwerth, A. C.; Coresh, J.; Culleton, B.; Hamm, L. L.; Parfrey, P. Kidney Disease as A Risk Factor for Development of Cardiovascular Disease: A Statement from The American Heart Association Councils on Kidney in Cardiovascular Disease, High Blood Pressure Research, Clinical Cardiology, and Epidemiology and Prevention. *Circulation*, **2003**, *108* (17), 2154–2169.
99. Serna-Saldivar, S. O.; Zorrilla, R.; De La Parra, C.; Stagnitti, G.; Abril, R. Effect of DHA Containing Oils and Powders on Baking Performance and Quality of White Pan Bread. *Plant Foods Human Nutr.* **2006**, *61* (3), 121–129.
100. Silva-Kazama, D. C. D.; Santos, G. T. D.; Pintro, P. T. M.; Visentainer, J. V.; Kazama, R.; Petit, H. V.; Marchi, F. E. D. Effect of Storage on Fatty Acid Profile of Butter from Cows Fed Whole or Ground Flaxseed with or without Monensin. *Revista Brasileira de Zootecnia* (Brazilian J. of Zootechnology), **2010**, *39* (10), 2297–2303.
101. Simopoulos, A. P. Genetic Variants in the Metabolism of Omega-6 and Omega-3 Fatty Acids: Their Role in The Determination of Nutritional Requirements and Chronic Disease Risk. *Exp. Biol. Med.* **2010**, *235* (7), 785–795.
102. Simopoulos, A. P.; Leaf, A.; Salem, N. Essentiality of and Recommended Dietary Intakes for Omega-6 and Omega-3 Fatty Acids. *Ann. Nutr. Metabol.* **1999**, *43*, 127– 131.

103. Sinclair, A. J.; Attar-Bashi, N. M.; Li, D. What Is the Role of Alpha linolenic Acid for Mammals? *Lipids* **2002**, *37* (12), 1113–1123.
104. Singh, K. K.; Mridula, D.; Rehal, J.; Barnwal, P. Flaxseed: A Potential Source of Food, Feed and Fiber. *Crit. Rev. Food Sci. Nutr.* **2011**, *51* (3), 210–222.
105. Singh, R. B.; Moshiri, M.; De Meester, F.; Juneja, L.; Muthusamy, V.; Manoharan, S. The Evolution of Low Ω-6/ Ω-3 Ratio Dietary Pattern and Risk of Cardiovascular Diseases and Diabetes. *J. Altern. Med. Res.* **2011**, *3*, 45–70.
106. Starling, S. Omega-3 in Infant Formula World first. 2008. www.nutraingredients.com/Industry/Omega-3-in-infant-formulaworld-first (accessed on July 12, 2021).
107. Stillwell, W.; Wassall, S. R. Docosahexaenoic Acid: Membrane Properties of a Unique Fatty Acid. *Chem. Phys. Lipids* **2003**, *126* (1), 1–27.
108. Tambade, P. B.; Sharma, M.; Singh, A. K.; Surendranath, B. Flaxseed Oil Microcapsules Prepared Using Soy Protein Isolate and Modified Starch: Process Optimization, Characterization and In Vitro Release Behavior. *Agric. Res.* **2020**, *2020*, 1–11.
109. Trautwein, E. A. n-3 Fatty Acids-Physiological and Technical Aspects for Their Use in Food. *Eur. J. Lipid Sci. Technol.* **2001**, *103*, 45–51.
110. Truan, J. S.; Chen, J. M.; Thompson, L. U. Flaxseed Oil Reduces the Growth of Human Breast Tumors (MCF7) at High Levels of Circulating Estrogen. *Mol. Nutr. Food Res.* **2010**, *54* (10), 1414–1421.
111. Ullah, R.; Nadeem, M.; Imran, M. Omega-3 Fatty Acids and Oxidative Stability of Ice Cream Supplemented with Olein Fraction of Chia (*Salvia Hispanica* L.) Oil. *Lipids Health Dis.* **2017**, *16* (1), 34–40.
112. Umesha, S. S.; Manohar, R. S.; Indiramma, A. R.; Akshitha, S.; Naidu, K. A. Enrichment of Biscuits with Microencapsulated Omega-3 Fatty Acid (Alpha-Linolenic Acid) Rich Garden Cress (*Lepidium Sativum*) Seed Oil: Physical, Sensory and Storage Quality Characteristics of Biscuits. *LWT-Food Sci. Technol.* **2015**, *62* (1), 654–661.
113. USDA Nutrient Data Laboratory. 2002. http://www.nalusda.gov/fnic/foodcomp./(accessed on August 03, 2021).
114. Veena, N.; Nath, B. S.; Srinivas, B.; Balasubramanyam, B. V. Quality Attributes of Dahi Prepared from Milk Fortified with Omega-3 Fatty Acids, Phytosterols and Polydetxrose. *J. Food Sci. Technol.* **2017**, *54* (7), 1765–1775.
115. Venegas-Calerón, M.; Sayanova, O.; Napier, J. A. An Alternative to Fish Oils: Metabolic Engineering of Oil-Seed Crops to Produce Omega-3 Long Chain Polyunsaturated Fatty Acids. *Progress Lipid Res.* **2010**, *49* (2), 108–119.
116. Vijaimohan, K.; Jainu, M.; Sabitha, K. E.; Subramaniyam, S.; Anandhan, C.; Devi, C. S. Beneficial Effects of Alpha Linolenic Acid Rich Flaxseed Oil on Growth Performance and Hepatic Cholesterol Metabolism in High Fat Diet Fed Rats. *Life Sci.* **2006**, *79* (5), 448–454.
117. Vuksan, V.; Whitham, D.; Sievenpiper, J. L.; Jenkins, A. L.; Rogovik, A. L.; Bazinet, R. P.; Hanna, A. Supplementation of Conventional Therapy with The Novel Grain Salba (*Salvia Hispanica* L.) Improves Major and Emerging Cardiovascular Risk Factors in Type 2 Diabetes: Results of A Randomized Controlled Trial. *Diab. Care* **2007**, *30* (11), 2804–2810.
118. Wainwright, P.E.; Bulman-Fleming, M.B.; Levesque, S.; Mustsaers, L.; Mutcheon, D. A Saturated Fat Diet During Development Alters Dendritic Growth in Mouse Brain. *Nutr. Neurosci.* **1998**, *1*, 49–58.
119. Ward, G. R.; Huang, Y. S.; Bobik, E.; Xing, H. C.; Mutsaers, L.; Auestad, N.; Wainwright, P. Long-Chain Polyunsaturated Fatty Acid Levels in Formulae Influence Deposition of

Docosahexaenoic Acid and Arachidonic Acid in Brain and Red Blood Cells of Artificially Reared Neonatal Rats. *J. Nutr.* **1998**, *128* (12), 2473–2487.

120. Wikipedia. Chemical Structure of Alpha Linolenic Acid. 2020. https://en.wikipedia.org/wiki/AlphaLinolenic_acid#/media/File:ALAnumbering.svg (accessed on May 30, 2021).
121. Wikipedia. Chemical Structure of Arachidonic Acid. 2020. https://en.wikipedia.org/wiki/Arachidonic_acid#/media/File:AAnumbering.png (accessed on May 30, 2021).
122. Wikipedia. Chemical Structure of Docosahexaenoic Acid. 2020. https://en.wikipedia.org/wiki/Docosahexaenoic_acid#/media/File:DHA_numbers.svg (accessed on May 30, 2021).
123. Wikipedia. Chemical Structure of Eicosapentaenoic Acid. 2020. https://en.wikipedia.org/wiki/Eicosapentaenoic_acid#/media/File:EPAnumbering.png (accessed on May 30, 2021).
124. Wikipedia. *Chemical Structure of Linoleic Acid.* 2020. https://en.wikipedia.org/wiki/Linoleic_acid#/media/File:LAnumbering.png (accessed on May 30, 2021).
125. Yep, Y. L.; Li, D.; Mann, N. J.; Bode, O.; Sinclair, A. J. Bread Enriched with Microencapsulated Tuna Oil Increases Plasma Docosahexaenoic Acid and Total Omega-3 Fatty Acids in Humans. *Asia Pac. J. Clin. Nutr.* **2002**, *11* (4), 285–291.

**CHAPTER 6**

# ASSESSMENT OF OMEGA-3 FATTY ACIDS AS A FUNCTIONAL COMPONENT IN FOOD PRODUCTS

PINAR GUMUS


## ABSTRACT

The demands of most consumers for secure and natural food products are increasing day by day. There is expanding trend for functional foodstuffs fortified by ω-3 fatty acids (FAs). This novel approach meets essential nutritionary needs with healthy eating habits. Since the human body cannot produce ω-3 FAs, they have been accepted as essential FAs. The ω-3 PUFAs, especially DHA and EPA can maintain a positive effect on health. Regular consumption of adequate quantities of ω-3 FAs with many health benefits (such as retarding autoimmune and inflammatory ailments, cancers, physiological, cognitive, and cardiovascular ailments) may provide an overall increase in the health and well-being of the people. This chapter presents an overview of ω-3 FAs, ω-3 fortification of foodstuffs, and also about the most recent research advances on the incorporation of ω-3 FAs into the food system. Future studies to discover new sources of ω-3 FAs and to raise the stability of ω-3 FAs, and also the usage of these ω-3 sources for food systems can be an important way for food fortification, human nutrition, and preventing diseases.


## 6.1 INTRODUCTION

There has been a global trend toward ω-3 fatty acids (FAs) in the food and agriculture industries due to consumer interest in recent years. Innovative

Plant-Based Bioactive Compounds and Food Ingredients: Encapsulation, Functional, and Safety Aspects.
Junaid Ahmad Malik, Megh R. Goyal, Preeti Birwal, & Ritesh B. Watharkar (Eds.)

foods are enriched by adding functional ingredients, such as vitamins, prebiotics, probiotics, minerals, essential FAs, carotenoids, polyphenols, anthocyanins, and also PUFAs.[12,35,39,48] FAs are categorized as saturated and unsaturated. ω-3 FAs, which belong to the PUFAs family, are very important in human nutrition due to their antimicrobial and antioxidant properties.[10,15,23,32]

There is an increasing demand for healthy and nutraceuticals fortified by minor compounds in various sectors of the food industry due to their diverse health advantages and ascending consumer awareness. The main aim is the consumption of functional foods containing antioxidant plant extracts, polyunsaturated FAs, probiotics, vitamins, and minerals. In the consumption of fortified foods, it is important to obtain stable products and to protect the functional attributes from production through consumption.[12,19,22,34] Consumption of polyunsaturated FAs that are basic nutrients for normal growth and development has been promoted since they can play a vital role in the inhibition and healing of various diseases, such as coronary artery ailment, hypertension, diabetes mellitus, arthritis, autoimmune ailments, and some types of cancer.[17]

This chapter explores an overview of ω-3 FAs and ω-3 fortification of foodstuffs and also of the most recent research works done on ω-3 FAs that incorporate into the food system.

## 6.2 WHAT ARE THE ω-3 FATTY ACIDS?

FAs are categorized into two main categories: saturated FAs and unsaturated FAs. Subcategories of unsaturated FAs are monounsaturated FAs and polyunsaturated FAs. Classification of FAs is shown in Figure 6.1. Suggested intakes of ω-3 FAs can cause an increase in the well-being of the population, can enable crucial health advantages, and also can prevent autoimmune and cancers, physiological, inflammatory, cognitive, and cardiovascular ailments[21,24,29,38,46]. ω-3 PUFAs are rich in double bonds. This property makes them greatly vulnerable to lipid oxidation. The high sensitivity to oxidative degradation is one of the main matters related to ω-3 FAs. Inhibition of lipid oxidation is one of the primary technological drawbacks for the successful improvement of ω-3-enriched foods.[19,24,42,44,47]

ω-3 supplementation attracted great interest in recent years. The ω-3 FAs acquired from many sources should be included in the daily diet to prevent many diseases. There are three kinds of ω-3 FAs; ALA (18:3), EPA (20:5), and DHA (22:6). ALA is derived from plant sources of ω-3, whereas EPA and DHA are accepted as marine ω-3 s. Some natural sources of ω-3 FAs are illustrated in Figure 6.2. Marine sources of ω-3 FAs are marine fish oil, marine mammals, krill, squid, and algae; plant sources are walnuts, hazelnuts, flax seeds, chia seeds, kiwifruit, soybean and hemp, rapeseed, and canola seed oils. These seeds and oils are the most largely used source of vegetarian ω-3 PUFAs in food enrichment.[3,6,27,42,46]

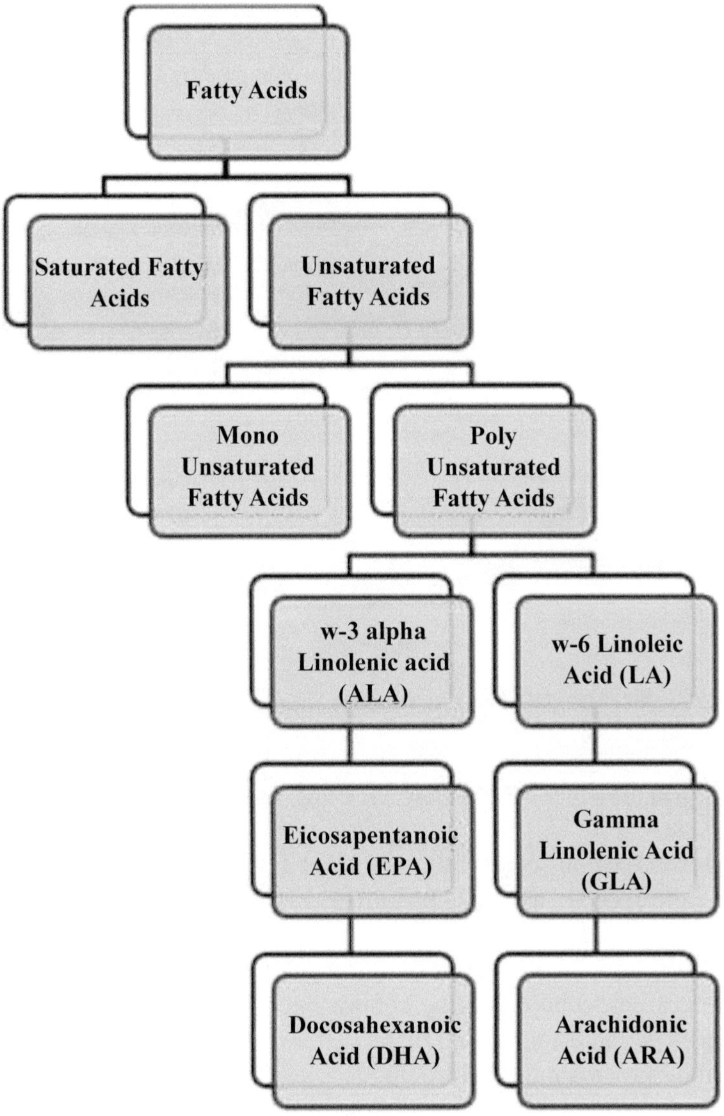


**FIGURE 6.1** Classification of fatty acids.

**FIGURE 6.2** Some natural sources of ω-3 FAs.

## 6.3 DEVELOPMENT OF FUNCTIONAL FOODS WITH ω-3 FAs

Consumer interest in secure and natural food products has increased day by day. The role of ω-3 in dietary supplements has gained importance. Functional foods can be considered to be those whole, fortified, enriched, or enhanced foods that improve the general condition and health. They can be produced by adding a particular bioactive component. Enriched foods are so-called because of the addition of one or more functional nutrients that were lost during food processing. Fortified foods are so-called because they have the addition of one or more functional nutrients that do not naturally oocur in the food. Enhanced foods are accepted as the addition of one or more functional nutrients into food either by modification or by indirect methods.[3,14,20,24,27,31]

Over recent years, there has been a significant concentration on novel approaches in the processing of food products. Functional foodstuffs fortified with LC ω-3 FAs have become more popular. The incorporation of different bioactive compounds for obtaining functional foods is one of the present approaches. The fortification of food products using long-chain ω-3 PUFAs is accepted as a creative way to ensure health benefits, to increase the nutritional value, to provide healthy eating and diets. Many functional components found in food products like vegetable oils have the potential to positively affect health benefits, body function, and nutrition values.[7,9,16,24,37,42]

Emulsions are a class of dispersed systems consisting of two immiscible liquids, which is termed the continuous phase. The process that transforms two separate immiscible liquids into an emulsion is known as emulsification. Because ω-3 oils do not mix with water, they must be emulsified in order to fortify beverages with ω-3 oils. The most commonly used FAs are the ω-3 FAs, ALA, DHA, and EPA. The emulsification of ω-3 FA prevents its oxidation and extends its shelf life, thus increasing its acceptability to consumers. Beverage is one of the most suitable food products for ω-3 enrichment. Today, consumers are looking for beverages enriched with ω-3 FAs that tastes better and meet their demands for natural, healthy, and nutritious components. Therefore, fruit juices can be the ideal medium for functional beverages because of their essential nutritional content. The fortification of beverages can be accomplished by adding the ingredients directly to the product or by adding the ingredients in encapsulated form. While incorporating these supplements into beverage formulations, the solubility, pH, temperature, light and general stability, color, and flavor of the resulting beverage are the important parameters considered during enriched beverage production.[3,27,40,44]

ω-3 PUFAs like fish oils are highly vulnerable to oxidation and degradation, which reduces their shelf life.[42] This is an undesirable operation in the food industry as it causes food spoilage. The main goal for the successful improvement of ω-3 enriched foods is the inhibition of lipid oxidation. Oxidation of ω-3 FAs can be detected by peroxide value and p-anisidine value, headspace propanil determination, and non-isothermal differential scanning calorimetry. Adding antioxidants is one of the most useful ways to preserve oil systems from oxidation. However, concerns are growing about the possible side effects of synthetic antioxidants. Therefore, there is an ascending demand for the use of naturally sourced antioxidants for better health.[8,25,50]

## 6.4 RECENT STUDIES ON FUNCTIONAL FOODS ENRICHED WITH ω-3 FATTY ACIDS

There are many examples of simple lipid emulsions of PUFAs in the scientific literature. The greater part of works has utilized fish oil as a source of ω-3 PUFAs when obtaining simple lipid emulsions. However, in recent years, there has been attempts to reveal the different sources of fish oil to integrate ω-3 FA into products. Eventually, more researches are emerging on simple PUFAs using other sources of ω-3 s.

Diguet et al.[13] studied that enriched breakfast cereals prepared by extrusion and the flavor is determined by a trained panel. The sensory features of products supplemented with microencapsulated ω-3 were determined to be better than nonencapsulated oil or non-cross-linked gelatin matrices.

Borneo et al.[6] designed a filling for sandwich cookies including 400 mg of eicosapentaenoic acid and docosahexaenoic acid encapsulated in a matrix of starch and gelatin. Cookies were stored at 18°C and 35°C and under two different packaging circumstances for 28 d. There were no important losses of EPA and DHA throughout storage under the study circumstances. The results showed that it is the potential to obtain shelf-stable enriched foods by high levels of ω-3 FA.

Goyal et al.[21] characterized flaxseed oil powder that could serve as a potential delivery system of ω-3 FAs in the vegan diet. The peroxide value showed high oxidative stability of the microencapsulated oil after 6 months of storage at $35 \pm 1$°C. It was revealed that flaxseed oil can be stabilized in higher density and applied as a supplement in milk to meet the nutritionary needs of ω-3 FAs in a vegan diet.

O/W nanoemulsion systems were formed by ultrasound with flaxseed and algae oil loads of up to 50%. Nanoemulsions with mean droplet size measuring 192 (flaxseed) and 182 nm (algae) using mixtures of the emulsifiers were made. This method could be used to form vegetarian LC-ω-3PUFA nanoemulsions proper for incorporation into fortified nutraceuticals with the possibility to enhance LC-ω-3 PUFA intake and bioavailability.[36]

Chen et al.[8] examined the effect of natural hydrophilic antioxidants, metal chelators, and their mixtures on the oxidative stability of algae oil and W/O (algae) emulsion. The results claimed that green tea extract and ascorbic acid had the superior defensive impact against algal oil oxidation and formed a 4-d lag phase, whereas rosmarinic acid, grape seed extract, deferoxamine, grape seed extract polymer and EDTA did not have important protective effects.

Komaiko et al.[33] used four types of sunflower phospholipids in fish oil nanoemulsion production by microfluidization. It was indicated that sunflower phospholipids are suitable and natural options for unnatural surfactants in nanoemulsion formation.

Karadag et al.[29] investigated the capability of Icelandic brown seaweed *Fucus vesiculosus* extracts to prevent lipid oxidation in granola bars enriched by fish oil emulsion in water, and whether the addition of seaweed extracts affects the physical microstructure of oil droplets in granola bars.

Comunian et al.,[11] on the other hand, encapsulated fish oil with a microfluidic device using different types of proteins and concentration combinations in O/W emulsions to preserve fish oil against oxidation. Fish oil productions in gelatin, casein, and hydrolyzed whey protein were measured by optical and cryotransmission electron microscopy. The particle size, zeta potential, structural stability under alternative stress circumstances, interfacial tension, and oxidative stability were also measured. The formulations were found to be stable under alternative stress circumstances; the mixture of 0.5% gelatin and casein (1:1) showed the best preservation against oxidation for fish oil.

In the study of Sivapratha and Sarkar,[45] plant-derived flaxseed oil rich in ω-3 FAs was distributed in the primary, secondary, and tertiary emulsion systems. The results indicated that secondary emulsions can resist changes in ambient stresses of salt and heat, and can also preserve the oil phase from decomposition better than other emulsions.

Alcântara et al.[2] evaluated the effect of homogenization method, ultrasound, or rotor–stator on the features of emulsions composed of chia oil and the stability of this powder oil. The utilization of ultrasound resulted in an important decrease in droplet size (4.62 vs 0.48 μm) and a significant raise in encapsulation efficiency (42 vs 96%). In this method, it has been shown that it is possible to obtain a chia oil microparticle with an industrially important oil concentration (30%) with high encapsulation efficiency and better oxidative stability by using green ultrasonic technology for the oil homogenization step.

In this work, O/W emulsions were prepared via a microplasticizer using 1% Tween 80, Quillaja Saponin, gum arabic surfactants, and 5% ω-3 FAs. pH was determined by the impact of temperature and ionic strength. It was detected that apricot nectar, which is a functional food enriched with ω-3 FAs, has good physical and oxidative stability, reduces % FFA release, and enhances β-carotene biotransmission. It has been determined that this product is at an acceptable level by consumers. It was predicted that by adding emulsions rich in ω-3 FAs to apricot nectar, oxidative deterioration can be prevented for a long time, thus, these functional products can be developed.[1]

Beikzadeh et al.[5] studied that flaxseed oil, as a possible source of polyunsaturated FAs, is vulnerable to oxidation. This study has examined the quality characteristics of bread including encapsulated and free flaxseed oils. Results showed that the addition of microencapsulated flaxseed oil into breads aids to protect the sensory attributes of the control sample when compared with the breads enriched with free flaxseed oil.

Gumus and Gharibzahedi[9] stated in their study that the most important sources of ω-3 FAs to strengthen yogurts are fish, flaxseed, and algae oils. It was claimed that yoghurt supplementation with microcapsules and emulsion systems that trap ω-3 FAs compared with free oils leads to better rheological, water-holding capacity, and textural properties with a lower rate of oxidation and syneresis.

Sánchez et al.[44] made the emulsions with krill oil and gum arabic in the ratio 1:4 and spray-dried. The impact of drying temperatures on antioxidant activity, astaxanthin retention, and microencapsulation efficiency was determined.

Kairam et al.,[28] in their study, enriched bread with encapsulated flaxseed oil, garlic oil, and a combination of flaxseed oil and garlic oil hybrid microcapsules and investigated their effects on oxidative stability and quality parameters.

In this study, the quality and chemical compositions, volume, total baking loss, and yield and sensory evaluation in wheat bread fortified by teff flour and chia seed were determined. It was determined that the addition of teff flour or chia seeds affected all the quality properties.[51]

## 6.5 SUMMARY

Functional foodstuffs supplemented with LC ω-3 FAs have become more popular. There are many examples of simple lipid emulsions of PUFAs in the scientific literature. ω-3 FAs have been supplemented to commercial foods, such as milk, cheese, yogurt, bread, and beverage. This chapter exhibited an overview of ω-3 FAs and ω-3 fortification of foodstuffs and also the most recent research works on essential FAs incorporated into the food system. Later studies were performed to find out the novel sources of ω-3 FAs and to enhance the stability of ω-3 FAs, and also the utilization of these ω-3 sources in the food products can be a significant way for food enrichment and treatment of diseases.

## KEYWORDS


- **omega-3 fatty acids**
- **functional food**
- **food fortification**
- **emulsion**


## REFERENCES


1. Abacı, Z. M. *Omega-3 yağ asitleri ile zenginleştirilmiş kayısı nektarının fiziksel ve oksidatif stabilitesinin belirlenmesi* (Determination of Physical and Oxidative Stability of Apricot Nectar Enriched with Fatty Acids). Master's Degree Thesis; Department of Food Engineering at Inonu University, Malatya, Turkey; 2020; p 95.
2. Alcântara, M. A.; de Lima, A. E. A.; Braga, A. L. M.; Tonon, R. V.; Galdeano, M. C.; da Costa Mattos, M.; Influence of the Emulsion Homogenization Method on the Stability of Chia Oil Microencapsulated by Spray Drying. *Powder Technol.* **2019**, *354*, 877–885.
3. Amanda F. L.; Wendy D. J. *Functional Foods*. FSHN12–17, one of a series of the Food Science and Human Nutrition Department, UF/IFAS Extension, 2012. http://edis.ifas.ufl.edu (accessed on 31 Jan 2022).
4. Anbudhasan, P.; Surendraraj, A.; Karkuzhali, S.; Ramasamy, D. Development of Omega-3 Fatty Acid Enriched Stable Functional Foods: Challenges and Options. *Int. J. Innov. Res. Rev.* **2014**, *2* (1), 1–13.
5. Beikzadeh, S., Shojaee-Aliabadi, S., Dadkhodazade, E., Sheidaei, Z., Abedi, A. S., Mirmoghtadaie, L.; Hosseini, S. M. Comparison of Properties of Breads Enriched with Omega-3 Oil Encapsulated in β-Glucan and *Saccharomyces cerevisiae* Yeast Cells. *Appl. Food Biotechnol.* **2019**, *7* (1), 11–20.
6. Borneo, R.; Kocer, D.; Ghai, G.; Tepper, B. J.; Karwe, M. V. Stability and Consumer Acceptance of Long-Chain Omega-3 Fatty Acids (Eicosapentaenoic Acid, 20: 5, N-3 And Docosahexaenoic Acid, 22: 6, N-3) in Cream-Filled Sandwich Cookies. *J. Food Sci.* **2007**, *72* (1), S049–S054.
7. Bush, L.; Stevenson, L.; Lane, K. E. The Oxidative Stability of Omega-3 Oil-in-Water Nanoemulsion Systems Suitable for Functional Food Enrichment: Systematic Review. *Crit. Rev. Food Sci. Nutr.* **2019**, *59* (7), 1154–1168.
8. Chen, B.; Rao, J.; Ding, Y.; McClements, D. J.; Decker, E. A. Lipid Oxidation in Base Algae Oil and Water-In-Algae Oil Emulsion: Impact of Natural Antioxidants and Emulsifiers. *Food Res. Int.* **2016**, *85*, 162–169.
9. Cholewski, M.; Tomczykowa, M.; Tomczyk, M. Comprehensive Review of Chemistry, Sources and Bioavailability of Omega-3 Fatty Acids. *Nutrients* **2018,** *10* (11), 1662–1668.
10. Comunian, T. A.; Favaro-Trindade, C. S. Microencapsulation Using Biopolymers as An Alternative to Produce Food Enhanced with Phytosterols and Omega-3 Fatty Acids: A Review. *Food Hydrocoll.* **2016,** *61*, 442–457.

11. Comunian, T. A.; Ravanfar, R.; Selig, M. J.; Abbaspourrad, A. Influence of the Protein Type on the Stability of Fish Oil-in-Water Emulsion by Glass Microfluidic Device. *Food Hydrocoll.* 2018, *77*, 96–106.
12. Dias, D. R.; Botrel, D. A.; Fernandes, R. V. D. B.; Borges, S. V. Encapsulation as a Tool for Bioprocessing of Functional Foods. *Curr. Opin. Food Sci.* **2017**, *13*, 31–37.
13. Diguet, A., Guillermic, R. M., Magome, N., Saint-Jalmes, A., Chen, Y., Yoshikawa, K., & Baigl, D. Photomanipulation of a Droplet by the Chromo Capillary Effect. *Angewandte Chemie* (Applied Chemistry) *Int. Ed.* **2009**, *48* (49), 9281–9284.
14. Drenjančević, I.; Kralik, G.; Kralik, Z.; Mihalj, M.; Stupin, A.; Novak, S.; Grčević, M. The Effect of Dietary Intake of Omega-3 Polyunsaturated Fatty Acids on Cardiovascular Health: Revealing Potentials of Functional Food. In *Superfood and Functional Food*; Shiomi, N., Waisundara, V., Eds.; Rijeka– Croatia: InTech Open; 2017; pp 207–232.
15. Encina, C.; Vergara, C.; Giménez, B.; Oyarzún-Ampuero, F.; Robert, P. Conventional Spray-Drying and Future Trends for the Microencapsulation of Fish Oil. *Trends Food Sci. Technol.* **2016**, *56*, 46–60.
16. Feizollahi, E; Hadian, Z; Honarvar, Z. Food Fortification with Omega-3 Fatty Acids: Microencapsulation as an Addition Method. *Curr. Nutr. Food Sci.* **2018**, *14* (2), 90–103.
17. García-Márquez, E.; Higuera-Ciapara, I.; Espinosa-Andrews, H. Design of Fish Oil-in-Water Nanoemulsion by Microfluidization. *Innov. Food Sci. Emerg. Technol.* **2017**, *40*, 87–91.
18. Genot, C.; Kabri, T. H.; Meynier, A. Stabilization of Omega-3 Oils and Enriched Foods Using Emulsifiers. In: *Food Enrichment with Omega-3 Fatty Acids*; Jacobsen, C., Nielsen, N. S., Horn, A. F., Sørensen, A. D. M., Eds.; Woodhead Publishing: Sawston, 2013; pp 150–193.
19. Geranpour, M.; Assadpour, E.; Jafari, S. M. Recent Advances in Spray Drying Encapsulation of Essential Fatty Acids and Functional Oils. *Trends Food Sci. Technol.* **2020,** *102*, 71–90.
20. Goksen, G.; Gumus, P. The Future Trend Natural Preservatives in the Food System: Essential Oils. *Eur. J. Sci. Technol.* **2021**, *28*, 441–443.
21. Goyal, A.; Sharma, V.; Sihag, M. K.; Tomar, S. K.; Arora, S.; Sabikhi, L.; Singh, A. K. Development and Physicochemical Characterization of Microencapsulated Flaxseed Oil Powder: A Functional Ingredient for Omega-3 Fortification. *Powder Technol.* **2015,** *286*, 527–537.
22. Gumus, C. E.; Gharibzahedi, S. M. T. Yogurts Supplemented with Lipid Emulsions Rich in Omega-3 Fatty Acids: New Insights into the Fortification, Microencapsulation, Quality Properties, and Health-Promoting Effects. *Trends Food Sci. Technol.* **2021**, *110*, 267–279.
23. Hashim, A. F.; Hamed, S. F.; Hamid, H. A. A.; Abd-Elsalam, K. A.; Golonka, I.; Musiał, W.; El-Sherbiny, I. M. Antioxidant and Antibacterial Activities of Omega-3 Rich Oils/ Curcumin Nanoemulsions Loaded in Chitosan and Alginate-Based Microbeads. *Int. J. Biol. Macromol.* **2019**, *140*, 682–696.
24. Hasler, C. M. Functional Foods: Benefits, Concerns and Challenges - A Position Paper from the American Council on Science and Health. *J. Nutr.* **2002**, *132* (12), 3772–3781.
25. Hermida, L. G.; Gallardo, G. Food Applications of Microencapsulated Omega-3 Oils. In *Microencapsulation and Microspheres for Food Applications*; Sagis, L. M. C., Ed.; Academic Press: Cambridge, 2015; pp 271–299.
26. Jamshidi, A.; Cao, H.; Xiao, J.; Simal-Gandara, J. Advantages of Techniques to Fortify Food Products with Benefits of Fish Oil. *Food Res. Int.* **2020**, *137*, 109–353.

27. Jankuloska, V.; Kalevska, T.; Nedelkoska, D. N. Components in Vegetable Oils as Ingredients of Functional Foods. *Food Knowl. Int. J.* **2020**, *40* (3), 579–584.
28. Kairam, N.; Kandi, S.; Sharma, M. Development of Functional Bread with Flaxseed Oil and Garlic Oil Hybrid Microcapsules. *LWT* **2021**, 136, E-article 110300.
29. Karadağ, A.; Hermund, D. B.; Jensen, L. H. S.; Andersen, U.; Jónsdóttir, R.; Kristinsson, H. G.; Jacobsen, C. Oxidative Stability and Microstructure of 5% Fish-Oil-Enriched Granola Bars Added Natural Antioxidants Derived from Brown Alga *Fucus vesiculosus*. *Eur. J. Lipid Sci. Technol.* **2017**, *119* (4), E-article 1500578.
30. Kasapoğlu, K. N.; Daşkaya-Dikmen, C.; Yavuz-Düzgün, M.; Karaça, A. C., Özçelik, B. Enrichment of Beverages with Health Beneficial Ingredients. In *Value-Added Ingredients and Enrichments of Beverages*; Grumezescu, A. M., Holban, A. M., Eds.; Academic Press: Cambridge, 2019; pp 63–99.
31. Kaur, N.; Chugh, V.; Gupta, A. K. Essential Fatty Acids as Functional Components of Foods: A Review. *J. Food Sci. Technol.* **2014**, *51* (10), 2289–2303.
32. Kaushik, P.; Dowling, K.; Barrow, C. J.; Adhikari, B. Microencapsulation of Omega-3 Fatty Acids: A Review of Microencapsulation and Characterization Methods. *J. Funct. Foods* **2015**, *19*, 868–881.
33. Komaiko, J.; Sastrosubroto, A.; McClements, D. J. Encapsulation of ω-3 Fatty Acids in Nanoemulsion-Based Delivery Systems Fabricated from Natural Emulsifiers: Sunflower Phospholipids. *Food Chem.* **2016**, *203*, 331–339.
34. Kralovec, J. A.; Zhang, S.; Zhang, W.; Barrow, C. J. Review of the Progress in Enzymatic Concentration and Microencapsulation of Omega-3 Rich Oil from Fish and Microbial Sources. *Food Chem.* **2012**, *131* (2), 639–644.
35. Lammari, N.; Louaer, O.; Abdeslam-Hassen, M.; Fessi, H.; Elaïssari, A. Plant Oils: From Chemical Composition to Encapsulated Form Use. *Int. J. Pharma.* **2021**, *601*, E-article 120538.
36. Lane, K. E.; Li, W.; Smith, C. J.; Derbyshire, E. J. The Development of Vegetarian Omega-3 Oil-in-Water Nanoemulsions for Integration into Functional Food Products. *J. Funct. Foods* **2016**, *23*, 306–314.
37. Lane, K. E.; Zhou, Q.; Robinson, S.; Li, W. The Composition and Oxidative Stability of Vegetarian Omega-3 Algal Oil Nanoemulsions Suitable for Functional Food Enrichment. *J. Sci. Food Agric.* **2020**, *100* (2), 695–704.
38. Ofosu, F. K.; Daliri, E. B.; Lee, B.; Yu, X.; Chuncheon-si, G. D. Current Trends and Future Perspectives on Omega-3 Fatty Acids. *Res. Rev. J. Biol.*, **2017**, *5*, 11–20.
39. Panse, M. L.; Phalke, S. D. Omega-3 Beverages. In *Value-Added Ingredients and Enrichments of Beverages*; Grumezescu, A. M., Holban, A. M., Eds.; Academic Press: Cambridge, 2019; pp 353–382.
40. Robins, M. M.; Watson, A. D.; Wilde, P. J. Emulsions: Creaming and Rheology. *Curr. Opin. Colloid Interface Sci.* **2002**, *7* (5–6), 419–425.
41. Rodriguez-Leyva, D.; Bassett, C. M.; McCullough, R.; Pierce, G. N. The Cardiovascular Effects of Flaxseed and Its Omega-3 Fatty Acid, Alpha-Linolenic Acid. *Can. J. Cardiol.* **2010**, *26* (9), 489–496.
42. Ruiz-Ruiz, J. C.; Ortiz-Vazquez, E. D. L. L.; Segura-Campos, M. R. Encapsulation of Vegetable Oils as Source of Omega-3 Fatty Acids for Enriched Functional Foods. *Crit. Rev. Food Sci. Nutr.* **2017**, *57* (7), 1423–1434.
43. Saini, R. K.; Keum, Y. S. Omega-3 and Omega-6 Polyunsaturated Fatty Acids: Dietary Sources, Metabolism, and Significance—A Review. *Life Sci.* **2018**, *203*, 255–266.

44. Sánchez, C. O.; Zavaleta, E. B.; García, G. U.; Solano, G. L.; Díaz, M. R. Krill Oil Microencapsulation: Antioxidant Activity, Astaxanthin Retention, Encapsulation Efficiency, Fatty Acids Profile, In Vitro Bio Accessibility and Storage Stability. *LWT*, **2021,** *147*, E-article 111476.
45. Sivapratha, S.; Sarkar, P. Oxidative Stability and Effect of Stress Factors on Flax Seed Oil-in-Water Emulsions Stabilized by Sodium Caseinate – Sodium Alginate – Chitosan Interfacial Membrane. *Chem. Papers*, **2018**, *72* (1), 1–14.
46. Surette, M. E. The Science Behind Dietary Omega-3 Fatty Acids. *Can. Med. Assoc. J.* **2008,** *178* (2), 177–180.
47. Taneja, A.; Singh, H. Challenges for the Delivery of Long-Chain Omega-3 Fatty Acids in Functional Foods. *Annu. Rev. Food Sci. Technol.* **2012**, *3*, 105–123.
48. Timilsena, Y. P.; Adhikari, R.; Barrow, C. J.; Adhikari, B. Microencapsulation of Chia Seed Oil Using Chia Seed Protein Isolate - Chia Seed Gum Complex Coacervates. *Int. J. Biol. Macromol.* **2016**, *91*, 347–357.
49. Tur, J. A.; Bibiloni, M. M.; Sureda, A.; Pons, A. Dietary Sources of Omega-3 Fatty Acids: Public Health Risks and Benefits. *Br. J. Nutr.* **2012**, *107* (S2), S23–S52.
50. Vellido-Perez, J. A.; Ochando-Pulido, J. M.; Brito-de la Fuente, E.; Martinez-Ferez, A. Novel Emulsions–Based Technological Approaches for the Protection of Omega–3 Polyunsaturated Fatty Acids Against Oxidation Processes: A Comprehensive Review. *Food Struct.* **2021**, *27*, E-article 100175.
51. Zięć, G.; Gambuś, H.; Lukasiewicz, M.; Gambuś, F. Wheat Bread Fortification: The Supplement of Teff Flour and Chia Seeds. *Appl. Sci.* **2021**, *11* (11), E-article 5238.
52. Jayasinghe, D. How to get Omega-3 As a Vegan: The Best Plant Sources of Omega-3. February 17, 2019. https://veganfitlyfe.wordpress.com/2019/02/17/how-to-get-omega-3-as-a-vegan-the-best-plant-sources-of-omega-3/

# CHAPTER 7

# ROLE OF NUTRACEUTICAL-BASED FUNCTIONAL FOODS IN HUMAN HEALTH

GARGI GHOSHAL

## ABSTRACT

The plant-based nutraceuticals provide numerous health benefits and play a significant role in the healing of various ailments and in the physiological functions of our body. This has led to the reduction in the use of synthetic drugs. This chapter presents delivery routes of nutraceuticals through oral or skin or eyes, etc. It is of utmost importance to design these natural bioactive compounds for their bioavailability and minimum physicochemical reactions in our body. Consumers demand safe and quality biomedicines, which require strict regulations for the manufacturing and delivery of these components.

## 7.1 INTRODUCTION

In 1989, Stephen De Felice introduced the term nutraceutical consisting of two words "*nutra*" (meaning nutrients/nutrition) and "*ceutical*" (meaning agent having disease-preventing effect).[42] Therefore, nutraceuticals provide not only nutrition values but also show disease-preventing effects (Table 7.1). Nutraceuticals are part of the natural food or plant part containing bioactive compounds (such as proteins, lipids, dietary fibers, carbohydrates, vitamins, minerals, antioxidants, therapeutic agents, or other nutrients) to prevent diseases or to incorporate immunity to fight against chronic diseases. Nutraceuticals differ from food supplements in terms of two features: (a) nutraceuticals are considered

Plant-Based Bioactive Compounds and Food Ingredients: Encapsulation, Functional, and Safety Aspects.
Junaid Ahmad Malik, Megh R. Goyal, Preeti Birwal, & Ritesh B. Watharkar (Eds.)

traditional food items and as the elite item of the diet; (b) nutraceuticals not only substitute diet but also aid in the treatment of diseases or disorders.

**TABLE 7.1** Nutraceuticals with Allied Terminology and Their Functions.

| Parameter | Function | References |
|---|---|---|
| Antioxidants | Agents that prevent oxidation in cells by free radical generation by deteriorating biomolecules, such as DNA, protein, lipids, and carbohydrate | [35] |
| Bioactivity | A particular agent's importance on live tissues | [42] |
| Botanicals | Plant parts or their extracts to provide health-beneficial effects | [42] |
| Designed foods | Food supplements or phytochemicals to provide health-beneficial effects and to reduce the risk of chronic diseases | [37] |
| Dietary supplements | These are anticipated for intake in the form of powder, capsule; but not indicated as normal food or main staple food of a meal or diet | [9, 29] |
| Functional foods | Foods include physiologically active ingredients with health-beneficial effects other than basic nutrition | [45] |
| Nutraceuticals | Either nutrients or non-nutrient composites having health-beneficial, illness defensive, or therapeutic features | [2] |
| Phytochemicals | Bioactive compounds have conveying human health benefits; and are used on the basis of individual dietary plan, and lifestyle | [25] |

Nutraceuticals are related to functional foods or biologically active phytochemicals with therapeutic characteristics. Dietary supplements have more distinct role in human health (such as nutritional substances) to increase ingestion of these components in the gastrointestinal tract (GIT). Pathogenic bacteria can be inhibited and the growth of beneficial probiotic bacteria is possible by using phytochemicals and their derivatives. Interactions among phytochemicals, probiotics, prebiotics, and intestinal microflora have health-beneficial effects on humans.[28,38]

Dietary supplements (such as green tea, vitamins D and E, selenium, soy proteins, carotenoids, and lycopenes) contribute to complimentary health benefits further than the essential nourishment with the introduction of functional foods and nutraceuticals.[33,35,89] Numerous food nutraceuticals have been considered in cancer therapies.

Nutraceuticals are dietary supplements when incorporated in food and are recognized as functional foods.[20] They are similar to usual foods and consumed with a normal diet item (Tables 7.2 and 7.3). Nonetheless, nutraceuticals are formed from food items, but are sold in the form of pills, powder, capsules,

**TABLE 7.2** Various Nutraceutical Substances Derived from Different Sources.

| Plant-derived Nutraceuticals | | Animal-derived Nutraceuticals | |
|---|---|---|---|
| **Name** | **Source** | **Name** | **Source** |
| Allicin, Adenosine | Onion, garlic | Magnesium | Meat, beef and dairy products |
| Anthocyanates | Red wine | Zinc | Red meats, some sea foods, fortified cereals |
| Ascorbic acid | Citrus fruits and vegetables | Eicosapentaenoic acid (EPA) | Fish oils |
| Capsaicin | Peppers | Manganese | Sea foods, meat |
| Cellulose | Cell-wall components of plants | Lecithin | Egg |
| Curcumin | Turmeric | Selenium | Organ meats, seafood, dairy products |
| Daidzein | Soybeans, legumes | Iron | Meat, fish, milk and dairy products |
| Ellagic acid | Fruits, such as grapes, strawberries, raspberries, and walnuts | Sodium | Processed and restaurant foods made with added salt |
| Gallic acid | Tea, clove, vinegar | Coenzyme (Q10) | Heart, liver, kidney, Pork, beef and chicken, herring, mackerel, sardine |
| Genestein | Soybeans and other legumes, apios | Folic acids | Chicken liver |
| Glutathione | Asparagus, potato, peppers, carrots, cabbage, cauliflower, Brussels sprout, broccoli, avocado, squash, spinach, and melons | Copper | Sea foods, nuts, |
| Indole, carbonol | Leafy vegetables, such as cabbage, broccoli, cauliflower, Brussels, kale, and sprouts | Creatine | Red meat, fish, raw beef, salmon |
| Luteolin, Lutein, zeaxanthin | Fruits and leaves, such as kale, collards, spinach, corn, and citrus | Chlorine | Fish, beef lever, poultry lever, meats, eggs |
| Lycopene | Tomatoes and tomato products | Iodine | Sea weeds, seafoods, dairy products, processed foods, iodized salt |

**TABLE 7.2** *(Continued)*

| Plant-derived Nutraceuticals | | Animal-derived Nutraceuticals | |
|---|---|---|---|
| **Name** | **Source** | **Name** | **Source** |
| Mono-unsaturated fatty acids | Tree nuts, olive oil | Calcium | Milk |
| Pectin | Fruits and vegetable peels; seeds | Minerals | – |
| Quercetin | Onion, red grapes, citrus fruit, broccoli, Italian, yellow squash | Spingolipids | Milk and dairy products, meat, fish, eggs |
| α-Glucan conjugated | Oat bran | Conjugated linoleic acid (CLA) | Meat and dairy products |
| α-Tocopherol | Fish, oil | Phosphorus | Milk and dairy products, meat, eggs. |
| β-Carotene | Carrots, papaya, mango, Rosemary | Potassium | Yogurt, milk, yellow fin tuna |
| β-Ionone | Berry, citrous fruits | Molybdenum | Organ meats, particularly lever and kidney |
| γ-Tocotrienol | Palm oil, rice bran oil | Docosa hexaenoic acid (DHA) | Fish oils |
| δ-Limonene | Oranges | Chromium | Meats, poultry, fish, |
| Inulin, Fructo- oligo-saccharides (FOS) | Whole grains, onions, garlic | – | |
| Isothiocyanates | Cruciferous vegetables | – | |
| Resveratrol | Grapes (skin), red wine | – | |
| Catechins | Teas, berries | | |
| Lignans | Flax, rye | | |
| Folic acid | Leafy vegetables, fortified cereals, enriched bread, multigrain breads | – | |

or syrup. Nutraceuticals possess anticancer, antidiabetic, anti-obesity properties, and these can help to reduce cardiovascular and chronic degenerative diseases, inflammatory disorders, and can boost immune systems.[10]

**TABLE 7.3** Fermented Dairy Products as Source of Nutraceuticals.

| Name | Health benefits |
|---|---|
| (A) Bioactive peptides | • Maintain immune system |
| • Glutamine peptide | • Regulate protein turnover |
| • Peptide FM | • Glycogen replenishment |
| • Casein phosphopeptides | • Inhibit dietary fat deposition |
| (B) Bioactive proteins | • Alter fat metabolism |
| • Lacto peroxidase | • Enhance calcium absorption |
| • Lactoferin | • Enhance iron absorption |
| | • Prevention of dental carries |
| | • Promote beneficial intestinal flora |
| | • Probiotic effect |
| | • Regulate iron transport and absorption |
| | • Promote beneficial intestinal flora |
| (C) Bioactive minerals Lactoval (natural milk calcium/mineral complex) | Enhance calcium absorption |

Availability of labor, capable researchers and scientists, cutting-edge research facilities and laboratories, plenty of raw material resources, etc. have contributed to India being a leading-edge country for the availability of nutraceuticals throughout the world.[73,88]

This chapter describes (1) the significance of nutraceuticals; (2) their classification, and applications; (3) routes of nutraceutical delivery; (4) formulation strategies; and (5) nanotechnology-based applications for the delivery of nutraceuticals and functional foods.

## 7.2 SIGNIFICANCE OF USING NUTRACEUTICALS AS POTENTIAL HEALTH-BENEFICIAL COMPONENTS

Following are the causes of favoring the use of nutraceuticals:

- Novel molecules are hard to identify, as well as the processes are costly than before; therefore, numerous pharmaceutical companies are searching for novel nutraceuticals, thus there is definitely a vast emergent market place.

- Faith on "food-like substances" among purchasers because of harmless nature or lowest toxicity as compared with conventional pharmaceuticals.
- Improper dietetic practices and eventually fatal deadly diseases leading to deaths due to cardiovascular diseases, particular category of cancers, etc. have led to the development of nutraceuticals to treat in specific circumstances have momentum.[22,36,71]
- Dietetic therapy and phytotherapy have appeared as novel perception and curing structures and have rapidly and extensively broadened in the modern era.[55,98]
- Long history to utilize and for better acceptance by patients and public approval.
- Nutraceuticals have already achieved recognition because the general public has started to accept nutraceuticals to protect them from side effects related to synthetic medicines.
- Renewable sources and local availability, farming, and delivery of nutraceuticals under eco-friendly conditions.
- Rising cost of treatments with traditional allopathic medicine, latest legislations, and novel inventions.[23,24,40]
- The manifestation of diet–disease interactions has directed us to hunt for precise components of minerals, extracted from animals or plants sources, which have favorable functions to maintaining our physical fitness.[14]
- Though public awareness in this particular subject is developing, yet, producers have hunted to accomplish the desire for this product to establish the market. There are more than 5,000 recognized nutraceutical products globally.

## 7.3 TYPES OF NUTRACEUTICALS

Nutraceuticals are categorized on the basis of the origin of the valuable component and/or the identification of their therapeutic benefits (Table 7.4) or on the basis of their chemical composition.[17,18,20,36,70] These advancements permit us to classify the nutraceuticals as molecules or elements in a food matrix.

**TABLE 7.4** Classification of Nutraceuticals and Their Functions.

| Nutraceuticals | Protection |
|---|---|
| Isoprenoid derivatives α-Carotene, lycopene | Cancer risk reduction, eye and gut health |
| Phenolic substances | Cognitive function |
| Fatty acids | Improve immunity |
| **Probiotics** | Maternal and infant health, gut health |
| I. *Lactobacilli L. rhamnosus*, *L. casei*, *L. reuteri*, | |
| II. *Bifidobacterium*, *B. infantis*, *B. breve*, *B. lactis*, *B. longum*, | |
| III. *Streptococcus* | |
| IV. *Lactococcus, L. reuteri, L. platinum, L. agilis* | |
| V. *Enterococus,* | |
| VI. *Bacillus,* | |
| VII. *Saccharomyces,* | |
| VIII. *Pediococcus*. | |
| **Prebiotics** | Respiratory and skin health |
| 1. Oligosaccharides | |
| 2. Inulins | |
| 3. Lactilol | |
| Proteins and amino acids | Weight management |
| Soy-based products | Women's health, Cardiovascular health |
| Dairy-based ingredients | Cancer risk reduction |
| Plant extracts and phytochemicals | Cognitive and mental function |
| **Antioxidants** | Bone and joint health |
| Vitamins C and E; α-Carotene, lycopene, lipoic acid, glutathione, dismastase, Coenzyme Q-10 | |
| Nutritional lipids and oils | Eye and gut health |
| Fibers | Immune system |
| Carbohydrates | Body building |
| Amino acid-based substances | Weight management |
| Microbes | Maintain body microflora balance in the GIT |
| Minerals: selenium, copper, manganese, zinc, super oxide | Enhance immunity |

## 7.4 APPLICATIONS OF NUTRACEUTICALS

Nutraceuticals are utilized as protective drugs; however, are regularly suggested as complementary or unconventional treatments to persistent diseases by changing the route of various infections and avoiding such situations prior to start. The simple dietetic changes have been thought to prevent the growth of many types of cancers, and utilization of antioxidants has been correlated to neurodefensive results both in vitro and in vivo research studies.[8,18,26,27,84] Consumption of highly nutritive proteins (such as soy-based foods) in humans has achieved attention to reduce weight, serum cholesterol, blood pressure, etc.[54] Additionally, a peptide (such as milk protein) is a new approach to reduce hypertension. Incidence of multiple sclerosis can be lowered by consuming a diet rich in vitamin D and ω-3 fatty acids that are available in fish oils.[77]

Procyanidin in grape seed extract may be able to reduce cholesterol, nausea and pancreas disease, and ache.[90] To treat diabetes, extracts of bitter melon and cinnamon have been successful. Research studies have documented that vitamin E enrichment resulted to lower AIDS in HIV-positive patients. The EuroSIDA study indicated the shortage of hydroxyvitamin D in such patients; whereas treatment with antiretroviral (ARV) was linked with high-risk patients with AIDS.[89] Therefore, it is concluded that an enrichment strategy with vital micronutrients is a crucial step to protect our body from unfavorable physiological activities.

The research studies have illustrated that usual antidepressant patients showed 40 to 60% improvement in neurological disorders in patients with neurotic confusion habits. Many researchers have studied how each nutraceutical has a multi-targeted function on the cell cycle and is easily accessible to the patient.[27]

Potential investigations should spotlight the formulation of co-drugs containing the nutraceuticals for improved efficiency, diminishing the dosage, and considerable reduction of side effects. The strategy to use soy-derived nutraproducts (such as vitamin D3 metabolites, calcitrol) has been successful in women with menopause conditions. The design and formulation of bioactive nutrients to develop the pharmacokinetics have posed several therapeutic issues.[73]

## 7.5 ROUTES OF NUTRACEUTICAL DELIVERY

### *7.5.1 ORAL DELIVERY*

Oral delivery of nutraceuticals is the most preferable route; however, it may influence the bioavailability of nutraceuticals. Oral pathway delivery of nutraceuticals includes chondroitin, lycopene, resveratrol, coenzyme Q-10, glucosamine, creatin, melatonin, green tea extract, acetyl L-carnitine, S-adenosyl methionine, lipoic acid, and dehydro-epiandro-sterone, vitamins, etc. Resveratrol in grapes and polyphenols in peanuts are quickly metabolized in the liver and GIT, thus improving the oral bioavailability. Fresh techniques and strategic design play a crucial role in oral delivery of nutraceutical agents by shielding them from quick removal and increasing the solubility in the GIT.[4]

### 7.5.2 DERMAL DELIVERY

The effect of nutraceutical-based antioxidants might be improved with the simultaneous oral and skin use of bioactives like lutein. Superior further reliable plasma level, the skin function produces nutraceuticals in the skin. Although the dermal delivery might help in prolonged period of storage, yet, wrinkle intensity in aged skin may be reduced due to the defensive action of the skin. Examples of nutraceuticals for use as dermal treatment include curcumin, coenzyme Q-10, genistein, gluconolactone, N-acetyl cysteine, and sulfated polysaccharides.[79,97]

### 7.5.3 NUTRACEUTICALS AS OCULAR THERAPY

When nutraceuticals cannot be delivered by normal oral pathways, nutraceutical delivery can be done by ocular therapy installation in the ocular cavity. Numerous nutraceuticals (e.g., Coenzyme Q-10, leutin, vitamin E) when routed through the eyes have disease-altering benefits for the eyes because of their antioxidants, and anti-inflammatory properties. Intra-ocular therapy of nutraceuticals relies on the residence time permeability in the form of drops or ointment. The bioavailability of bioactive component is generally low.[13,48,96]

## 7.6 FORMULATION STRATEGIES FOR NUTRACEUTICALS[57–59]

### 7.6.1 LIPOSOME CARRIER SYSTEM

All delivery systems are summarized in Table 7.5. Liposomes are vesicles made from phospholipids; liposomes have been exploited to modify the pharmacokinetics of numerous nutraceuticals, such as herbals, vitamins, enzymes, and minerals.[41,96] The key factors for applying liposomes include the composition and size of nutraceuticals, surface charge, lipid concentration, vesicle size, etc. Liposomes are mainly used in cosmetic delivery systems and for skin treatment. Liposomes are active in carrying bioactive agents in the stratum cornea and can increase the accumulation of vitamins in the skin.[48]

### 7.6.2 ELECTROSPUN FIBER MATS

The delivery of diverse nutraceuticals is possible by using mats of electrospun fiber carrier.[81] Scientists have created nanofiber mats using cellulose acetate (CA) for the delivery of vitamins to the skin.[79–81] Vitamin E, Vitamin A (as α-tocopherol), and trans-retinoic acid entrapped in spun fiber mats were able to enhance the cumulative discharge of vitamins during the designed assessment periods. The delivery of curcumin using cellulose acetate (CA) has also been observed. In this particular study, the author exhibited that the system was nontoxic and delivery of curcumin to the skin was successful as a potential topical treatment with antioxidants, antitumor, anti-inflammatory activities. The *asiaticoside*-loaded CA fiber mats that were extracted from *Centela asiatica* L. were used successfully for transdermal release; and 26% of the maximum release of the drug was obtained through different pig-skin models.

Another research result reveals that when AP content was increased, then the diameter of the fiber mat was decreased. These results also highlighted the antibacterial property of PCL/AP mats against the skin pathogen *S. aureus*. It was found that the plain PCL alone was unable to stop the growth of pathogenic bacteria, but PCL/AP mats provided antioxidant and antibacterial actions, and a large surface area for functioning via the skin; and these activities were comparable with silver nanoparticles-coated PCL/AP mats.

### 7.6.3 MICROSPONGE AND NANOSPONGE

The microsponge and nanosponge delivery systems provide enormous beneficial effects due to their large inner, exterior, and high bioactive entrapment capability. Researchers generated genetically modified melanin and entrapped in a porous melano-sponge-α structure to use melanin on the skin surface as UV-A and UV-B sun-protecting agent. Microsponges can also decrease allergenic actions when cinnamic aldehydes were incorporated into this arrangement. In another study, resveratrol-entrapped cyclodexdrin-based nanosponges were prepared for the enhancement of solubility and stability of resveratrol. These nanosponges are hypercross-linked cyclodextrin polymers that form three-dimensional networks. The nanosponge composite enhanced the in vitro discharge, penetration, effectiveness, and steadiness of resveratrol, and was considered appropriate formulation for external use and rapid delivery of the resveratrol.[7,85]

### *7.6.4 CYCLO DEXTRIN COMPLEXATION*

Cyclodextrin (CD) and derivatives have been extensively used as delivery media to develop the natural solubility, strength, penetration, and bioavailability of nutraceuticals.[7,19,47,85] Such cyclodextrin molecule has cage-like structure so that many vitamins in contact with the biological membrane can complex for the improvement of physicochemical properties. β-Cyclodextrin is the generally used cyclodextrin because of its accurate cavity size. The 7-dehydrocholesterol is a water-insoluble vitamin that is used in cosmetics and pharmaceutical goods; however, it is difficult to incorporate it in skin products.[81]

Silymarin has poor aqueous solubility, bioavailability, and also requires a new formulation approach for its efficient oral delivery.[19,53] Usually, silymarin β-CD complexes are created by co-precipitation process to increase the suspension rates and for more sustained release. Other investigators also used β-CD to produce a nutraceutical garlic oil insertion. Garlic oil has strong antioxidant and antimicrobial features but is inadequate as a well-designed element due to its intrinsic volatility and poor physicochemical steadiness. DSC studies revealed that the nutraceutical was effectively included within the CD voids and prevented the oxidation with an eventual enhancement in the steadiness of complex. In vitro release studies revealed that the GO/β-CD complex has excellent water solubility and a controlled discharge rate was also observed. It was found that hydroxyl propyl-β-CD has a synergistic effect on vitamin A as it enhances it solubility 35,000 folds, therefore causing the improvement in oral delivery of vitamins A, D, E, and K. Hydroxyl propyl-β-CD and their resultant astaxanthin insertion complexes were produced successfully by improving the solubility and adopting an approach for controlled release of nutraceuticals.

Curcumin (anticancer and antiviral nutraceutical) is not soluble in aqueous media at neutral or low acidic pH; but its solubility was improved when complexed with cyclodextrin (CD). *Kaempferia parviflora* (KP) plant extract was added in 2-hydroxypropyl CD and there was 3.5 times higher permeability and 21–34 times higher bioavailability.[11,95]

### 7.6.5 BIODEGRADABLE HYDROGELS

Biodegradable hydrogels have been extensively investigated to hold, guard, alter the release of large range of pharmaceutical composites containing nutraceuticals.[1,54,78] Many pharmaceutical polymers can disintegrate based on varying environments and altering pH, hydration, thermal, and stimulus conditions. And most of the absorption of nutritive molecules takes place in the small intestine, and is used to ease the practice by defending nutraceuticals from disintegration or degradation and by making a possible controlled release.

Soy gel carrier[54] provides the extra beneficial effect of additive nourishment effect, given by the soy carrier, lower need for inorganic solvents on the other hand that may be well-suited with food-based complement and the bioactive is untreated during gelation procedure which would have degraded the susceptible bioactive, the benefits of this hydrogel medium may be comprehensive for the release of nutraceutical products.

Milk protein hydrogels have perfect structural and physicochemical characteristics as carrier media for bioactives in the intestine environment. Milk proteins are expensive, easily obtainable, and nontoxic and have already been utilized as a delivery vehicle to transport a multitude of the drug via the GIT. The pH status can trigger the release of nutraceuticals, and it was observed in milk protein hydrogels.[54] Transglutaminase of microbial origin was incorporated in casein hydrogel matrix and was found suitable for nutraceutical encapsulation.

Similarly, vitamin B12 has been entrapped in casein hydrogel to allow easy release at a controlled rate in the body.[12] A different pH-sensitive delivery matrix for riboflavin containing poly(glycidyl methacrylate) was embedded in sodium alginate hydrogel. It was reported that about 30% of vitamin after 24 h could be observed in simulated gastric fluid (SGF) when poly(glycidyl methacrylate) was grafted with sodium alginate polymer as carrier. There was only 18% use of vitamin in SGF when plain SA polymer beads were cross-linked with calcium.

Alginate (the common polysaccharide) is commonly used for gel formation and it can react with polyvalent cations easily. Along with alginate, antimicrobial and anti-inflammatory *Aloe vera* is also used very frequently as an herbal hydrogel that is formed by solvent casting method for therapeutic use and wound dressing.

Another novel deacetylated poly N-acetyl glucoseamine nanofiber-based hydrogel formulation having therapeutic properties is commonly used to treat degenerative disc disease. Other formulation of cationic polymer chitosans-based hydrogel containing the complexed chondroitin sulfate is applied in the treatment of joint pain due to arthritis, the hydrogel follows pH control release and preferably at basic pH.[52,54,78]

## 7.7 NANOTECHNOLOGY-BASED APPLICATIONS FOR NUTRACEUTICAL DELIVERY

Recently, nanotechnology has gained recognition. It is applied to improve the solubility of the insoluble nutraceuticals to regulate the diversity of dosage structure that is also assured to attain excellent bioavailability and site-specific release of nutraceuticals. Thus, the application of nanotechnology is recommended to enhance the industrial perspective of a massive amount of nutraceuticals.[21,49,58,64,67,74,86,92]

### *7.7.1 NANOSUSPENSION AND NANOEMULSION*

The main factors to control solubility rate and maximum solubility affect the bioavailability by the diffusion of nutraceutical either in oral or external delivery respectively. Hydrophobic insoluble nutraceuticals can be made deliverable in the form of nanosuspension and nanoemulsion to improve and enhance their release. Lutin, widely used antioxidant was entrapped by making 400 nm-sized particles within a nanosuspension prepared by a high-pressure homogenization process followed by lyophilization and included in creams and gels prepared for skin care. Another formulation includes Lutin converted into pellets to load in gelatin capsules in hard form for oral delivery. The nanoformulation offered enhanced dispersion because of enhanced solubility and a bigger exterior. It was reported that the penetration of the lutin nanocrystal formulation was 14 times superior as compared with a crude lutin powder when passed through a 0.1 μm artificial cellulose nitrate membrane. Therefore, for oral formulation, better the dissolution, consequently enhanced bioavailability.

Similarly, another hydrophobic phytochemical isolated from sweet orange peels is 5-hydroxy 6,7,8,4-tetramethoxyflavon (PMF), though this herbal nutraceutical is having a higher melting point, lower aqueous solubility, and bioavailability. Thus, it was included in different nanoemulsions to deliver effectively in the nutritional supplement.

For carotenoid and fat-soluble vitamins (A, D, E, and K) for oral delivery, nanoemulsions have been used in liposomes. The researchers reported that when hydrophobic β-carotene is incorporated in nanoemulsion formulation, it provides excellent resistance against mechanical, chemical, and biological deterioration in neutral or lower pH gastric environmental conditions.[32] Further, it was reported that the bioavailability of vitamin A in nanoemulsion was superior in corn oil as compared with the orange oil as corn oil contains long-chain whereas orange oil contains medium-chain fatty acids. For transdermal delivery of hyaluronic acid (HA), lipophilic bioactive can be delivered using nanoemulsion of oil in water. The optimized formulation of HA along with vitamin E applied as model bioactive for proficient partition distribution is useful to enhance stratum corneum permeability into the profound external skin stratum than only HA nanoemulsion system.[24,50,59]

### *7.7.2 NANO STRUCTURE LIPID CARRIER (NLC)*

Nanostructure lipid carrier (NLC) is comparatively a novel invention of nano-sized lipid matrix structures for optimal nutraceutical release via the dermal, oral, and external route. Cetylpalmitate and fatty acids like caprylic/capric were incorporated in NLC for the dermal delivery of CoQ10 effectively as compared with other formulations. It was reported that the above-mentioned formulation offers excellent physical strength and entrapment efficiency, and the release pattern follows the original quick diffusion of CoQ10 into the skin, later, a maintained continuation phase exists ensuring the continuous release of the vitamin. An isometric shape and approximately 200 nm particle size were determined by scanning electron microscopy analysis. Further, under storage of NLC with varying temperatures – low (4°C), room (25°C), and high (40°C), the loaded retention of CoQ10 was more than 90% in nano-sized range for 1 year at all the temperature ranges proving the NLC as steadiness enhancer and also an enhancer to improve the strength of nutraceuticals for the successful dermal delivery system.

Another group of researchers has revealed that for better lutin oral delivery, fish oil-based NLC's was made with dual benefits as omega-3 fatty acid is present in fish oil. Therefore, this system works as an enhanced delivery mechanism system.

NLCs are also suitable for intravenous release of silybin, which is a liver protecting nutraceutical that prevents hepatotoxicity. It was proved that the above-mentioned formulation is suitable for site-specific biphasic delivery of silybin and prolonged residence time in serum for enhanced action.[47,51]

### 7.7.3 NANOMICELLES

Quercetin, an important phytochemical, has potential application in the treatment of cancer, neurological, and cardiovascular diseases, but due to its hydrophobic nature, it poses problem to develop clinical trial of any formulation. Therefore, it was entrapped in nanomiceller delivery system of copolymer polyethylene glycol (PEG) derivatized phosphatidyl ethanolamine (PE). These nontoxic formulations provided higher solubility and were steady at pH range 1.2–7 and were stable in their in vivo preclinical studies. Dramatic enhancement in the anticancer activity was observed when 30 mg/kg quercetin dose was used and applied in mice compared with control standard quercetin suspension in ethanol. Vitamin $D_2$ nanoencapsulation was done using casein micelles for protection against UV degradation. Such micellar systems can be recognized as means with the resultant perspective for delivery-responsive nutraceutical with the barrier to commercialization.[82]

### 7.7.4 NANOPARTICLE NANOCAPSULES AND NANOENCAPSULATION

Numerous phytonutrients have restrictions as additional nutraceutical ingredients because of high vulnerability to deprivation or poor water solubility as phytonutrients are sensitive to temperature, pH, and oxygen and in the presence of their chances of degradation. A very important example is epigalocatechin-3-gallate present in green tea, which has excellent antioxidants, neuroprotection, antitumor, cardiovascular protection, as well as numerous other health benefits, but it is not stable. It shows rapid degradation when present in soft drinks and formation of intermediate or degradation products that cause the development of off-flavor and discoloration of the drinks and loss of activity of the nutraceuticals during long storage times. In this regard, nanoencapsulation causes retention of color and flavor during prolonged storage and enhanced delivery of EGCG along with other antioxidant polyphenols.

In another study, it was found that epigalocatechin-3-gallate (EGCG) complexed with thermally modified β-lactoglobulin (βlg) resulting excellent resistance to EGCG against oxidation. In this combination, the size range of nanoparticles of βlg-EGCG was smaller than 50 nm, incorporating perfect precision and appropriateness of the nanosystem for incorporation in cola drinks. Another important encapsulant is milk protein utilized to nanoencapsulate β-carotene, including fatty acids and vitamin D to convey nutraceuticals site specifically. The poor water-soluble β-carotene vitamin A precursor was nanoencapsulated using casein and dextrans producing spherical nanoparticles of casein and dextran as shell and β-carotene core.

Another exciting research was done in making solid lipid nanoparticles (SLN) along with surfactant for oral and dermal formulations. Resveratrol SLNs were developed by scientists having report about the continuous release of the drug and reported the quick penetration of the nutraceutical through the skin and confers resveratrol cytotoxicity outcome is appropriate for the healing of skin cancer. Vitamin E-loaded nanocapsules were developed in lab-scale and trial was performed in pilot scale followed by storage study at $40 \pm 2^{\circ}C$ temperature and RH of $75 \pm 5\%$ for 3 and 6 months which resulted in particle sizes of 185 nm and 253 nm, respectively; superior chemical, physical properties, and firmness of the nanocapsules authenticated that this dosage form was extremely appropriate for the protection of the consistency of nutraceuticals like vitamins that have shown thermal and UV degradation vulnerability. Therefore, nanoencapsulation of complex bioactive nutrients improves their solubility and preserves their activity and prevents their degradation by providing stability.

Jain et al.[30] fabricated and studied the functional quality of lipidic nano-constructs of lycopene which provides improved cytotoxicity in the cell line of MCF-7 studied for breast cancer. They investigated the anticancer efficiency of lycopene loaded solid lipid nano particles (LYC-SLNs). They involved homogenization–evaporation technique using varying proportion of biocompatible gelucire and compritol ATO 888. Micromeritics properties, cytotoxicity, apoptosis, and cellular uptake in vitro discharge persuaded in MCF-7 cell lines were evaluated.[30]

In another study, the same group formulated whey protein isolate nanoparticles and lycopene was incorporated (LYC-WPI-NPs) to augment the bioavailability of lycopene as well as its anticancer activity. The nanoformulations were identified as sphere-shaped of size ranging 100–350 nm and the entrapment efficiency of lycopene in nanoparticles was found to be in the range of 50 and 65%, and the study also demonstrated subsequently the sustained release of the drug which is 75% following 16th h. The cytotoxicity study of nanoparticles was carried out using MCF-7 breast cancer cell line. It was concluded that protein nanoparticles with entrapped lycopene capable of recovering the biopharmaceutical feature of lycopene confirming that anticancer action, reduced tumor propagation, reduced the inactivation rate in treated mice, therefore demonstrating the possible efficacy of cancer therapeutic interference.[33]

Jain et al.[32] prepared β-carotene-encapsulated SLN using a diverse amalgamation of glyceryl monostearate and gelucire. β-Carotene has anticancer activity but is highly sensitive to oxidation, eventually reduced the bioavailability and spoilage through oral administration. It is rigorously lowered. The categorization studies were carried out for particle size, morphology, stability, and release behavior. Pharmacokinetic studies of BC-SLNs using Wistar rats and in vitro cytotoxicity using MCF-7 breast cancer cell lines were performed, and it was recognized that entrapment of BC in SLNs influences its anticancer effectiveness. A superior anticancer activity, bioavailability of BC on oral administration were observed in BC incorporated SLN as compared with the free BC. They reported that SLN could be an efficient, promising approach to enhance the biopharmaceutical characteristics of carotenoids for their anticancer efficacy.[32]

### 7.7.5 SOLID DISPERSIONS

Other important method, solid dispersion (SD) is applied to improve the bioavailability of oral nutraceuticals. In this method, a nutraceutical is suspended in a water-soluble polymer matrix and solubility was increased, and consequently, the rate of solubilization of the nutraceutical enhanced. Hydrophobic nutraceuticals have slow suspension rates and eventually less bioavailability than enviable giving dosed through GIT. It was found that optimized SD ratio of 1.5:6 of poloxamer 407 and Aerosil 200 used as the carrier for CoQ10 achieved 85% dissolution and 4 times more solubility as compared with only 5% dissolution in control; additional stability of CoQ10 was estimated and established to enhance the quantity of poloxamer 407.

It was reported that when resveratrol is entrapped in SLN with changing types and strength of organization connecting polymer and bioactive, only accurate polymer combination was chosen having the adequate intermolecular bond between polymer and bioactive to hinder recrystallization and for the successful production of stable solid dispersions. Resveratrol combined with eudragit 100 and poly(vinyl pyrrolidine) K29/32 provided the maximum stable structure as solid dispersion to improve the oral bioavailability of many essential nutraceuticals.[91]

## 7.8 SELF-EMULSIFYING DRUG DELIVERY SYSTEM (SEDDS)

Novel SEDDS is being accepted by several research groups because of enhanced oral bioavailability of nutraceuticals. In this system, quick self-emulsification, consequently diffusion in the arrangement is possible in water solution, which directs to an extremely quick enhancement in dissolution. Among other delivery methods for CoQ10, SEDDS provides five-fold higher enhancement for oral bioavailability. The entrapment, stability, and enhanced bioavailability of Tocotrienols were also studied in SEDDS. Further, it was found that 2.5–4.5 times larger $C_{max}$, a higher AUC, and a diminished lag time to incorporation were observed in high-fat meals under SEDDS compared with the non-SEDDS experiments. Similarly, silymarin in SEEDS resulted 99% release in 7 min and favored amalgamation into the lymphatic arrangement compared with only 13.9% release with the dry herbal extract.

Another system of drug delivery is the self-nano emulsifying DDS (SNEDDS) for nutraceutical delivery. Leaf extract of Persimmon containing quercetin and kaempferol when incorporated in SNEDDS, 1.5- and 1.6-times higher bioavailability was obtained. An antimicrobial, aphrodisiac, and anti-inflammatory methoxyflavon-incorporated SMEDDS accomplished marked enhancement in permeation across CaCo-2 cells and bioavailability of the compound; and the site-specific release in the liver was most appropriate for plant extract to prevent first-pass effects and finally high plasma intensity.[18,62]

## 7.9 MICROPARTICULATE SYSTEMS: MICROPARTICLES, MICROSPHERES AND MICROCAPSULES

Jain et al.[35] formulated the entrapment of β-Carotene in microencapsulate by complex coacervation technique using whey protein isolates (WPI) and Gum Acacia (GA) as coating materials. They aimed to study its controlled release behavior. In vitro discharge study revealed first burst release, eventually followed by sustained release of β-carotene.[35]

Also, Jain et al.[31] produced microcapsulated β-carotene using casein and gum tragacanth and this release study resulted in optimum conditions, such as Pr:Ps=2:1, 4.35 pH, encapsule size 159.71±2.16 μm, yield 82.51±0.412%, entrapment efficiency 79.36±0.541%; and maximum electrostatic interaction. Steadiness of encapsulated β-carotene was estimated at selected temperatures (lower (5°C), room (25°C) and high (40°C)) for 3 months. Encapsulation robustly enhanced the steadiness of micronutrients. Their study promotes the use of microcapsules as a carrier for the protection and constant release of micronutrients.[31]

In another study, Jain et al. formulated casein and anionic polysaccharides electrostatic complexes incorporated with lycopene by coacervation method. The functional attributes of electrostatic complex were studied in vivo rat model.[34] They also studied the encapsulation of β-carotene to improve the stability and bioavailability following complex coacervation method with casein (CA) and guar gum (GG). The 65.95±5.33% encapsulation efficiency, a uniform surface morphology were obtained with maximum 98% discharge following zero-order kinetics effectively using Korsmeyer-Peppas model.[86]

## 7.10 PARTICLE COATING FOR PROTECTION, SITE-SPECIFIC DELIVERY AND ENHANCED EFFICACY

Combination of two different techniques is suitable for several release arrangements to guarantee the achievement of effectiveness of nutraceuticals. An attractive approach for entrapping and release of gastric responsive nutraceuticals orally for the production of β-lactoglobulin coating was documented compared with chitosan nanoparticles alone. Although, chitosan having nutraceutical carrier properties is eco-friendly, biocompatible, muco-adhesive, and versatile. Yet, it is unstable at low gastric pH and can degrade quickly, thus it predisposes the nutraceuticals before reaching the targeted site and there is a loss of integrity of nutraceuticals. Therefore, chitosan can be protected by applying coating of β-lactoglobulin to stop the degradation of nutraceuticals at low gastric pH. This is due to the fact that β-lactoglobulin protein is highly resistant to pepsin degradation.

Researchers made an innovative technique to release CoQ10 efficiently as therapeutic agent to treat cataract. Their release system consisted of CoQ10-loaded soy phospatidyl cholin (SPC) liposome (an outstanding drug carrier) and coating material trimethyl chitosan polymer of irregular molecular weights. Liposomes can provide adequate ophthalmic drug formulation, because inertness does not offer any interference with the vision.[72]

The study was carried out to improve bioflavonoid release using the combined tools by producing hesperidin-loaded NLCs coated with methoxyl pectin, alginate, or chitosan.

## 7.11 NUTRACEUTICAL DERIVATIVES FOR IMPROVED EFFICACY AND DELIVERY MECHANISMS

A recent research study has used derivatives of EGCG (such as esterified with docosapentanoic acid and other fatty acids) for improved lipophilicity to give better antiviral, anti-inflammatory, antioxidants activities and to embrace usefulness as a favored substitute to indigenous EGCG molecules.

Vitamins after chemical modification are converted to a new molecule, which can be used efficiently to treat many chronic diseases. Researchers created six novel derivatives of tocopherol that were confirmed using human MCF-7 and MDA-MB-231 breast cancer cell lines to exhibit anticancer activity.

Other vitamin derivatives were developed to deliver the nutraceutical more efficiently. D-α-tocopheryl polyethene glycol succinate (vitamin E—TPGS) is an esterified vitamin E derivative that has been used as water-soluble form of vitamin E and might be applied to make micelles, liposomes, nanoparticles, and many TPGS-based delivery systems for enhanced nutraceutical delivery.[12,97] Table 7.5 indicates delivery systems for nutraceutical composites.

## 7.12 SUMMARY

Nutraceutical-based balancing medicines, vitamins, and herbal formulations are gradually being prepared due to the increasing recognition that nutraceuticals offer several advantages in curing or avoidance of diseases. Therefore, several release mechanisms have been developed to enhance the bioaccessibility of nutrients at diverse target sites during the utilization of a new formulation strategy. Numerous nutraceutical composites may work optimally when combined with multicomponent products. The potential strategy might entail designs, which permit each bioactive nutraceutical to be independently confined, wrapped up, and united with another one. Combination of technologies will facilitate the binding of the strength of distinct components, that in turn will offer progressing in the efficiency of multicomponent nutraceutical products.

**TABLE 7.5** Nutraceutical Delivery Systems.

| Nutraceuticals | Carrier or delivery system | Beneficial effects of the delivery system |
|---|---|---|
| Vitamin E acetate | Liposomal carrier system | Superior deposition in skin[63] |
| Grape seed extract | | better solubility and commencement of action[41] |
| CoenzymeQ10 | | better skin diffusion[48] |
| Melatonin | | Improved oral bioavailability[41] |
| Resveratrol | | Improved stability and sustained liberation formulation[5] |
| Zinc gluconate | | Enhanced AUC of oral formulations.[41] |
| a-Tocopherol/vitamin A | Electrospun fiber mats | CA nanofibers for steady discharge[81] |
| Curcumin | | Ultra-fine fibrous mats for skin liberation[79] |
| Asiaticoside | | Herb-loaded CA mats as an injury dressing[80] |
| Ascorbyl palmitate | | PCL nanofibrous mats for enhanced constancy and antibacterial features |
| Melanin | Microsponges and nanosponges | Melanosponge for skin use; and defense from UV-A and UV-B |
| Resveratrol | | Cyclodextrin (CD)-based nanosponges for better solubility, stability, penetration for topical and buccal delivery[7] |
| Acemannan | | Sponges for renewal of periodontal tissues[11] |
| 7-Dehydrocholesterol | Cyclodextrin (CD) complexation | Hydroxypropyl-b-CD compound with better solubility[44] |
| Silymarin | | b-CD complex with persistent discharge[19] |
| Astaxanthin | | Controlled discharge and better solubility[95] |
| *Kaempferia parviflora* | | Better bioavailability after oral consumption[57] |
| Curcumin | | Composites with enhanced strength and solubility[87] |
| Riboflavin | Biodegradable hydrogels | Provides gut protection and pH-sensitive release[61] |
| Ginseng | | Antioxidant[61] |
| Vitamin $B_{12}$ | | Controlled release[15,78] |
| *Aloe vera* | | Wound healing dressing[67] |
| Riboflavin | | pH controlled release and better vitamin entrapment[1] |

**TABLE 7.5** *(Continued)*

| Nutraceuticals | Carrier or delivery system | Beneficial effects of the delivery system |
|---|---|---|
| Chondroitin sulfate | | Controlled release of Chondroitin[68] |
| Poly N-acetyl glucosamine | | Used degenerative disc disease[26] |
| Lutein | Nanotechnology | Nanocrystal suspensions: increased oral bioavailability[59] |
| 5-Hydroxy-6,7,8,4-tetramethoxyflavone (PMF) | | Nanoemulsions for efficient delivery[50] |
| Vitamins A, D, E, K | | Nanoemulsions with improved absorption[24] |
| β-carotene | | Nanoemulsions with higher strength and bio-accessibility[69] |
| Vitamin E | | Hyaluronic acid nanoemulsion better partition and distribution into stratum corneum[46] |
| CoenzymeQ$_{10}$ | | NLC's with steadiness, high entrapment, biphasic discharge[83] |
| Lutein | | NLC's with controlled release and gut defence[51] |
| Silybin | | NLC's with high AUC and biphasic release[39] |
| Melatonin | | SLNs attain constant plasma intensity[39] |
| Resveratrol | | SLNs for rapid dermal delivery with better efficiency[86] |
| Quercetin | | Nano micelles with better solubility and strength[82] |
| Vitamin D$_2$ | | Nano micelles offer UV guard[75] |
| Vitamin E | | Nanocapsules with firmness and thermal safeguard[43] |
| Epigallocatechin-3-gallate (ECGC) | | Safeguard and discharge in food purposes[76] |
| Lycopene | | WPI nanoparticles for solubility and activity[36] |
| Vitamin C | | Chitosan nanoparticles with improved shelf life and controlled release[3] |
| Thymol | | Better solubility[93] |
| Carvacrol | | Enhanced solubility[93] |
| Vitamin A | Solid dispersions | Quick release[65] |

**TABLE 7.5** *(Continued)*

| Nutraceuticals | Carrier or delivery system | Beneficial effects of the delivery system |
|---|---|---|
| Coenzyme Q10 | | For better solubility, stability, and dissolution[60] |
| Resveratrol | | Better stability[91] |
| Coenzyme $Q_{10}$ | Self-emulsifying drug delivery system (SEDD) | Five-fold increase in bioavailability and rapid suspension[62] |
| Vitamin E | | 2.5–4.5 times higher $C_{max}$ and diminished lag time to absorption[94] |
| Milk thistle | | Better delivery to lymphatic system[53] |
| Persimmon leaf | | 1.5–1.6 times higher delivery than control[50] |
| *Kaempferia parviflora* | | Eelevated permeation through Caco-2 cells[57] |
| *Fadogiaan cylantha* (Makoni tea) *Melissa officinalis* (lemon-balm) *Tussilago farfara* (colts foot) | Microparticulate systems | enhanced solubility, shelf life, steadiness and masking properties on disagreeable odor[72] |
| Vitamin A palmitate | | improved encapsulation efficiency and release profile[56] |
| Lycopene | | Microcapsules with improved stability[30,31] |
| Vitamin E, Ascorbic acid | | longer discharge and guarding the vitamins[65] |
| Coenzyme $Q_{10}$ | Particle coatings | better maintenance time in cornea and steadiness[96] |
| Vitamin E | | Controlled release of bioactives[66] |
| Gastric-sensitive nutraceuticals, e.g.: Hesperetin | | Provides safety from gastric pH and pepsin; and improved absorption better release and steadiness[16] |
| Nutraceutical derivatives | | Better bioactivities: anti-inflammatory, antioxidant, and antiviral enhanced delivery and applications[12,97] |

## KEYWORDS


- **bioavailability**
- **delivery**
- **functional food**
- **nutraceutical**
- **plant bioactive**
- **therapeutic food**


## REFERENCES


1. Abd El-Ghaffar, M. A.; Hashem, M. S.; Mostafa, K. E.; Awady, A. M. R. The pH-Sensitive Sodium Alginate Hydrogels for Riboflavin Controlled Release. *Carbohyd. Polym.* **2012**, *89* (2), 667–675.
2. Ali, R.; Athar, M.; Abdullah, U.; Abidi, S. A.; Qayyum, M. Nutraceuticals as Natural Healers: Emerging Evidences. *Afr. J. Biotechnol.* **2009**, *8* (6), 891–898.
3. Alishahi, A.; Mirvaghefi, M. R.; Tehrani, H.; Maher, F.; Elsabee, Z. Shelf-Life and Delivery Enhancement of Vitamin C Using Chitosan Nanoparticles. *Food Chem.* **2011**, *126* (3), 935–940.
4. Alves-Rodrigues, A.; Shao, A. Science Behind Rutin. *Toxicol. Lett.* **2004**, *150*, 57–83.
5. Amri, A.; Chaumeil, J. C.; Sfar, S.; Charrueau, C. Administration of Resveratrol: What Formulation Solutions to Bioavailability Limitations? *J. Control Release* **2012**, *158*, 182–193.
6. Ansari, K. A.; Azhar, M.; Altamush Siddiqui, M. Energy Analysis of Single-Effect Vapor Absorption System Using Design Parameters. *J. Energy Resour. Technol.* **2021**, *143* (6), 37–45.
7. Ansari, K. A.; Vavia, P. R.; Trotta, F.; Cavalli, R. Cyclodextrin-Based Nanosponges for Delivery of Resveratrol: In-vitro Characterization, Stability, Cytotoxicity and Permeation Study. *AAPS Pharm SciTech.* **2011**, *12*, 279–286.
8. Baylock, R. L.; Maroon, J. Natural Plant Products and Extracts that Reduce Immune Excitotoxicity Associated Neuro-degeneration and Promote Repair Within the Central Nervous System. *Surg. Neurol. Int.* **2012**, *3*, 19–23.
9. Brower, V. Nutraceuticals: Poised for a Healthy Slice of the Healthcare Market? *Nat. Biotechnol.* **1998**, *16*, 728–731.
10. Cencic, A.; Chingwaru, W. The Role of Functional Foods, Nutraceuticals, and Food Supplements in Intestinal Health. *Nutrients* **2010**, *2* (6), 611–625.
11. Chantarawaratit, P.; Sangvanich, W.; Banlunara, K.; Soontornvipart, *P. Thunyakitpisal acemannan* Sponges Stimulate Alveolar Bone, Cementum and Periodontal Ligament Regeneration in a Canine Class Furcation Defect Model. *J. Periodontal Res.* **2013**, *49* (2), 164–178.

12. Chen, W.; Park, S.K; Yu, W.; Xiong,A.; Kline, K. Synthesis and Screening of Novel Vitamin E Derivatives for Anticancer Functions. *Eur. J. Med. Chem.* **2012**, *58*, 72–83.
13. Choonara, Y. E.; Pillay, V.; Carmichael, T. R.; Leith, C. R.; Meyer, L. C.; Toit, D.; Naylor, S.; Wanblad, C. *In Vivo* Evaluation of a Biodegradable Donut-Shaped Minitablet for Prolonged Posterior Segment Drug Delivery in the Rabbit Eye Model. *J. Pharma. Sci.* **2012,** *100* (5), 1819–1832.
14. Dhami, V. P. S.; Chopra, K. The World of Nutraceuticals. *Eastern Pharmacist* **2000,** *2000*, 29–30.
15. Elzoghby, A. O.; El-Fotoh, A. W. S.; Elgindy, N. A. Casein-Based Formulations as Promising Controlled Release Drug Delivery Systems. *J. Controlled Release* **2011**, *153* (3), 206–216.
16. Fathi, M.; Varshosaz, J. Novel Hesperidin Loaded Nano-Carriers for Food Fortification: Production and Characterization. *J. Funct. Foods* **2013**, *5* (3), 1382–1391.
17. Frank, J.; Chin, X. W. D.; Schrader, C.; Eckert, G. P.; Gerald-Rimbach, G. Do Tocotrienols have Potential as Neuroprotective Dietary Factors? *Age. Res. Rev.* **2012,** *11* (1), 163–180.
18. Gholsa Y. N.; Yadav, S. R. Development of Nutraceutical for Chemoprevention: A Review. *Int. J. Pharm. Technol.* **2012**, *4*, 1950–1973.
19. Ghosh, A.; Biswas, S.; Ghosh, T. Preparation and Evaluation of Silymarin β-Cyclodextrin Molecular Inclusion Complexes, *J. Young Pharmacists* **2011**, *3* (3), 205–210.
20. Ghoshal, G. Recent Trends of Innovations in Nutraceutical Research. In *Recent Innovations in Food and Nutritional Sciences*; Giri, A., Ed.; Levant Books Publishers: Kolkata, India, 2020; pp 64–90.
21. Ghoshal, G.; Jain, A.; Katare, O. P. Harnessing Nanotechnology Using Nutraceuticals for Cancer Therapeutics and Intervention. In *Nano Nutraceutricals*; Singh, B., Ed., Chapter 4; CRC Press: Boca Raton, 2018; pp 57–70.
22. Ghoshal, G. Beverage: A Potential Delivery System for Nutraceuticals. In *Nutrient in Beverages*; Grumezescu, A., Holban, A-M., Eds., 1st ed., Chapter 4; Academic Press, 2019; pp 111–142.
23. Ghoshal, G; Kansal, S. K. The Emerging Trends in Functional and Medicinal Beverages and Its Health Implication. In *Functional and Medicinal Beverages*; Grumezescu, A., Holban, A-M., Eds., 1st ed.; Academic Press: New York, 2019; pp 41–71.
24. Gonnet, M.; Lethuaut, L.; Boury, F. New Trends in Encapsulation of Liposoluble Vitamins. *J. Controlled Release* **2010**,*146* (3), 276–290.
25. González-Sarrías, A.; Larrosa, M.; García-Conesa, M. T.; Francisco, A.; Barberán, T.; Espín, J. C. Nutraceuticals for Older People: Facts, Fictions and Gaps in Knowledge. *Maturitas* **2013,** *75*, 313–334.
26. Gorapalli, D.; Vournakis, J.; Seth,A.; Whyne, C. Evaluation of a Novel Poly N-acetyl Glucosamine (pGlcNAc) Hydrogel for Treatment of the Degenerating Intervertebral Disc. *Life Sci.* **2012**, *91*, 25–26.
27. Gupta, S. C.; Kim, J. H.; Prasad, S.; Aggarwal, B. B. Regulation of Survival, Proliferation, Invasion, Angiogenesis, and Metastasis of Tumor Cells Through Modulation of Inflammatory Pathways by Nutraceuticals. *Cancer Metastasis Rev.* **2010**, *29* (3), 405–434.
28. Gupta, S.; Chauhan, D.; Mehla, K.; Sood, P.; Nair, A. An Overview of Nutraceuticals: Current Scenario. *J. Basic Clin. Pharm.* **2010**, *1* (2), 55–62.
29. Halsted, C. H. Dietary Supplements and Functional Foods: Two Sides of a Coin? *Am. J. Clin. Nutr.* **2003,** *77*, 1001–1007.

30. Jain, A.; Sharma, G.; Kuswah, V.; Ghoshal, G.; Singh, B.; Jain, S.; Shivhare, U. S.; Katare, O. P. Fabrication and Functional Attributes of Lipidic Nanoconstructs of Lycopene: An Innovative Endeavor for Enhanced Cytotoxicity in MCF-7 Breast Cancer Cells. *Colloids Surfaces B: Biointerfaces* **2017**, *152*, 482–491.
31. Jain, A.; Thakur, D.; Ghoshal, G.; Katare, O. P.; Singh, B.; Shivhare, U. S. Formation and Functional Attributes of Electrostatic Complexes Involving Casein and Anionic Polysaccharides: An Approach to Enhance Oral Absorption of Lycopene in Rats In Vivo. *Int. J. Biol. Macromol.* **2016**, *93*, 746–756.
32. Jain, A.; Sharma, G.; Ghoshal, G.; Singh, B.; Shivhare, U. S.; Katare, O. P. Beta Carotene-Encapsulated Solid Lipid Nanoparticle (BC-SLNs) as Promising Vehicle for Cancer: An Investigative Assessment. *AAPS-Pharma Sci. Tech.* **2019**, *20* (3), 100–117.
33. Jain, A.; Sharma, G.; Ghoshal, G.; Kesharwani, P.; Singh, B.; Shivhare, U. S.; Katare, O. P. Lycopene Loaded Whey Protein Isolate Nanoparticles: An Innovative Endeavor for Enhanced Bioavailability of Lycopene and Anti-cancer Activity. *Int. J. Pharma.* **2018,** *546* (1–2), 97 105,
34. Jain, A.; Thakur, D.; Ghoshal G.; Katare, O. P.; Shivhare, U. S. Characterization of Microcapsulated β-Carotene Formed by Complex Coacervation Using Casein And Gum Tragacanth. *Int. J. Biol. Macromol.* **2016**, *87*, 101–113.
35. Jain, A.; Thakur, D; Ghoshal, G.;Katare, O. P.; Shivhare, U. S. Microencapsulation by Complex Coacervation Using Whey Protein Isolates and Gum Acacia: An Approach to Preserve the Functionality and Controlled Release of β- Carotene. *Food Bioprocess Technol.* **2015**, *8*, 1635–1644.
36. Jain, A.; Sharm, G.; Kushwah, V.; Ghoshal, G.; Jain, A.; Singh, B.; Shivhare, U. S.; Jain, S.; Katare, O. P. Beta-Carotene-loaded Zein Nanoparticles to Improve the Biopharmaceutical Attributes and to Abolish the Toxicity of Methotrexate: A Preclinical Study for Breast Cancer. *Artif. Cells Nanomed. Biotechnol.* **2018**, *46* (1), 402–412.
37. Jain, P.; Pundir, R. K. Nutraceuticals: Recent Developments and Future Prospective. In: *Recent Trends in Biotechnology and Therapeutic Applications of Medicinal Plants*; Shahid, M.; Shahzad, A.; Malik, A., Sahai, A., Eds.; Springer Science+Business Media: Dordrecht; 2013; DOI 10.1007/978-94-007-6603-7_9.
38. Kalra, E. K. Nutraceutical–Definition and Introduction. *AAPS Pharm. Sci.*, **2003**, *5* (2), E25–E30.
39. Kandimalla, K. K.; Babu, R. J.; Singh, M. Biphasic Flux Profiles of Melatonin: The Yin-Yang of Transdermal Permeation Enhancement Mediated by Fatty Alcohol Enhancers. *J. Pharma. Sci.* **2010**, *99* (1), 209–218.
40. Kaur, P.; Ghoshal, G.; Banerjee, U. C. Traditional Bio-preservation in Beverages: Fermented Beverages. In *Preservatives and Preservation Approaches in Beverages*; Grumezescu, A., Holban, A-M., Eds., 1st ed.; Academic Press: San Diego, 2019; pp. 69–113.
41. Keller, B. C. Liposomes in Nutrition. *Trends Food Sci. Technol.* **2001**, *12* (1), 25–31.
42. Keservani, R. K.; Kesharwani, R. K.; Vyas, N.; Jain, S.; Raghuvanshi, R.; Sharma, A. K. Nutraceutical and Functional Food as Future Food: A Review. *Der Pharmacia Lettre*, **2010**, *2* (1), 106–116.
43. Khayata, N.; Abdelwahed, W.; Chehna, M. F.; Charcosset, C. Stability Study and Lyophilization of Vitamin E-Loaded Nanocapsules Prepared by Membrane Contactor. *Int. J. Pharma.* **2012**, *439* (1–2), 254–259.

44. Kim, S. H.; Youn, J. Y.; Kim, K. M.; Kang, K. C.; Lee, S. J. Characterization of an Inclusion Complex of, 7-Dehydrocholesterol and Cyclodextrin. *J. Ind. Eng. Chem.* **2010**, *16* (1), 119–121.
45. Koletzko, B.; Aggett, P. J.; Bindels, J. G. Growth, Development and Differentiation: A Functional Food Science Approach. *Br. J. Nutr.* **1998**, *80*, 5–45.
46. Kong, M.; Chen, X. G. Investigations on Skin Permeation of Hyaluronic Acid Based Nanoemulsion as Transdermal Carrier. *Carbohyd. Polym.* **2011**, *86* (2), 837–843.
47. Lacatusu, I.; Mitrea, E.; Badea, N.; Stan, R.; Meghea, A. Lipid Nanoparticles Based on Omega-3 Fatty Acids as Effective Carriers for Lutein Delivery. Preparation and *In Vitro* Characterization Studies. *J. Funct. Foods* **2013**, *5* (3), 1260–1269.
48. Lee, W.; Tsai, T. Preparation and Characterization of Liposomal Coenzyme Q10 for In Vivo Topical Application. *Int. J. Pharma.* **2010**, *395*, 1–2.
49. Jia, L.; Zhang, D.; Li, Z.; Feng, F.; Wang, Y.; Dai, W. T.; Duan, C.; Zhang, Q. Preparation and Characterization of Silybin-loaded Nanostructured Lipid Carriers. *Drug Deliv.* **2010**, *17* (1), 11–18.
50. Li, Y.; Zheng, J.; Xiao, H.; McClements, D. J. Nanoemulsion-Based Delivery Systems for Poorly Water-Soluble Bioactive Compounds: Influence of Formulation Parameters on Polymethoxyflavone Crystallization. *Food Hydrocoll.* **2012**, *27* (2), 517–528.
51. Liu, C.; Wu, C. Optimization of Nanostructured Lipid Carriers for Lutein Delivery. *Colloids Surfaces A* **2010**, *353* (2–3), 149–156.
52. Livney, Y. D. Milk Proteins as Vehicles for Bioactives. *Curr. Opin. Colloid Interface Sci.* **2010**, *15* (1–2), 73–83.
53. Losio, T.; Voinovich, D.; Serdoz, F. Oral Bioavailability of Silymarin Phytocomplex Formulated as Self-Emulsifying Pellets, Phytomedicine. *Int. J. Phytother. Phytopharmacol.* **2010**, *18* (6), 505–512.
54. Maltais, A.; Gabriel, E.; Muriel, S. Tableted Soy Protein Cold-set Hydrogels as Carriers of Nutraceutical Substances. *Food Hydrocoll.* **2010**, *24* (5) 518–524.
55. Mani, T. T.; Mukherjee, P. K.; Manimaran, S.; Suresh, B. Indian Herbal Drug Development—Problems and Prospects. *Pharma Times* **2002**, *34*, 13–14.
56. Martínez-Sancho, C.; Herrero-Vanrell, R.; Negro S. Vitamin A Palmitate and Aciclovir Biodegradable Microspheres for Intraocular Sustained Release, *Int. J. Pharma.* **2006**, *326* (1–2), 100–106.
57. Mekjaruskul, C.; Yang, Y. T.; Leed, M. G.; Sadgrove, M. P.; Jay, M.; Sripanidkulchai, B. Novel Formulation Strategies for Enhancing Oral Delivery of Methoxyflavones in *Kaempferia parviflora* by SMEDDS or Complexation with 2 Hydroxypropyl-β-Cyclodextrin. *Int. J. Pharma.* **2013**, *445* (1), 1–11.
58. Braithwaite, M. C.; Tyagi, C.; Tomar, L.; Kumar, P.; Yahya, E.; Choonara, Y. E.; Pillay, V. Nutraceutical-Based Therapeutics and Formulation Strategies Augmenting Their Efficiency to Complement Modern Medicine: An Overview. *J. Funct. Foods* **2014**, *6*, 82–99.
59. Mitri, K.; Shegokar, R.; Gohla, S.; Anselmi, C.; Rainer, H.; Müller, R. H. Lutein Nanocrystals as Antioxidant Formulation for Oral and Dermal Delivery. *Int. J. Pharma.* **2011**, *420* (1), 141–146.
60. Nepal, P. R.; Han, H.; Choi, H. Enhancement of Solubility and Dissolution of Coenzyme $Q_{10}$ Using Solid Dispersion Formulation. *Int. J. Pharma.* **2010**, *383* (1–2), 147–153.

61. Norajit, K.; Kim, K. M.; Ryu, G. Comparative Studies on the Characterization and Antioxidant Properties of Biodegradable Alginate Films Containing Ginseng Extract. *J. Food Eng.* **2010**, *98* (3), 377–384.
62. Onoue, S.; Uchida, A.; Kuriyama, K. Novel Solid Self Emulsifying Drug Delivery System of Coenzyme $Q_{10}$ with Improved Photochemical and Pharmacokinetic Behaviors. *Eur. J. Pharma. Sci.* **2012**, *46*, 492–499.
63. Padamwar, M.; Pokharkar, V. Development of Vitamin Loaded Topical Liposomal Formulations Using Factorial Design Approach: Drug Design and Stability. *Int. J. Pharma.* **2006**, *5*, 11–20.
64. Paneva, D.; Manolova, N.; Argirova, M.; Rashkov, I. Antibacterial Electrospun Poly(ε Caprolactone)/ Ascorbyl Palmitate Nanofibrous Materials. *Int. J. Pharma.* **2011**, *416*(1), 346–55.
65. Patravale, V. B.; Mandawgade, S. D. Novel Cosmetic Delivery Systems: An Application Update. *Int. J. Cosmetic Sci.* **2008**, *30* (1) 19–33.
66. Peng, C. C.; Kim, J.; Chauhan, A. Extended Delivery of Hydrophilic Drugs from Silicone-hydrogel Contact Lenses Containing Vitamin E Diffusion Barriers. *Biomaterials* **2010**, *31*(14), 4032–4047.
67. Pereira, R.; Mendes, A.; Bártolo, P. Alginate/Aloe vera Hydrogel Films for Biomedical Applications. *Procedia CIRP* **2013**, *5*, 210—215.
68. Piai, J. F.; Lopes, L. C.; Fajardo, A. R.; Rubira, A. F.; Muniz, E. C. Kinetic Study of Chondroitin Sulphate Release from Chondroitin Sulphate/Chitosan Complex Hydrogel. *J. Mol. Liquids* **2010**, *156* (1), 28–32.
69. Qian, C.; Decker, E. A.; Xiao, H.; McClements, D. J. Nanoemulsion Delivery Systems: Influence of Carrier Oil on β-carotene Bioaccessibility, *Food Chem.* **2012**, *135*, 1440–1447.
70. Rajasekaran, A.; Sivagnanam, G.; Xavier, R. Nutraceuticals as Therapeutic Agents: A Review. *Res. J. Pharm. Technol.* **2008**, *1* (4), 28–40.
71. Rishi, R. K. Nutraceuticals: Borderline Between Food and Drug? *Pharma Rev.* **2006,** *2006,* 51–53.
72. Sansone, F.; Mencherini, T.; Picerno, P.; Rita, M.; Aquino, P.; Rosaria-Lauro, M. Maltodextrin/Pectin Microparticles by Spray Drying as Carrier for Nutraceutical Extracts. *J. Food Eng.* **2011**, *105* (3), 468–476.
73. Schwartz, C. G. Vitamin D and Intervention Trials in Prostate Cancer: From Theory to Therapy. *Ann. Epidemiol.* **2009**, *19*, 96–102.
74. Schwartz, C. J.; Baisaeng, N.; Hoppel, M.; Löw, M.; Valenta, C. Ultra-small NLC for Improved Dermal Delivery of Coenyzme $Q_{10}$. *Int. J. Pharma.* **2013**, *447* (1–2), 213–217.
75. Semo, E.; Kesselman, E.; Danino, D.; Yoav, D. Livney, Y. D. Casein Micelle as a Natural Nano-capsular Vehicle for Nutraceuticals. *Food Hydrocoll.* **2007**, *21* (5–6), 936–942.
76. Shpigelman, A.; Israeli, G.; Livney, Y. D. Thermally-Induced Protein–Polyphenol Co-assemblies: Beta Lactoglobulin-based Nanocomplexes as Protective Nanovehicles for EGCG. *Food Hydrocol.* **2010**, *24* (8), 735–743.
77. Sirtoti, C. R.; Gali, C.; Anderson, J. W.; Arnold, A. Nutritional and Nutraceutical Approaches to Dyslipidemia and Atherosclerosis Prevention: Focus on Dietary Proteins. *Atherosclerosis* **2009**, *203* (1), 8–17.
78. Song, F.; Zhang, L.; Shi, J.; Li, N. Novel Casein Hydrogels: Formation, Structure and Controlled Drug Release. *Colloids Surface B* **2010**, *79* (1), 142–148.

79. Suwantong, O.; Opanasopit, P.; Ruktanonchai, U.; Supaphol, P. Electrospun Cellulose Acetate Fiber Mats Containing Curcumin and Release Characteristic of the Herbal Substance. *Polymer* **2007**, *48*, 7546–7557.
80. Suwantong, O.; Ruktanonchai, U.; Supapho, P. Electrospun Cellulose Acetate fiber Mats Containing Asiaticoside (*Centella asiatica*) Crude Extract and the Release Characteristics of Asiaticoside. *Polymer* **2008**, *49* (19), 4239–4247.
81. Taepaiboon, P.; Rungsardthong, U.; Supaphol, P. Vitamin-loaded Electrospun Cellulose Acetate Nanofiber Mats as Transdermal and Dermal Therapeutic Agents of Vitamin A Acid and Vitamin E. *Eur. J. Pharma. Biopharma.* **2007**, *67* (2), 387–97.
82. Tan, B.J; Liu, Y.; Chang, K. L.; Lim, B. K. W.; Chiu, G. N. C. Perorally Active Nanomicellar Formulation of Quercetin in the Treatment of Lung Cancer. *Int. J. Nanomed.* **2012**, *7*, 651–661.
83. Teeranachaideekul, V.; Souto, E. B.; Junyaprasert, V. B.; Muller, R. H. Cetyl Palmitate-Based NLC for Topical Delivery of Coenzyme $Q_{10}$—Development, Physicochemical Characterization and In-vitro Release Studies. *Eur. J. Pharma. Biopharma.* **2007**, *67* (1), 141–148.
84. Telrandhe, U. B.; Kurmi, R.; Uplanchiwar,V.; Mansoori, M. H.; Jain, V.; Jain, R. K.; Jain, S. K. Nutraceutical: A Resource in Modern Medicine. *Int. J. Univ. Pharm. Life Sci.* **2012**, *2* (1), 179–195.
85. Teskac, K; Kristl, J. The Evidence for Solid Lipid Nanoparticles Mediated Cell Uptake of Resveratrol. *Natl. J. Pharma.* **2010**, *390* (1), 61–69.
86. Thakur, D.; Ghoshal, G.; Jain, A.; Shivhare, U. S.; Katare, O. P. Microencapsulation of β-carotene Based on Casein/Guar Gum Blend Using A Zeta Potential-Yield Stress Phenomenon: An Approach to Enhance Photo-stability and Retention of Functionality. *AAPS Pharm. Sci. Tech.* **2017**, *18*, 5–10.
87. Tonnesen, H. H.; Másson, M.; Loftsson, T. Studies of Curcumin and Curcuminoids. XXVII. Cyclodextrin Complexation: Solubility, Chemical and Photochemical Stability. *Int. J. Pharma.* **2002**, *244*, 127–135.
88. Verma, S. An Empirical Analysis of Consumer's Attitude Towards OTC Health Supplements in India. *Int. J. Ind. Culture Busi. Manage.* **2009**, *2* (1), 1–10.
89. Viard, J.; Souberbielle, J.; Kirk, O.; Reekie, J.; Knysz, B.; Losso, M.; Gatell, J.; Pedersen, C.; Bogner, J. R.; Lundgren, J. D. Vitamin D and Clinical Disease Progression in HIV Infection: Results from the EuroSIDA Study. *AIDS* **2011**, *25* (10), 1305–1315.
90. Weber, H. A.; Guthrie, J. R.; Hodges, A. E.; O'Brien, B. M. Comparison of Pro-anthocyanidins in Commercial Antioxidants: Grape Seed and Pine Bark Extracts. *J. Agric. Food Chem.* **2007,** *55* (1), 148–56.
91. Wegiel, L.; Edgar, K. J.; Mauer, L. J.; Taylor, L. S. Scope of Nutraceuticals. *J. Pharma. Sci.* **2013**, *102*, 171–181.
92. Wijesinghe, W. A. J. P.; Jeon, Y. Biological Activities and Potential Industrial Applications of Fucose Rich Sulfated Polysaccharides and Fucoidans Isolated from Brown Seaweeds: A Review. *Carbohyd. Polym.* **2012**, *88* (1), 13–20.
93. Wu, Y.; Luo, Y.; Wang, Q. Antioxidant and Antimicrobial Properties of Essential Oils Encapsulated in Zein Nanoparticles Prepared by Liquid–Liquid Dispersion Method. *LWT Food Sci. Technol.* **2012**, *48* (2), 283–290.
94. Yap, S. P.; Yuen, K. H. Influence of Lipolysis and Droplet Size on Tocotrienol Absorption from Self-Emulsifying Formulations. *Int. J. Pharma.* **2004**, *281* (1–2), 67–78.

95. Yuan, C.; Xu, X.; Jin, Z. Inclusion Complex of Astaxanthin with Hydroxypropyl-β-cyclodextrin: UV, FTIR, NMR and Molecular Modeling Studies. *Carbohydrate Polymers*, **2012**, *89* (2), 492–496.
96. Zhan, J; Wang, S. Topical Use of Coenzyme $Q_{10}$-Loaded Liposomes Coated with Trimethyl Chitosan: Tolerance, Precorneal Retention and Anti-Cataract Effect. *Int. J. Pharma.* **2009**, *372* (1–2), 66–75.
97. Zhang, Z.; Tan, S.; Feng, S. Vitamin E TPGS as a Molecular Biomaterial for Drug Delivery. *Biomaterials* **2012**, *33* (19), 4889–4906.
98. Zhao, J. Nutraceuticals, Nutritional Therapy, Phytonutrients, and Phytotherapy for Improvement of Human Health: A Perspective on Plant Biotechnology Application. *Recent Patents Biotechnol.* **2007**, *1*, 75–97.

# CHAPTER 8

# POTENTIAL OF BDELLIUM TREE (*COMMIPHORA WIGHTII*) FOR NUTRACEUTICALS

ANUPAM JAISWAL, MEGHA VERMA, SHREYA CHITNAVIS, STUTI PANDEY, RAKSHA AGRAWAL, KAMAL SHAH, DURGESH NANDINI CHAUHAN, and NAGENDRA SINGH CHAUHAN

## ABSTRACT

To overcome the side effects of synthetic drugs, herbal nutraceuticals have been suggested. Extracts of these nutraceuticals have therapeutic action to reduce the development of several diseases. *Commiphora wightii* produces oleo-gum resin guggul that contains guggulsterone as the chief compound with antiarthritic, anti-inflammatory, and hypolipidemic properties. Guggulsterone has two isomers E-guggulsterone and Z-guggulsterone, which show their antagonistic property against farnesoid-X receptor which have anti-inflammatory activities. This chapter suggests that guggul can be used as a nutraceutical with promising therapeutic activity against many chronic diseases.

## 8.1 INTRODUCTION

Evaluation studies have shown that plant-based edibles (such as fruits, nuts, vegetables) have a direct impact on human health and disease balance. As our lifestyle has evolved, these age-related chronic diseases (such as diabetes, arthritis, CVS disorders) get related to dietary habits. This factuality has

Plant-Based Bioactive Compounds and Food Ingredients: Encapsulation, Functional, and Safety Aspects.
Junaid Ahmad Malik, Megh R. Goyal, Preeti Birwal, & Ritesh B. Watharkar (Eds.)

uplifted the thought of various health organizations around the world to suggest an increased intake of plant-derived products to enhance human health and slow down the development of diseases. These products can be used as health supplements and are termed nutraceuticals. Additionally, they will ease the consumer disappointment on expensiveness, as these plant-derived products are low in cost as compared with synthetic drugs. "*Let food be the medicine and medicine be the food*" has been rightly quoted by Hippocrates.[8]

The term "*nutraceuticals*" was coined by "Stephen De Felice" and can be written as "nutrition" and "pharmaceutical." Stephen De Felice suggests that nutraceuticals can be described as "*a food that provides medical or health benefits including the prevention and/or treatment of diseases.*" Whereas Canada Health Organization describes it as "a *product prepared from food but sold in the form of pills.*"[2,3] The Food and Drug Administration (FDA) regulates nutraceuticals under different sets of regulations.

Many compounds fall under the nutraceuticals category, such as phyto-constituents (such as phytosterol, flavonoid, glycosides), probiotics, vitamins with different medicinal properties to enhance human health. Phytosterols have the main therapeutic property to lower reactive oxygen species, and this property can be used to form nutraceuticals against inflammatory diseases.[4] *Commiphora wightii*, an Indian bdellium tree, is a flowering plant to produce oleo-gum resin "*guggul*" with significant medicinal applications. The main chemical constituent in *Commiphora wightii* is "guggulsterone" (GS), which shows promising effectiveness against osteoarthritis, rheumatoid arthritis, obesity, urinary disorder.[5] This property of guggul can be used to form ayurvedic formulations as nutraceuticals.

This chapter explores the potential of *Commiphora wightii* as nutraceutical by studying its morphology, distribution, and uses of its chemical constituents.

## 8.2 NATURAL DISTRIBUTION AND MORPHOLOGY OF *COMMIPHORA WIGHTII* (INDIAN BDELLIUM TREE)

*Commiphora wightii* is a type of flowering plant belonging to the family Burseraceae. There are four species of the genus *Commiphora*, such as *C. wightii*, *C. caudate, C. agallocha*, and *C. stocksiana*, and all four species are distributed in India.[5] Out of these four species, *C. wightii* is well-distributed in states, such as Maharashtra, Rajasthan, Gujarat, Karnataka, and Madhya

Pradesh of India. *Commiphora wightii* is well-known for its fragrant oleo-gum resin "guggul" with pharmacological actions.

Guggul (Figure 8.1) has pale brown color and it exudes out from the bark of the tree during November through February.[5] *Commiphora wightii* is a highly branched, small shrub with a sharp ending at the end of the branches. *C. wightii* plant bears unisexual flowers with brown-red color, whose size is relatively small. Different species of *Commiphora* have different flower arrangements: either they are found in groups or found in a single arrangement. Its fruit is ovoid with green to red color.[6]

**FIGURE 8.1** Guggul gum.

## 8.3 CHEMICAL CONSTITUENTS OF *COMMIPHORA WIGHTII*

Oleo-gum resin of *Commiphora wightii* (Figure 8.2) has shown many therapeutic effects. The chief chemical compound in guggul is guggulsterone (GS), which has two stereoisomers, they are E-guggulsterone and Z-guggulsterone. Other than these, guggul gum resin contains 0.37% essential oil, which comprises myrecene, dimyrcene, polymyrcene, and guggulsterone-I, II, III.[9] The extraction process uses ethyl acetate as a solvent. The insoluble fraction of guggul gum in ethyl acetate solvent is characterized as a carbohydrate.

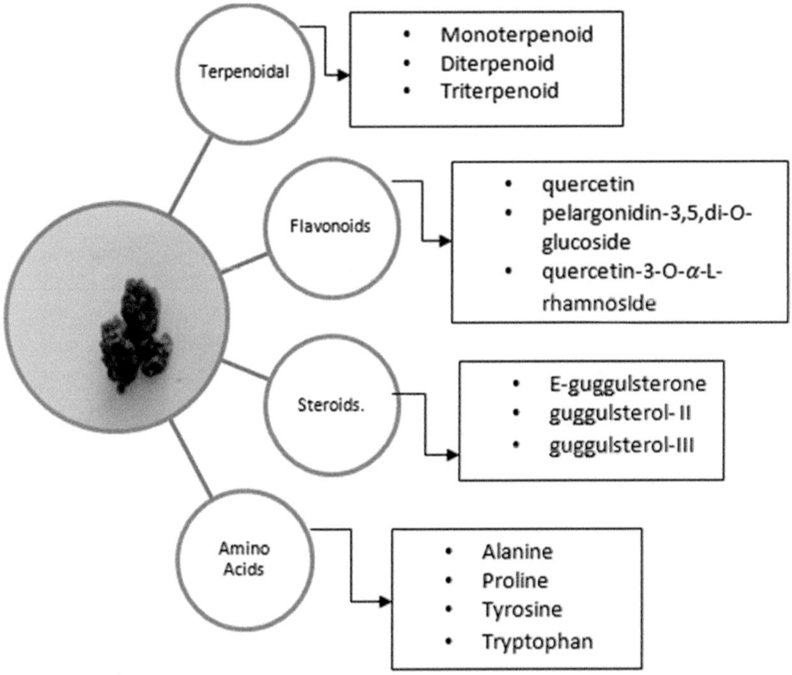


**FIGURE 8.2** Chemical constituents of guggul.

Guggulsterone (GS) is very important with a specific property, and GS is the only plant product which is antagonistic to farnesoid-X receptor (bile acid receptor). This receptor mainly regulates the bile acid synthesis and lipid metabolism.[15] Guggulsterone-E and guggulsterone-Z are known to lower inflammation by inhibiting NF-κB.

Different types of formulations are available, which contain gum guggul with different therapeutic actions, such as lowering cholesterol, lowering arthritic pain, and much more. Guggulsterone (GS) has been recommended to standardize the formulation so that nutraceuticals must contain a sufficient quantity of guggulsterone in it.

## 8.4 THERAPEUTIC ACTIVITIES OF GUGGUL

### *8.4.1 ANTI-INFLAMMATORY AND ANTIARTHRITIC ACTIVITY*

Inflammation caused by stress, poor diet, and poor lifestyle may result in many disorders. Arthritis is an inflammatory disorder of joints, which can cause severe pain in joints, such as the knee, hip, shoulder. Both arthritis and general inflammation can affect an individual's lifestyle. To overcome this, guggul-based nutraceuticals can be used. Guggulsterone, the chief compound of guggul, has shown a reduction in swelling around joints and can reduce pain.[16] Guggulsterone also inhibits NF-κB receptor which helps in reducing the inflammation. It has been documented that guggul is highly effective against osteoarthritis (OA).[18] Guggul extract shows an effective decrease in knee pain and swelling of the knee joint. It also increases knee flexion and helps in increasing walking distance. Guggul has been reported to have a very high efficacy against arthritis in the preclinical trials.[11]

### 8.4.2 HYPOLIPIDEMIC ACTIVITY

Hypolipidemic causes various disorders such as atherosclerosis in which cholesterol is deposited in arterial walls resulting in difficulty in the blood flow. The first animal study was conducted during 1964–1966 using the rabbit model, and it was demonstrated that guggulsterone is effective in lowering serum cholesterol levels in rabbit when hyperlipidemia is induced in rabbit.[18] Guggul resin is an antagonist of farnesoid-X receptor, it shows hypolipidemic activity. The extract of guggul resins blocks the HMG-CoA in the HMG pathway due to which cholesterol biosynthesis is blocked, which lowers the LDL level in serum.[18]

### 8.4.3 THYROID STIMULATING ACTIVITY

Various experimental studies have suggested that mainly Z-guggulsterone can enhance thyroid activity by increasing the iodine intake by the thyroid.[19] Guggulsterone also increases the production of T3 and T4 hormones, which can also help in lowering lipid content in serum.

### 8.4.4 CARDIOPROTECTIVE ACTIVITY

Guggul extract has shown cardioprotective activity by preventing platelet aggregation, as platelet aggregation is the main factor for coronary artery diseases. Gum guggul also treats ischemia, which helps in treating heart disorders.[18]

### 8.4.5 CHRONIC KIDNEY DISEASE (CKD)

Guggul has been reported to show reduced inflammation in chronic kidney disease patients by modulating the prostaglandin E2 (PGE2)[17] and due to an increase in inflammatory cytokines levels by the intake of guggul.[13]

### 8.4.6 ASTHMA

Guggul has shown an enhanced prognosis against several signs and symptoms of bronchial asthma in various patients. On undergoing treatment with guggul, the signs and symptoms, such as rhondi, dyspnea, and asthma attacks were reduced.[7]

### 8.4.7 DIABETES MELLITUS

The intake of guggul has shown effective results in decreasing the risk factors related to *diabetes mellitus*. With the help of guggul, patients were able to maintain the fructosamine levels and hepatic enzyme activities. Guggul also helps to maintain the lipid profiles near normal levels in diabetic patients.[1]

### 8.4.8 ECZEMA AND PSORIASIS

Guggul shows greater efficacy in the treatment of eczema and psoriasis. It diminishes the grade of erythema and skin superficial symptoms and it has also been shown to decrease the usage of topical corticosteroids. It shows a great improvement in the scales of psoriasis and eczema. It also improves both erythema and itching.[20]

### 8.4.9 BREAST FIBROADENOMAS

Guggul is found to be beneficial in the treatment of breast fibroadenomas. Guggul reduces fibroadenoma size and volume in young women without having any harmful effects.[14]

### 8.4.10 FASCIOLIASIS

Guggul is found to be safe, well-tolerated, and effective in the management of fascioliasis, which is a parasitic worm infection. The patients reportedly show significant improvement in the general condition in which there is a decrease in the fasciola egg count. It also shows a great improvement in signs and symptoms of the disease without having any adverse side effects after the treatment.[12]

### *8.4.11 GINGIVITIS*

Guggul has been found to be very effective against gingivitis. Guggul has been reported to cause a significant decline in the inflammatory index of the patient after treatment.[10]

## 8.5 AVAILABLE GUGGUL FORMULATIONS

### *8.5.1 AMRITA GUGGUL*

The ingredients of Amrita guggul include Guduchi, Guggul, Haritaki, Amalaki, Dantimula, Trivritimula, Pippali, Bibhitaki, Vidangna. Amrita guggul is used for treating leprosy, malignant jaundice, loss of appetite, enlargement of the spleen, etc.[18]

### *8.5.2 ABHAGUGGUL*

It contains ingredients, such as Babool, Haritaki, Bibhitaki, Amalaki, Maricha, Shunthi, Shudhaguggulu. It can cure the conditions of osteoporosis and low bone density and increases bone strength.[18]

### *8.5.3 ABHAYADI GUGGUL*

It contains ingredients, such as Haritaki, Munakha, Dalchini, Lvang, Harida, Sunthi, Pippali, etc. It is used to treat constipation, indigestion, and gout and can cure mania.[18]

### *8.5.4 DASHYANG GUGGUL*

It contains the following ingredients: Sunthi, Maricha, Pippali, Haritaki, Bibhitaki, Amalaki, Musta, etc. It can cure obesity, nervous problems, and rheumatic infection and is used as an alternative, tonic, and stimulant.[18]

### 8.5.5 *LAKSHAGUGGULU*

It contains various ingredients: Laksha, Asthisamhrita, Arjun, Ashwagandha, Nagbala, Shudhaguggulu. It is mainly used to treat bone fracture and osteoporosis.[18]

### 8.5.6 *NAVAKA GUGGUL*

It has the following ingredients: pippali, sunthi, maricha, haritaki, bibhitaki, amalaki, vidang, musta, etc. It can be used in the treatment of PCOD-related obesity, diabetes, and also it relieves rheumatoid arthritis pain.[18]

## 8.6 SUMMARY

This chapter explains the potential therapeutic effect of guggul which can be utilized in nutraceuticals. *Commiphora wightii* (Indian bdellium tree), is a flowering plant which secretes oleo-gum resin, guggul from its branches which have many pharmacological properties. The chief constituent in guggul was found to be guggulsterone, which has two isomers E-guggulsterone and Z-guggulsterone. Guggulsterone has many pharmacological effects and has anti-inflammatory, antiarthritic, antioxidant, and cardioprotective properties. This property of guggul can be utilized in nutraceuticals and can help individuals suffering from chronic health diseases. This chapter suggests that guggul can be used to formulate nutraceuticals to treat chronic diseases.

## KEYWORDS

- **Commiphora**
- **guggul**
- **guggulsterone**
- **nutraceuticals**
- **pharmacological**

## REFERENCES


1. Ahangarpour, A.; Heidari, H.; Fatemeh, R. A.; Pakmehr, M.; Shahbazian, H.; Ahmadi, I. Effect of *B. serrate* Supplementation on Blood Lipid, Hepatic Enzymes and Fructosamine Levels in Type 2 Diabetic Patients. *J. Diab. Metabol. Disorders* **2014,** *13* (1), 29–36.
2. Behera, L.; Mehta, A.; Dholariya, C.; Sukhadiya, M.; Gunaga, R.; Patel, S.; Vegetative Propagation of Guggul (*Commiphora wightii* (Arn.) Bhan.): A Commercially Important and Threatened Medicinal Plant Species. *e-Planet* **2020,** *18* (2), 164–169.
3. Bhatia, A.; Bharti, S.K.; Tripathi, T.; Mishra, A.; Sidhu, O. P.; Roy, R.; Nautiyal, C.S.; Metabolic Profiling of *Commiphora wightii* (guggul) Reveals a Potential Source for Pharmaceuticals and Nutraceuticals. *Phytochemistry* **2015,** *110*, 29–36.
4. Chaudary, G. Pharmacological Properties of *Commiphora wightii*: An Overview. *Int. J. Pharm. Pharma Sci.* **2012,** *4*, 73–75.
5. Das, L.; Bhaumik, E.; Raychaudhuri, U.; Chakraborty, R. Role of Nutraceuticals in Human Health. *J. Food Sci. Technol.* **2012,** *49,* 173–183.
6. Espín, J.C.; García-Conesa, M.T.; Tomás-Barberán, F. A. Nutraceuticals: Facts and Fiction. *Phytochemistry* **2007,** *68* (22–24), 2986–3008.
7. Gupta, I.; Gupta, V.; Parihar, A.; Gupta, S.; Lüdtke, R.; Safayhi, H. Effects of *B. serrata* Gum Resin in Patients with Bronchial Asthma. *Eur. PubMed Central* **1998,** *3* (11), 511–514.
8. Jain, A.; Gupta, V.B. Chemistry and Pharmacological Profile of Guggul: A Review. *Ind. J. Trad. Knowl.* **2006,** *5* (4), 478–483.
9. Kamat, S. D. An Unpublished Medical Treatise Dravyaratnākara-Nighaṇṭu. *Ann. Bhandarkar Orient. Res. Inst.* **2006,** *87*, 75–91.
10. Khosravi Samani, M.; Mahmoodian, H.; Moghadamnia, A.; Poorsattar Bejeh Mir, A.; Chitsazan, M. The Effect of Frankincense in the Treatment of Moderate Plaque-Induced Gingivitis. *DARU J. Pharma. Sci.* **2011,** *19* (4), 288–294.
11. Kimmatkar, N.; Thawani, V.; Hingorani, L.; Khiyani, R.; Efficacy and Tolerability of *B. serrata* Extract in Treatment of Osteoarthritis of Knee. *Phytomedicine* **2003,** *10* (1), 3–7.
12. Massoud, A.; El Sisi, S.; Salama, O.; Massoud, A. Preliminary Study of Therapeutic Efficacy of a New Fasciolicidal Drug Derived from *Commiphora molmol* (myrrh). *Am. J. Trop. Med. Hygiene* **2001,** *65* (2), 96–99.
13. Moreillon, J. J.; Bowden, R. G.; Deike, E.; Griggs, J.; Wilson, R.; Shelmadine, B. The Use of an Anti-Inflammatory Supplement in Patients with Chronic Kidney Disease. *J. Complement. Integr. Med.* **2013,** *10* (1), 143–152.
14. Pasta, V.; Dinicola, S.; Giuliani, A.; Harrath, A. H.; Alwasel, S. H.; Tartaglia, F. A Randomized Trial of Boswellia in Association with Betaine and Myoinositol in the Management of Breast Fibroadenomas. *Eur. Rev. Med. Pharmacol. Sci.* **2016,** *20,* 1860–1865.
15. Poonia, P.; Mittal, S. K.; Gupta, V. K. Gum guggul: An Ayurvedic Boom. *Int. J. Pharmacogn. Phytochem. Res.* **2014,** *6* (2), 347–352.
16. Sairkar, P.K.; Sharma, A.; Shukla, N. P. Estimation of Guggulsterone -E and -Z in the Guggul-Based Commercial Formulations Using High-Performance Thin-Layer Chromatography. *J. Pharm. Bioallied Sci.* **2017,** *9* (1), 1–7.
17. Shelmadine, B. D.; Bowden, R. G.; Moreillon, J. J.; Cooke, M. B.; Yang, P.; Deike, E.; Giggs, J. O.; Wilson, R. L. Effects of an Anti-Inflammatory Supplement on Eicosanoid

Derivatives in Patients with Chronic Kidney Disease. *J. Altern. Complement. Med.* **2017,** *23* (8), 632–638.

18. Singh, H.; Satija, S.; Kaur, H.; Khurana, N.; Sharma, N.; Vyas, M.; Singh, T.G.; Mahajan, S.; Mehta, M. Novel Drug Delivery Approaches for Guggul. *Plant Arch.* **2019,** *19* (2), 983–993.
19. Thosar, S. L.; Yende, M. R.; Cultivation and Conservation of Guggulu (*Commiphora mukul*). *Ancient Sci. Life* **2009,** *29* (1), 22–25.
20. Togni, S.; Maramaldi, G.; Di Pierro, F.; Biondi, M. A Cosmeceutical Formulation Based on Boswellic Acids for the Treatment of Erythematous Eczema and Psoriasis. *Clin. Cosmetic Investig. Dermatol.* **2014,** *7*, 321–327.

# PART III

# HEALTH BENEFITS OF PLANT-BASED FOODS

## CHAPTER 9

# BIOACTIVE COMPOUNDS AND PHYTONUTRIENTS FROM CEREALS

SWATI SETHI, LOVEJOT KAUR, PRERNA NATH, and DEEP N. YADAV

### ABSTRACT

Apart from providing the basic nutrients required for growth, maintenance, and normal functioning of body, cereal grains contain enormous amount of phytonutrients that can promote health. Being plant-based, they cause no side effects that are typically accompanied with chemical or radiation treatments. In the past, phytonutrients in cereal grains have been used in traditional preparations as aqueous extracts for managing several chronic ailments. This chapter discusses: (1) potential sources of phytochemicals from whole grain cereals, pseudo-cereals, specialty cereals (pigmented grains), and their by-products. and (2) the suitable technologies for extraction of phytochemicals from cereals.

### 9.1 INTRODUCTION

In addition to fulfilling the basic nutritional requirements (such as carbohydrates, proteins, minerals, and vitamins) of human body, cereal grains are an indispensable source of phytochemicals. Nutritional phytochemicals are non-nutrient, bioactive plant components that have been linked to a lower risk of chronic illness.[103] The consumption of cereal grains has been documented to reduce the risk of developing several chronic ailments due to the presence of phytochemicals and bioactive compounds. Recently, there has been increased awareness among consumers toward health and the minimal use of synthetic additives.

Plant-Based Bioactive Compounds and Food Ingredients: Encapsulation, Functional, and Safety Aspects.
Junaid Ahmad Malik, Megh R. Goyal, Preeti Birwal, & Ritesh B. Watharkar (Eds.)

Phytochemicals have demonstrated important role in prevention of cancer and heart diseases, induced apoptosis, diabetes, osteoporosis, and aging due to their anti-oxidants, anti-inflammatory, chemo-preventive, hepatoprotective, neuroprotective, hypolipidemic, anti-allergic, and anti-microbial activities. The phytochemicals present in cereal grains, either alone or in combination with other components (synergistic effect) have remarkable therapeutic prospective in curing various illnesses. Phytonutrients or bioactives can be supplemented with processed food products to provide many functional benefits. Also, the phytonutrients extracted from their natural sources using conventional or advanced extraction technologies can be added in processed food products to intensify the clinical functions of the food products. Such food products are known as functional foods or health foods.

Quantitatively, the phytochemical content in respective food grains can be enhanced through appropriate genetic engineering techniques and food processing technologies (such as germination, fermentation, and extrusion), which portray a substantial role in augmenting the phytonutrient content. This kind of processing treatments affects the cell-wall structure, which ultimately, facilitates the transformation of bound form of phytochemicals into their free forms.[70]. Phytochemicals (Table 9.1) are classified into four categories, such as: (1) terpenoids; (2) phenolic metabolites; (3) alkaloids; and (4) other nitrogen-containing plant constituents.

Cereals contain a range of phytonutrients, such as carotenoids, flavonoids, tocopherols, lipoic acids, polyphenols, and glucosinolates. More precisely, whole grains are rich sources of phytonutrients, as these phytonutrients are particularly confined to the bran (a multi-layered outer covering of the grain) fraction of grain followed by germ. Non-conventional or under-explored varieties of pigmented cereal grains (such as pigmented rice (red and black), pigmented maize (blue, purple), pigmented wheat (red and purple), pigmented sorghum (black), and pigmented barley (black)) are excellent sources of bioactive compounds.

Substantial research studies have been piloted on the extraction of phytochemicals from cereal grains using conventional extraction technologies. Correspondingly, fruits and vegetable wastes have been extensively exploited for the extraction of phytochemicals using advanced or novel non-thermal processing technologies at laboratory scale, while the cereal by-products are minimally or insignificantly utilized.

This chapter focuses on potential of whole grain cereals (including non-conventional genotypes) and their by-products, as sources of phytochemicals.

**TABLE 9.1** Classification of Plant-Based Phytochemicals.

| Terpenoids | Phenolic metabolites | Alkaloids | Nitrogen-containing plant constituents | Prebiotics |
|---|---|---|---|---|
| Bufadienolides | Anthochlors | Amaryllidaceae | Cyanogenic glycosides | Dietary fiber |
| Cardenolides | Anthocyanins | Betalain | Glucosinolates | Oligosaccharides |
| Carotenoids | Coumarins | Diterpenoid | Nonprotein amino acids Amines | Resistant starch |
| Cucurbitacins | Benzofurans | Indole | Purines | |
| Diterpenoids | Flavonoids | Isoquinoline | Pyrimidines | |
| Iridoids | Chromones | Lycopodium, | | |
| Lactones | Flavonones | Monoterpene | | |
| Monoterpenoids | Flavonols | Sesquiterpene | | |
| Nortriterpenoids | Isoflavonoids | Peptide | | |
| Phytosterols | Phenolic acids | Pyrrolidine | | |
| Saponins | Lignans | Piperidine | | |
| Sesquiterpenoids Sesquiterpene | Phenols | Pyrrolizidine | | |
| Triterpenoid | Phenylpropanoids | Quinoline, | | |
| Other triterpenoids | Phenolic Ketones | Quinolizidine | | |
| | Stilbenoids | Steroidal | | |
| | Quinonoids | Tropane compounds | | |
| | Xanthones, Tannins | | | |

## 9.2 STRUCTURAL OVERVIEW OF PHYTOCHEMICALS IN CEREALS

The major phytochemicals in cereal grains are phenolic acids, flavonoids, carotenoids, phytosterols, tocols, lignin, ϒ-amino butyric acid, and dietary fibers. Structurally, phenolic acid comprises of a phenol ring that possesses a minimum of one carboxylic acid functionality. In plant metabolites, these refer to two diverse groups of organic acids (hydroxybenzoic and hydroxycinnamic acid) that exist with unique carbon structures (Figure 9.1). Structurally, flavonoids involve two aromatic rings (A and B rings) that are connected with an oxygenated heterocycle ring that is known as C ring (Figure 9.2). The presence of diverse structural groups in the C ring of flavonoids categorizes them into isoflavonoids, flavonols, flavanols, flavones, flavanones, and anthocyanins. Anthocyanins are predominant groups of water-soluble flavonoids.

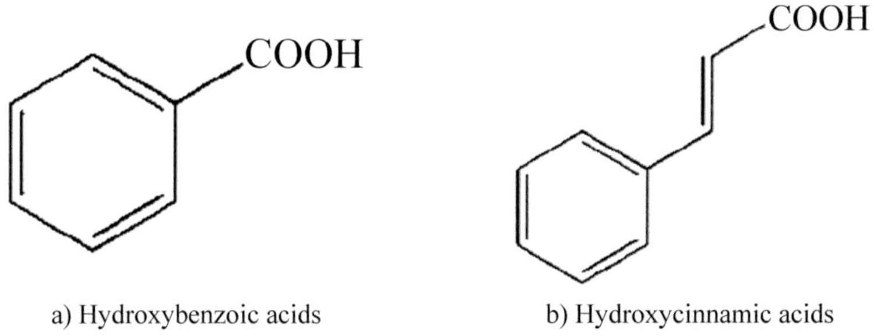


**FIGURE 9.1** Chemical structure of (a) hydroxybenzoic acids and (b) hydroxycinnamic acids.

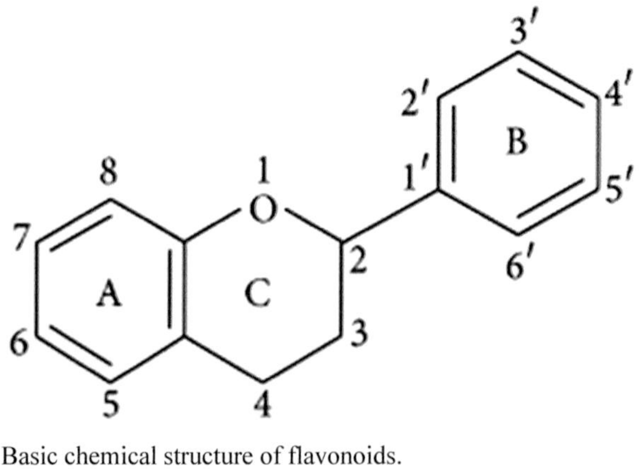


**FIGURE 9.2** Basic chemical structure of flavonoids.

Tetraterpenes having conjugated double bonds and delocalized π-electrons are known as carotenoids. Carotenoids are composed of 40-carbon basic structure of isoprene units that are cyclized either at one or both ends. Commonly, carotenoids exist in nature in their trans-arrangements (Figure 9.3). Due to the elongated sequence of conjugated double bonds, carotenoids display singlet oxygen quenching and light-absorbing ability.

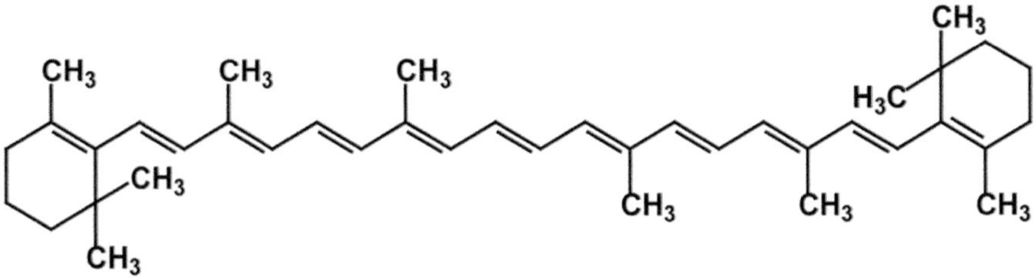


**FIGURE 9.3** Chemical structure of β-carotene.

Phytosterols are group of plant sterols and stanols, which are analogous to cholesterol, with only difference in the side chains (Figure 9.4). Tocopherols and tocotrienols exist as eight vitamers (isomers), such as

- α-tocotrienol;
- β-tocotrienol;
- Υ-tocotrienol;
- α-tocopherol;
- β-tocopherol;
- Υ-tocopherol, and
- δ-tocopherol with a common arrangement composed of a phytol side chain composed of isoprenoid units and 6- hydroxychroman group (Figure 9.5).

Tocopherols possess saturated phytol side chains, whereas tocotrienols consist of unsaturated phytol side chain. Due to structural and functional resemblance to 17-β-estradiol, lignans are grouped under phytoestrogens (Figure 9.6). Υ-Aminobutyric acid is a non-protein amino acid and functions as a major inhibitory neurotransmitter. It is principally synthesized from the

ϒ-decarboxylation of L-glutamic acid and the reaction is catalyzed by the enzyme glutamate decarboxylase.

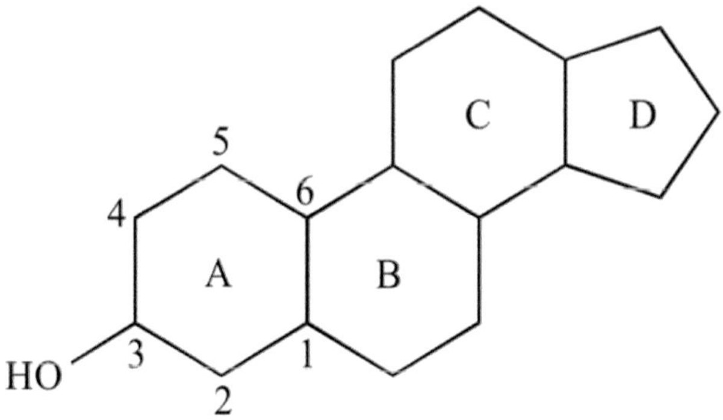


**FIGURE 9.4** Basic chemical structure of sterol.

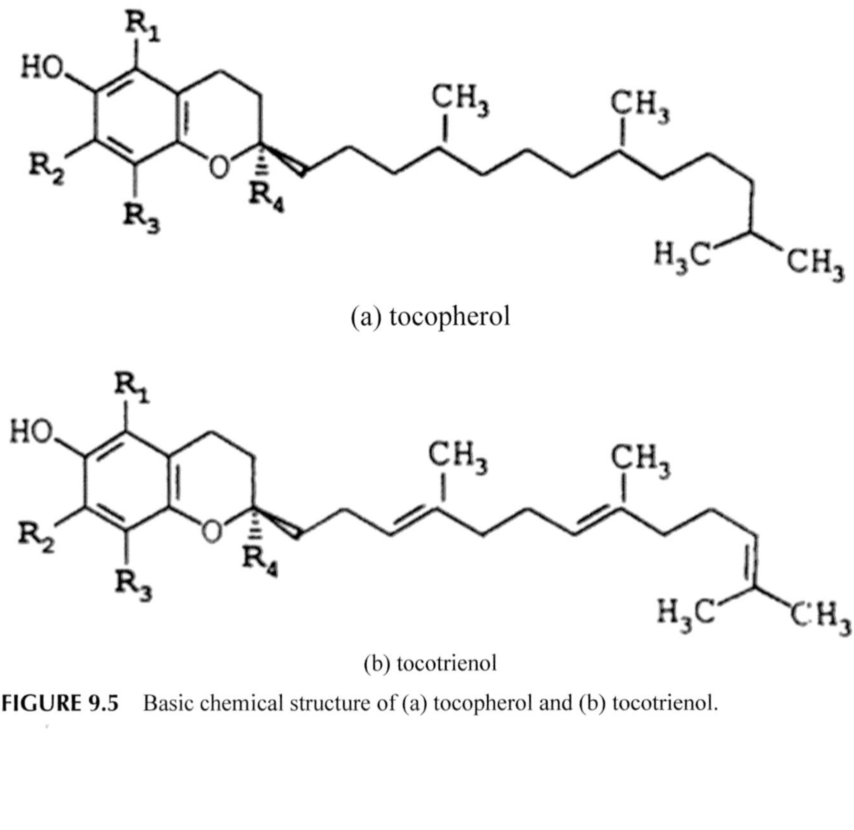


**FIGURE 9.5** Basic chemical structure of (a) tocopherol and (b) tocotrienol.

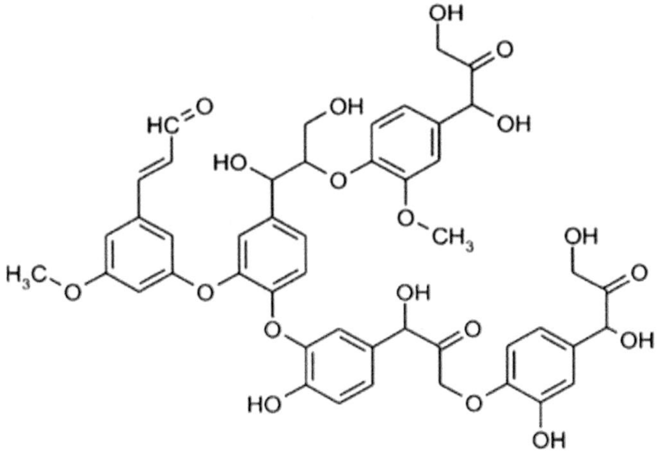


**FIGURE 9.6** Chemical structure of lignin.

## 9.3 PHYTONUTRIENTS AND SECONDARY METABOLITES IN CEREALS

### *9.3.1 WHEAT*

Structurally, the wheat grain consists of germ (or embryo); endosperm (the starchy portion, which also delivers nutrition during germination); the heavy cell-walled layer of aleurone (outer layer of endosperm) and the pericarp. The outer bran fraction of wheat is composed of the pericarp, testa, aleurone, and hyaline layers. On percent weight basis, the grain of wheat is composed of 14–16% husk, 2–3% germ or embryo, and 81–84% starchy endosperm. In addition to the basic essential nutrients required for growth and maintenance, wheat is also a source of health promoting phytonutrients. Several researchers have indicated the antioxidant properties and associated benefits of wheat and wheat-based products.

The major phytochemicals present in wheat grain are phenolics, carotenoids, and fiber. The phytochemicals are mainly confined to the outer bran layer of grain, which is also a by-product of wheat milling industry. Therefore, the bran can be incorporated as such in numerous processed foods to increase their nutritional or functional value; and the bran can be utilized for the extraction of phytochemicals by using suitable extraction technologies.

The phytochemical profile in grain and their concentration are dependent on various factors, such as variety or genotype, growing conditions, and the interface between environmental factors and genotype. Furthermore, food medium and cooking or processing conditions influence the bioavailability of phytochemicals.

Different studies have revealed that the phenolic acid and alkylresorcinols do exist in the germ/bran fractions of whole wheat grain and constitute 83% of total phenolic content, which is subsequently supported by higher antioxidant activity in bran fraction compared to other milled fractions.[105] The functional or physiological effect of wheat bran is divided into three categories: (1) nutritional effect (due to presence of essential nutrients); (2) mechanical effects (due to fiber content and its effect on gastrointestinal tract); and (3) antioxidant effects (attributed by the phytonutrients such as phenolic acid and alkylresorcinols).[185]

The phenolic acids and alkylresorcinols are linked with reduced risk of cardiovascular diseases (CVD). The phenolic antioxidants of wheat bran have displayed impediment in oxidation of low-density lipoproteins by binding with apolipoprotein-B[216]; while alkylresorcinols stimulate thromboxane production, inhibition of platelet binding to fibrinogen and hinder the formation of triglyceride, which in turn reduces the risk of CVDs. These antioxidant phytochemicals have also been identified to protect physiologically essential components from oxidative damage, such as DNA, proteins, and membrane lipids.[229]

Ferulic acid (a derivative of cinnamic acid (phenolic acid), 72% of which exist in bound form) contributes to approximately 60% of antioxidant activity in bran fraction.[185] Phenolic compounds (such as feruloyl oligosaccharides) have shown to protect human erythrocytes against free radical-induced oxidative damage.[204]

Apart from the phenolic compounds, the other important phytonutrients are carotenoids which function as antioxidants. The carotenoids such as lutein, zeaxanthin, and cryptoxanthin were ranged 25–145, 8.5–27, and 1–13.5 μg/100 g, respectively, in wheat varieties of different experimental lines.[75,153] The presence of carotene in wheat grain has also been reported.[75] Carotenoids act as radical scavengers and their mode of action is depicted by two proposed mechanisms[129]: (1) firstly, they act as free radical scavengers and (2) secondly, they also perform their role by taking part in gap junctional communication. This enables for cell-to-cell communication, which has the benefit of slowing tumor-causing cells' proliferation. Tocopherols are another important phytonutrients in wheat grain and act as antioxidants.

Furthermore, wheat grain is an imperative source of fiber, which is mainly confined to the outer bran layers. The whole wheat grain is comprised of fiber in a range of 11.6–12.7% on dry weight basis.[33] Since, the major portion of fiber is located in the outer layer and the bran layer is one of the most important sources of fiber and contains 46% of non-starch polysaccharide. The important non-starch polysaccharide present in wheat bran such as arabinoxylan, cellulose, and β-glucan were 70, 24, and 6%, respectively, of the total non-starch polysaccharide.[114]

The favorable health effect of consumption of fibers is to reduced risk of particular chronic ailments such as cardiovascular disease (CVD), certain cancers, type 2 diabetes, and metabolic syndrome.[46,121] Researchers also advocated the role of multidimensional arrangement of various bioconstituents in the wholegrain matrix that might work collectively to divulge proposed health benefits.[54]

Fibers present in whole grain function as prebiotics and are usually defined as non-digestible food constituents that work by selectively encouraging the growth and/or activity of one or a narrow range of favorable bacteria (probiotics) in the colon to divulge progressive.[62] Prebiotics are intended for their beneficial health benefits on bowel health, which in turn are linked with the reduced risk of colon cancer and CVDs.[40] The defensive mechanism of prebiotic constituents of dietary fiber in wheat bran (arabinoxylan, β-glucans, etc.) against colon cancer works in three ways[110]:

- The first mechanism takes into account of recognized impact on the dilution of possible carcinogens and backer of carcinogens by production of bulky stool that lowers the access to cells lining the colon.
- The second mechanism describes the accelerated transit of fecal material through the colon, leading to reduced access to fecal constituents of the colonic epithelial cells.
- The third mechanism is interpreted by fermentation of the wheat bran fibers to short chain fatty acids throughout the colon that might restrain carcinogenesis by affecting proliferation, differentiation, and apoptosis of colonocytes,[197] and stimulation of the immune system as well.[168]

The presence of fibers in wheat grain is also linked to the increased antioxidant activity.[102] This effect was attributed to the release of bound phenolic acids during colonic fermentation. Ferulic acid exists in bound form in whole wheat grain and is typically bound to arabinoxylans and other non-digestible polysaccharides.[118] During fermentation of non-digestible polysaccharides, the gastrointestinal enzyme esterase releases bound ferulic acid, which later contributes to the antioxidant activities.[10]

Additionally, wholegrain wheat contains higher levels of sulphur containing amino acids such as methionine and cysteine (0.6%).[54] These are identified as predecessor of glutathione (intracellular antioxidant) synthesis and contribute in regulating the oxidative status in cell.[123]

### 9.3.2 RICE

Rice is consumed in the form of whole rice (with bran), polished rice (without bran), and numerous processed products (idli, dosa, flaked rice, parched rice, weaning foods, extruded products, noodles, pasta, pudding, and fermented alcoholic beverages). The Rice processing industry is a well-established industry and manufactures a number of primary, secondary, and tertiary processed products along with a huge aggregate of by-products (mainly bran and husk). Conventionally, these by-products were mainly utilized for animal feeding, but with technological advancements through scientific interventions, increased awareness about health among consumers and economic utilization of all fractions generated during processing operations (milling), the by-products are used to formulate high-value products for food and pharmaceutical use.

Rice bran, a nutrient-dense portion of rice grain, is a valuable by-product of the rice milling industry that may be used to extract high-value components like phytonutrients and oil. Pericarp, aleurone, sub-aleurone layer, and germ are all components of rice bran. The genotype, environmental or growing conditions, and rice milling methods all affect the composition and phytochemical profile of rice bran.[67]

Rice (*Oryza sativa* L.) occurs in nature in diverse shades, such as white, brown, red, black, and purple with white (polished) being the most commonly cultivated. The colored rice genotypes are superior to white rice varieties in terms of their phytochemical and nutritional profiles. The rice bran represents 5–8% of the total grain and it contains an array of phytochemicals, such as, gamma oryzanol, tocopherols, and tocotrienols.[124,200]

The level of gamma oryzanol and tocols in rice bran ranged from 0.9 to 2.9% and from 0.10 to 0.14%, respectively, and are essentially reliant on environmental and genetic factors. Due to exclusively high amount of natural antioxidants, rice bran is the only by-product among agro-industrial by-products that find applications in pharmacy, cosmetics, and nutrition.[44] Rice bran is a good source of 18–22% edible oil with an adequate fatty acid profile contributing to its nutraceutical value.

The bioactive chemicals present in rice bran are phytosterols, tocotrienols, oryzanols, polycosanols, squalene, ferulic acid, phytic acid, gamma amino butyric acid, and inositol hexaphosphate, etc.[13] Oryzanols and Tocols are the two important groups of natural antioxidants in the rice bran with maximum antioxidant activity displayed by tocotrienols, oryzanols, and tocopherols[2] and their concentration differs with the variety of rice.[78]

Sterols exist in free and esterified form in plant products. The ferulic acid esters of phytosterols are referred as oryzanol. The biological and physiological effects of oryzanols are antioxidant[142] and anti-blood cholesterol lowering agents.[169]

Inhibition of tumor promotion, growth, promotion, and stimulation of the hypothalamus, treatment of nerve imbalance and menopausal disorders, protection of cells from oxidative deterioration of plasma very low-density lipoprotein, hypolipidemic effect, anti-cancer activity, and reduced risk of coronary heart disease are some of the other health benefits associated with dietary consumption of ϒ-oryzanol.[6,210,212] Gamma-oryzanol is also found to have beneficial effects on the management of type 2 *diabetes mellitus* as it normalizes the blood glucose level by raising glucokinase activity and its generation in the liver.[85] ϒ-Oryzanol exhibited inhibition of tumor elevation in tumor-bearing mice by inactivating macrophages, stimulating natural killer cells activity, and inhibiting angiogenesis.[91]

δ-Tocotrienol also showed anti-diabetic effects as it supports in improving the insulin sensitivity and utilization of body glucose and by operating as a peroxisome proliferator-activated receptor in diabetic mice.[53]

Pigmented rice varieties are enriched rice due to their edge on phyto-chemical and nutritional profiles in comparison to white rice. Pigmented or colored rice is considered as specialty rice due to their uniqueness in terms beneficial health benefits owing to the occurrence of anthocyanins and other phenolic compounds.[165] Among the pigmented rice varieties, red rice and black rice are the most commonly consumed varieties and are mainly cultivated in Asian countries.

Black rice is known to have an exceptionally high content of anthocyanins, which are types of flavonoids that act as antioxidants and are mainly located in the bran layer of the rice kernel. The antioxidant activity of anthocyanins is attributed to their ability to impede the development or to moderate the levels of free radicals that impair the reactive cells.[6] The anthocyanin fractions in black rice bran are about 26.3% with cyanidin-3-O-glucoside and peonidin-3-O-glucoside accounting for about 90% of total anthocyanins.[34] The reported health benefits with the consumption of black rice are[133,213,218,226]:

- anti-diabetic effect (due to exhibition of aldose reductase inhibitory activities by cyanidin-3- glucoside and pelargonidin-3-glucoside),
- hypocholesterolemic effect (due to reduction in cholesterol, LDL cholesterol, and concentration of triacylglycerol by anthocyanins),
- anti-oxidative effect (due to exhibition of scavenging and quenching activity for reactive oxygen species by phenolics),
- anti-carcinogenic effect,
- anti-atherosclerosis effect (because of a decrease in the extent of atherosclerotic plaques related to increased glutathione peroxidase),
- anti-allergic activities (due to inhibited histamine release), and
- gastric acid secretion and platelet aggregation are both inhibited.

The American Health Association and the American Cancer Society have suggested an increased ingestion of black rice to avert the risk of heart-related ailments and specific types of cancers. Correspondingly, the red rice varieties are also found to be a good source of natural antioxidants; however, the phytochemical profiles of red and black rice vary considerably.

The predominant phenolic acids in red rice cultivars are ferulic, protocatechuic, and vanillic acid, while protocatechuic acid predominates in black varieties followed by vanillic and ferulic acids. Red rice varieties exhibited superior DPPH and ABTS+antioxidant activities compared to black varieties due to presence of pro-anthocyanidins.[144]

The total anthocyanin content was estimated to be much higher in black rice cultivars (79.5–473.7 mg/100 g) than the red rice cultivars (7.9–34.4 mg/100 g); whereas, a narrow difference was observed in the total phenolic contents of red (460.32–725.69 mg/100 g) and black (417.11–687.24 mg/100 g) rice. In terms of oxygen radical absorbing capacity, the different rice cultivars are ranked as follows: red (69.91–130.32 μmol Trolox/g)>black (55.49–64.85 μmol Trolox/g)>green (35.32 μmol Trolox/g)>white (21.81 μmol Trolox/g) rice. Antioxidant capacity of different rice cultivars estimated as ferric reducing antioxidant power (FRAP) varied within 0.9–8.1 mmol Fe (II)/100 g of dry matter.[180]

The fiber fractions (hemicellulose rich) from defatted rice bran have shown positive health effects on diabetes management by slowing down the glucose absorption and by enhancing the utilization of glucose by the colonic fermentation products of fiber. Due to these health benefits, defatted rice bran has been extensively exploited as a raw material for fiber extraction and also as a functional constituent in the preparation of numerous processed food products, such as bakery products, pasta products, meat emulsions, and batter mixes.

### 9.3.3 MAIZE

Beyond the role of maize as basic nutrition, it provides desirable health benefits. The maize grain is composed of endosperm, germ, and bran with each part having different phytochemical profile. As in other grains, the phytochemicals in grain are mainly confined to bran and germ fractions. Likewise, the phytochemical profile depends on the variety, and climatic conditions, etc.

The phytochemicals in all types of maize grain are flavonoids and phenolic acids, sterols, lignins, and bound phytochemicals. A range of total phenolics and phenolic acids in maize bran are flavonoids, anthocyanins, ferulic acid, vannilic acid, coumaric acid, syringic acid, and caffeic acid. Of the whole grain, 87% of these phenolic compounds are located in bran and germ fractions. The health benefits linked with these phytochemicals in maize include lowered risk of cardiovascular disease,[196] type 2 diabetes,[214] obesity,[120] certain type of cancers,[131] and improvement of digestive tract health.

Ferulic acid is the predominant phenolic acid in the maize grain and is principally found in bound esterified form associated with structural cell-wall constituents, such as proteins, celluloses, and lignins. Consequently, to increase the bioavailability of ferulic acid for appropriate antioxidant action, processing technologies (such as thermal processing, fermentation, and freezing) have been found to intensify the total antioxidant activity by 44% owing to release of bound phenolic acids.[47]

Pigmented corn varieties (such as blue, red, and purple) exhibit enhanced concentration of anthocyanidins (325 mg/100 g on dry basis) including peonidin derivatives (15–20%), cyanidin derivatives (75–90%), and pelargonidin derivatives (5–10%). Yellow corn is a rich source of carotenoids (823 g/100 g of DW) together with zeaxanthin (40%), lutein (50%), β-carotene (4%), β-cryptoxanthin (3%), and α-carotene (2%). Highest total antioxidant activity (181.4±0.86 mol of ascorbic acid equivalent/g) was exhibited by maize between common staple grains, such as rice and wheat.[6] Foremost contributors to the antioxidant properties of corn are phytochemicals. The approximately 87% of the ferulic acids and flavonoids (major constituents of total phenolics of corn) are present in bound form; and these directly contribute to the total antioxidant activity.[6]

Dewanto et al.[47] observed that higher antioxidant activity was in heat processed sweet corn in comparison to raw sweet corn. Total phenolic content varies in different varieties of corn, and the range is from 243.8±4.6 to 320.1±7.6 mg of gallic acid equivalent/100 g of dry weight of corn. Lignins (matairesinol, lariciresinol, secoisolariciresinol, and pinoresinol) are also present in maize grain in significant amounts contributing to anticancer and antioxidant activities.

Predominant group of phenolic phytochemicals are flavonoids in maize. Anthocyanins get accumulated in the bran portion of maize grain and are accountable for the color of the grain. The majority of the anthocyanins (up to 50%) are present in the pericarp portion of maize grain. Depending on the concentration of anthocyanins, the maize grains exist in different colors (such as blue, red, and purple) with purplish red corn kernel having the rich amount of anthocyanins (141.7 mg anthocyanins/100 g of flour); and blue kernels possess lowest anthocyanins (62.7 mg/100 g of flour).[128] Beneficial health benefits based on epidemiological studies include reduced risk of diabetes, cardiovascular disease, and cancers.[103] Anthocyanins are recognized to divulge a good effect on digestive health with their anti-inflammatory activities.

Wu et al.[208] suggested the beneficial effect of anthocyanins consumption on reduced chances of colon cancer by escalating fecal content of butyric acid and also with reducing the body weight by 16.6% at treatment dose of 200 mg/kg. Luna-Vital et al.[108] suggested the biological effect through another mechanism, "anthocyanins restrain the production of triacylglycerol and fatty acids by holding back the level of mRNAs enzymes, thus diminishing the buildup of triacylglycerol in adipose tissue and liver."

Carotenoids are also present in maize grain. The concentration of carotenoids classifies the grain as yellow, orange, and red maize. Carotenoids play two important roles in human body, such as (1) they act as pro-vitamin A by converting carotenoids into retinol (Vitamin A) and (2) as antioxidants that can scavenge reactive singlet oxygen radicals.

Mechanical homogenization and thermal processing technologies have been found to escalate the bioavailability of carotenoids. Xanthophylls (lutein and zeaxanthin) in yellow maize also play an important role in the eye health by preventing oxidative damage and eye-stress. Xanthophylls are selectively occupied by the macula area in eye, where they absorb 90% of blue light (450–470 nm) consequently decreasing the short wave length of light from arriving the vital portion of eye.[163] The average xanthophyll concentration in maize grain is about 21.97 ug/g of corn.

The vitamin E functions as a natural antioxidant and also maintains the membrane integrity in the human body. They cause declining risk of certain cancers, cardiovascular diseases,[159] and also enhancing function of immune system and repairing DNA damage. All vitamers of vitamin E are available in corn except β-tocotrienol. The total average vitamin E concentration in yellow corn is 66.9 mg/kg of DW (dry weight) and is mainly restricted to the germ fraction of corn.[66]

Phytosterols are abundant in corn oil, with a concentration of 5.13–9.79 g/kg of plant sterols as compared to 1–5 g/kg in vegetable oils.[152] The majority of oil content is mainly confined to the germ portion of maize (24.2–30.7%) followed by pericarp and endosperm fractions (0.4–1.2%). Sitosterol is the main phytosterol available in corn that accounts for 77–87% of total phytosterols in corn and 13–23% of campestanol (while the endosperm portion of grain is rich in phytostanols).[71]

Rich intakes of plant sterols (1.6 g/day) have been accompanied with the reduction in cholesterol and serum LDL levels without disturbing HDL cholesterol levels. This effect might be attributed to the competition between phytosterols and cholesterol in the intestine, consequently impeding the cholesterol absorption.[136]

Additionally, the maize grain is found to contain type 2 resistant starch, which is complex to digest due to its native granule structure. This type of starch is present in higher amount in high amylose maize grain. Due to the slow digestion properties exhibited by such type of starch granules, RS2 showed beneficial effect on the management of diabetes and colon health.[132]

### *9.3.4 BARLEY*

Barley (*Hordeum vulgare* L.) is underutilized by the food processing industries due to its lower palatability (typical flavor and darker color or appearance). Out of total cultivated barley, 65% is utilized as animal feed,

33% is used for malting and only 2% is used as direct human consumption.[186] Barley is an important cereal crop due to its analogous nutritional value and associated functional properties.

The phytochemicals in barley are phenolic acids, flavonoids, lignans, tocols, phytosterols, and folates.[57] The concentration or level of phytonutrients is generally dependent on genotypic and/or environmental factors. The phytochemicals present in barley exist in free, bound, or conjugated form. The pigmented varieties (purple and black) of barley are rich in pigment anthocyanins and are nutritionally superior to the conventional/non-pigmented varieties.

Phenolic acids in barley are mainly confined to the outer portions of the grain and are generally present in bound form followed by free and conjugated forms. The benefits of barley phenolic acids are linked to the reduced risk of some chronic illness.[51] The predominant phenolic acids present in barley grain are ferulic acid (27%), vanillic acid (28%), syringic acid (17%), and p-coumaric acid (22%).[59] The free, conjugated, and bound forms of phenolic acid concentration in barley ranged between 4.6–23, 86–198, and 133–523 mg/g, respectively, whereas the total phenolic acids varied between 604 and 1346 mg/g.[1]

Flavonoids in barley grain are located in the pericarp portion of grain. Among flavonoids, flavanols, anthocyanins, and proanthocyanidins are predominant flavonoids in barley grain. The major anthocyanins in pigmented barley grain are cyanidin-3-glucoside, peonidin-3-glucoside, delphinidin-3-glucoside, pelargonidin, pelargonidin glycosides, and petunidin 3-glucoside. The degree of color depth in pigmented barley grain is linked to the concentration of anthocyanins. As is the case in other cereals, the occurrence of anthocyanins in barley grain is also limited to the bran fraction of the grain. The chief anthocyanins (cyanidin 3-glucosode, peonidin 3- glucoside, and pelargonidin 3-glucoside) accounted for 50–70% of the total anthocyanins in barley grain. The flavonoid concentration in barley grain of different genotypes (127 lines) ranged from 62 to 301 mg/g.[89] The associated health benefits of barley flavonoids are reduced risk of cancer and coronary heart diseases.

Lignans have biological properties, such as antioxidant, protective effects against coronary heart disease, antitumor, antimicrobial, estrogenic, and antiestrogenic activities.[183] Different types of lignans are reported in barley grain by Smeds et al.[178]

Barley is paramount source of tocols among cereal grains due to presence of all eight biologically active vitamers in considerably higher concentration and balanced ratios.[126] Tocols are acknowledged for antioxidant properties,[31] anticancer and cancer suppression effects,[187] immune system, moderating the risk of CVD, and apoptosis induction.[21,135] Also, a remarkable effect of tocols in clearing blockages of atherosclerosis in the carotid artery was observed, which in turn reduces the risk of stroke.[199]

The level of total tocotrienols and phytosterols in the oils extracted from different barley fractions ranged from 2911 to 6126 mg/g. Moreau et al.[127] demonstrated the concentration of tocols in whole grain barley in the following order: α-tocotrienol > α-tocopherol > ϒ –tocotrienol > ϒ-tocopherol > β-tocotrienol > ϒ-tocotrienol. The tocotrienol content in barley varied from 70.6 to 76.8% with majority of tocotrienols present in endosperm and pericarp portions of grain, while the tocopherols are confined to bran fraction of grain. A significant portion of tocols is present in the hulls of barley grain than the hull less barley. This indicates that barley can be an important source of tocotrienol among all cereals.

Barley is an effective source of phytosterol that showed positive effects in reducing cholesterol levels, reduced risks of CVD, and prevention of colon cancer.[82,126] Barley phytosterols are present in free and bound forms, which are often esterified to fatty acids, steryl glucosides, phenolic acids, or acylated steryl glycosides with different levels of esterification depending on the variety or genotype. The phytosterol level among different varieties of barley ranged from 820 to 1153 mg/g with sitosterol as the most abundant phytosterol followed by campesterol.[9] Like other phytochemicals, phytosterols are also limited to the outer grain layers.

Barley is a good source of β-glucan, a kind of soluble fiber. Lower plasma cholesterol, decreased glycemic response, weight control, and prebiotic effects are some of the favorable impacts of β-glucan on human health.[79] The amount of β-glucan in barley varies from 2.5 to 11.3%. Whole grain barley products with β-glucan levels of 0.75 g per serving or 3 g per day can now claim to lower the risk of coronary heart disease, according to the US Food and Drug Administration (FDA). Due to these stated health benefits, barley is being utilized as a commercial source of raw material for isolation and purification of β-glucan that can be utilized in preparation of functional foods. In addition to pure form of β-glucan, numerous functional food products prepared using β-glucan are available in market.

### *9.3.5 SORGHUM*

Phytochemicals such as phenolics, plant sterols, and policosanols are abundant in sorghum. Under the phenolic compounds category, the two major groups in sorghum are flavonoids and phenolic acids. Benzoic or cinnamic acid derivatives, such as ferulic acid, protocatechuic acid, syringic acid, caffeic acid, p-coumaric acid, and sinapic acid, are the most abundant phenolic acids in sorghum.[205] The bulk of phenolic acids found in the bran layer of sorghum grains are esterified to cell wall polymers in bound form. Ferulic acid is the most abundant phenolic acid in sorghum grain, and it contributes to the grain's strong antioxidant capabilities and health benefits.

The main flavonoids in sorghum grain are tannins and anthocyanins.[94] Sorghum grains contain significant amount of tannins and are classified into three categories on the basis of extractable tannin content as: type I (no significant levels of tannins extracted by 1% acidified methanol), type II (tannins that are extractable in 1% acidified methanol and not alone methanol), and type III (tannins that are extractable in both acidified form of methanol and methanol alone).[155]

Since this classification does not explain the difference in levels of anthocyanins and other phenolic constituents, an additional classification was suggested on the basis of grain appearance/color and total extractable phenols. On this basis, the sorghum grain is classified as:

- White sorghum (with no detectable anthocyanins or tannins and very low levels of extractable phenol);
- Red sorghum (lack tannins, a red pericarp with rich amount of extractable phenols);
- Black sorghum (pericarp is black and has high levels of anthocyanins);
- Brown sorghum (pigmented testa with significant levels of tannins and varying degrees of pericarp pigmentation).

Tannins in sorghum grain are mostly polymerized derivatives of flavan-3-ols and/or flavan-3,4-diols and are found as condensed tannins. The predominant anthocyanin in sorghum grain is 3-deoxyanthocyanidins, which is a rare type of anthocyanin found in food grains.[65] Structurally, these types of anthocyanins do not have a hydroxyl group at the third carbon and occur in nature distinctly as aglycones[184] with better stability in acidic solutions as compared to other anthocyanins.[188] Sorghum 3-deoxyanthocyanidins have a potential as a natural food colorant owing to their high stability.[17] The consumption of these anthocyanins is linked to the antioxidant and anticancer effects in human population.

Phytosterols present in sorghum are analogous in composition to those exist in maize. The sorghum phytosterols mainly include stanols or free sterols and their fatty acid or ferulate esters.[15] Sterols and stanols have almost similar structure with only difference of occurrence of a double bond at fifth position in sterols. The free phytosterols identified in sorghum grain consist of sitosterol, campesterol, and stigmasterol.[15] A level of 0.5 mg/g of total phytosterols has been reported in sorghum grain by Singh et al.[177] However, only limited information is available on sorghum phytosterols in the scientific literature.

### 9.3.6 MILLETS

Owing to the advantages of being drought-resistant, pest resistance, disease resistance, and short growing season, millet grains are receiving enormous scientific attention, particularly in the developing countries. Millets are of the following types:

- Finger millet (*Eleusine coracana*),
- Pearl millet (*Pennisetum glaucum*),
- Kodo millet (*Paspalum setaceum*),
- Barnyard millet (*Echinochloa utilis*),
- Proso millet (*Penicum miliaceum*),
- Foxtail millet (*Setaria italic*) and
- Little millet (*Panicum sumatrense*).

Apart from food uses, the millet grains are utilized for bioethanol and biofilm production in developed countries. Millets are high energy, nutrient dense grains that are being utilized in developing countries to combat malnutrition and also in developed countries to prevent or treat diseases like obesity, diabetes, CVD, etc. Millet grains are rich sources of phytochemicals and micronutrients. The major categories of phytochemicals in millets are phenolic acids, lignans, β-glucan, inulin, resistant starch, phytates, sterols, tocopherol, and carotenoids. The phenolic compounds in millets include phenolic acids and tannins with small amounts of flavonoids. These compounds contribute to the antioxidant properties and also impart important functions in inducing the immune system.

Phenolics are mostly found in the outer layers of the bran layer, which include the testa, aleurone layer, and pericarp. Approximately 60% of total polyphenols in millet grain are located in the seed coat tissue, which amounts for about 12% of the total seed mass. Millet grains contain phenolic chemicals in the form of soluble conjugates and free and insoluble bound forms.

According to Rao and Muralikrishna,[160] the major bound phenolic acid is ferulic acid (18.60 mg/100 g), the major free phenolic acid is protocatechuic acid (45.0 mg/100 g) in the millet grain. The chief bound forms of phenolics present in finger millets, that is, ferulic acid and p-coumaric acid accounts for 64–96 and 50–99%, respectively, of total phenolic acid of millet grains.[174] Clinical studies on diabetic rats suggested improved antioxidant status and better control on blood glucose level with a speedy dermal wound healing process from finger millet.[157] Dietary polyphenols have received tremendous attention due to their health promoting properties.

Phenolic extracts of several millet varieties (such as finger, foxtail, kodo, proso, pearl, and little millets) expressed significant antioxidant, metal chelating, and reducing powers. It is estimated that fox-tail millets contain 3.34 mg tocopherol/100 g (wet basis) and 47 mg polyphenolics/100 g; however, proso millet contains 2.22 mg tocopherol/100 g (wet basis) and 29 mg polyphenolics/100 g. Finger millets are superior source of proanthocyanidins, also referred as condensed tannins[50] with high antioxidant properties.

Antioxidant activities of finger, foxtail, proso, and little millets were estimated to be more and total carotenoids levels ranged from 78 to 366 mg/100 g in different millet varieties. The level of total tocopherols in proso and finger millet varieties was superior (3.6–4.0 mg/100 g) to that of little millet and foxtail varieties (approximately 1.3 mg/100 g). However, HPLC estimation of β-carotene demonstrated that it lacks in the millets and vitamin E signified higher levels of and α- and γ-tocopherols; however, it exhibited lower concentrations of tocotrienols in millets. Millets are rich in phytic acid, which is known for its cholesterol lowering properties.[41]

Pearl millet is rich source of soluble and insoluble dietary fibers, resistant starch, and antioxidants. Little millet and kodo millet were also reported to have dietary fiber content in a range between 37 and 38%, which is the highest among the cereals.[72] Finger millet carbohydrates are slowly digested and absorbed in comparison to those present in other cereals.[86]

Regular consumption of finger millet is associated with reduced chances of *diabetes mellitus*[63] and disorders of gastrointestinal tract[198] owing to their superior profile of polyphenols and dietary fibers.[36] The valuable outcome of phenolics on diabetes management was attributed to the limited inhibition of α-glucosidase and amylase during enzyme hydrolysis of complex carbohydrates and delayed glucose absorption that eventually regulates the blood glucose concentrations.[174]

Among carbohydrates, each category of dietary fibers has their own individual therapeutic effects. Water-soluble fiber includes arabinoxylan and β-glucan while water-insoluble fiber fractions include hemicelluloses, lignin and cellulose and water-unextractable arabinoxylan.[23,30] The health benefits linked to high fiber intakes are delayed nutrient absorption, increased fecal bulk, increased fecal transit time and fermentability characteristics, mobility of intestinal contents, lowering of blood lipids, and prevention of colon cancer.[193] The soluble fiber fractions showed physiological advantages in the form of hypoglycemic and hypocholesterolemic properties.[173]

Soluble fractions of dietary fibers are essential as they entrap fatty constituents in the gastro-intestinal tract and consequently, lessen the blood cholesterol levels and further worsen the risk of heart-related ailments. Some fibers can absorb mutagenic agents and are eliminated in the form of faces.[194] These types of fibers are about completely fermented by the large intestine microflora, leading to numerous favorable metabolic effects of fiber.[106,140] Resistant starch, a functional fiber present in millets, prevents enzymatic digestion and produces short-chain fatty acids in colon, particularly butyrate, which appears to stabilize the colonic cell proliferation as a defence system against colon cancer.[12]

### 9.3.7 PSEUDOCEREALS

The pseudocereals are substantial sources of several bioactive constituents, such as polyphenols, phytosterols, tocols, and saponins. Pseudocereals of importance include amaranth (*Amaranthus caudatus*, A. cruentus), buckwheat (*Fagopyrum esculentum*), and quinoa (*Chenopodium quinoa*).

Pseudocereals are alternative for gluten containing cereal grains due to their comparable functional and nutritional properties. Pseudocereal buckwheat is a good source of phytochemicals, such as phenolic acids, tocols, 3-flavanols, and rutin.[141] The phytochemical rutin plays an important role in controlling high blood pressure.[119]

Buckwheat expressed superior antioxidant properties than wheat, barley, oat, and rye.[230] Lyophilized extract of dehulled buckwheat and buckwheat hulls expressed 90.7 and 381.9 μg/mg of total phenolic contents, respectively. The major phenolic acids present in buckwheat grain are caffeic, o-coumaric, p-coumaric, ferulic, gallic, p-hydroxybenzoic, syringic, and vanillic. Buckwheat seeds and hulls contain 3870 and 13,140 mg/kg flavonoids, respectively.[141] Rutin was estimated as 470 and 770 mg/kg in the buckwheat seeds and hulls, respectively. Buckwheat tocopherols are comparable to those of common cereals and ranged from 14 to 22 mg/kg.[90]

Quinoa and amaranth are excellent sources of antioxidants. Amaranth has potential cholesterol- and lipid-lowering activities.[42] β-sitosterol is the main phytosterol in amaranth grain with total phytosterol content ranging from 561 to 834 μg/100 g.[116] Various bioactive phytochemicals in quinoa and amaranth seeds (such as phenolics, betanins, and carotenoids) are recognized for their health benefits, such as reduced cholesterol, antidiabetic effects, antioxidant, and anti-inflammatory effects.[3] Quinoa had total polyphenolic, anthocyanins, and flavonoids contents of 600 μg/g gallic acid equivalence, 96 mg cyanidin-3-glucoside/100 g, and 102 mg catechin equivalence/100 g, respectively.[64]

Betanin (a natural food colorant) acts as a competent scavenger of reactive oxygen species and hence functions as an antioxidant. Betacyanins, betanin, and isobetanin were found to be present in the red and black quinoa seeds.[52] Carotenoids, lutein (3.96–12.01 mg/kg), zeaxanthin (0.31–5.37 mg/kg), and β-carotene (0.26–1.07 mg/kg) were also observed in quinoa grains.[190]

## 9.4 GENETIC APPROACH TO IMPROVE THE AGGREGATION OF SECONDARY METABOLITES

Use of genetic approaches enables scientists to develop new cultivars that accumulate higher concentration of a phytonutrient and that can be stably fixed. Genetic improvement of crops has evolved from classical breeding to molecular-assisted approaches. Genes controlling the accumulation of these secondary metabolites can be elucidated by the global analysis of the metabolic pathways, which lead to the accumulation of a phytonutrient and by identifying the relation between the activities of key enzymes and phytochemical content in the regulation of a given pathway. After identification of the genes, these might be used as markers to select new improved genotypes by molecular approaches.

Functional as well as positional techniques can be used to authenticate the candidate gene. Scanning genome and candidate gene analyses are included in the position-based approach. This approach targets at nearby areas of a known quantitative trait loci (QTLs) and then the candidate genes are recognized from genes present in the chromosomal location in question. Then this candidate gene is validated by explaining that the variants of this gene are responsible for the variability of a trait. However, the major limitation of this approach is that some of the statistically significant detectable genes could not be confirmed to be localized near or within the known QTL region. Additionally, there are also several QTLs, which are false-positive and it is not necessary that an identified QTL always represents a single gene.

In the functional approach, most significant connections are between DNA and regulatory elements. Various databases and repositories can assist the functional approach for the identification of candidate genes. Databases contain information regarding metabolite and plant metabolic pathways. Therefore, plant databases having transcriptome data are now available and offer appropriate information about expressed genes under any growth stage, condition, tissues, or organs. Nowadays, different projects are sequencing cultivated species as well as their wild relatives. This varied information can help to select the candidate gene and it can further be validated through different biochemical and molecular approaches.

Key tools of the molecular breeding are marker-aided selection and genetic transformation. The method to be used genome editing depends on the amount of genetic variations of the desired trait. By making use of molecular breeding approaches, genes controlling the accumulation of phytonutrients can be used for choosing the genotypes in which the trait is fixed. Variances in the functioning of an enzyme may be due to variation in the quantity of enzyme or because of variable variants of an enzyme. A promising allele that is linked to a competent enzyme variant in a biosynthetic pathway can be a potential applicant that is associated with a larger buildup of a preferred phytochemical.

## 9.5 PROCESSING STRATEGIES TO AUGMENT THE BIOACCESSIBILITY AND BIOAVAILABILITY OF PHYTONUTRIENTS

Processing induced chemical or structural changes in food constituents can enhance the bioavailability or bioaccessibility of phytonutrients. To employ beneficial effects of phytonutrients through their respective biological activity, these must be free from the surrounding constituents during digestion in a bioaccessible form that is ultimately absorbed and conveyed to the bloodstream. Food processing technologies functions in following four possible ways to exert release of phytochemicals from the food matrix[162]:

- Chemical alterations into more bioaccessible and bioavailable forms;
- Cleavage of covalent or hydrogen bonds or hydrophobic forces that linked phytochemicals to surrounding macromolecules;
- Damaging microstructural hurdles to facilitate the release of phytonutrients from the matrix; and
- Create microstructures that protect phytochemicals from harsh environment until their absorption.

However, the processing parameters need to be suitably modulated, which otherwise might show detrimental effects on phytochemicals thus reducing its final concentrations in the food matrix.

Mechanical disruption, grinding, or comminution operations in solid plant-based foods favor the release of phytochemicals from food matrix due to cell-wall damage and enhance their bioaccessibility. Similarly, increased phenolic acids bioaccessibility in whole wheat bread with reduced bran particle size were observed by Hemery et al.[73] during baking.

Germination is a multifaceted functional process, where enormous endo-enzymes are stimulated and released, leading to disintegration and recombination of grain constituents, which might impact phytochemical activities in a progressive way. Germinated cereal grains are supposed to have a superior nutritive and physiological significance than their raw counterparts.[176] Abundant studies have supported the beneficial alterations in nutritional and chemical profile of germinated cereal grains. Researchers have found a significant increase in polyphenols content in germinated millet and brown rice. In brown rice, germination (47 h) increased the levels of free phenol and bound phenol by 76.67 and 44.64%, respectively.[139,150]

A significant increment in ferulic acid, luteolin, and p-coumaric acid was observed in germinated wheat in comparison to non-germinated one.[49] Conversely, few studies have encountered reduction in polyphenol content in grains, which might be attributed to the release of soluble polyphenols obtained from grains into the soaking medium (water) in the course of germination methodology or through the oxidation and breakdown of phenolics by enzymatic activities (esterase and polyphenols oxidase). Oh[137] described a substantial rise in the concentration of GABA in brown rice during germination. A considerable increase of GABA was also reported in barley, rye, sorghum, buckwheat, oat, wheat, and brown rice.[49]

Extrusion is comparatively a new technology that employs a combination of three unit operations, such as heating, cooking, and molding. Extrusion is a hydrothermal treatment that breaks the cell-wall structure, supports the conversion of bound polyphenols to free phenols, and assists in the extraction of free phenols. However, processing conditions and feed material composition play important roles in getting beneficial outcomes of extrusion on phytochemical activities. During extrusion processing of food material, if the moisture content of the feed is $>18\%$, then the polymerization of polyphenols has been promoted resulting in reduced extraction rate and antioxidant activity; while a feed material with $<15\%$ moisture content showed accelerated depolymerization of condensed tannin and their conversion into low molecular weight oligomers that are easy to extract. However, heat of compression directs to limited destruction of polyphenols with a definite magnitude of polymerizations and structural changes.[195]

Fermentation generally results in the formation of many small molecules with biological activity. Wheat, corn, brown rice, millets, and oat showed a significant increase in total phenol after the solid fermentation.[219] Solid-state fermentation for 5 days significantly increased the amount of total phenol and free phenol and ABTS antioxidant activity in maize by 20.05 and 36.73%, respectively.[166] Also, similar effects in total phenols and ABTS antioxidant activity were observed in sorghum slurry and porridge *via* liquid fermentation.[48] Food processing technologies play crucial role in augmenting the phytochemical activities and resultant biological effects if controlled and monitored accurately. Therefore, the processing conditions need to be optimized accordingly, keeping in view the sensitivity of respective phytochemical constituent.

## 9.6 EXTRACTION TECHNIQUES

Conventional extraction techniques employ organic solvents, moderate to high temperature and frequent agitation for extraction of phytochemicals. The efficiency of extraction of conventional method principally hinges on the solvents used (with different degree of polarity), raw material composition, and extraction conditions (time and temperature of extraction). Diverse structure, functionalities, site of action, specificity, and efficacy may build complications in the extraction of phytochemicals. The principal extraction techniques for extraction of bioactive compounds are solvent extraction (Soxhlet technique), maceration, and hydrodistillation.[182]

Soxhlet extraction is a solid–liquid extraction, where the process is executed frequently until the extraction is complete. Conversely, extensive requirement of solvent and longer extraction duration limits the application of this method.

Another important method of extraction is maceration, where the sample with reduced particle size is mixed with solvent under agitation to extract the required bioactive. The agitation increases the rate of diffusion and also removes the concentrated solution from the surface of the sample, which facilitates the extraction procedure. The extensive extraction duration restricts the application of this technique.

Hydrodistillation is an alternative conventional technique of extraction, where the extraction is accomplished with distilled water. Three major physicochemical processes are combined in this method, such as hydrodiffusion, hydrolysis, and heat breakdown. However, the high extraction temperatures limit its applications. To a definite magnitude, these constraints were conquered by the enzyme-assisted aqueous extraction technique.

Enzymatic extractions are dependent on the enzyme substrate interactions that are affected by size of raw material, enzyme concentration, reaction time, temperature and pH of medium, and solid-to-liquid ratio. The addition of enzyme heightens the reaction rate till the substrate concentration becomes limiting.[115] However, limited selectivity of enzymes and cost of enzymes confines the application of this technique. Limitations such as solvent toxicity, solvent residues in extracts, long extraction times, low yield, low extraction selectivity, poor stability, and thermal damage to heat labile bioactives have prompted the development of novel extraction technologies that can reduce or eliminate the use of organic solvents. These new techniques are classified as non-conventional extraction techniques. The extraction of highly valuable phytochemicals from plants has greatly benefitted from non-conventional methods. Substituting old technologies by these methods has various pros like decreased in energy consumption and time of processing, and reduction in the use of deleterious and expensive solvents. Extraction yields are also enhanced by using non- conventional methods. These new techniques are often used for the extraction of phytochemicals and are discussed in this section. However, the scientific literature on extraction applications of these technologies is generally available for bioactive compounds of fruits and vegetables sources, and very limited scientific literatures are available on these advanced extraction technologies in case of cereal-based sources.

### 9.6.1 SUPER CRITICAL FLUID EXTRACTION

The supercritical extraction method involves alterations in temperature and pressure; and these changes convert the gas in the supercritical fluid. This process is based on the principle of using a supercritical fluid, which has properties in-between of gas and liquid.[55,201] Supercritical fluid has distinct properties like low viscosity, high diffusivity, and small surface tension due to temperature and pressure that are above the critical values.[111] As a result of this, the fluid diffusal through natural solid matrix is enhanced, and the extraction of natural compounds is more efficient than that of conventional liquid solvents. On reduction of operating pressure and temperature, these distinct features are lost. Therefore, the extraction of solute along with solvent occurs at supercritical conditions; and later, when pressure and temperature are decreased below its critical condition, solute can be separated. Many diverse supercritical fluids are being used, namely, $CO_2$, propane, LPG, Sulphur hexafluoride ethane, nitrous oxide ethane, methanol, and water.

The $CO_2$–SF extraction is desirable for the extraction of heat-labile compounds because reduced adopted temperature and quicker extraction results in reduced contact time with heat and pressure. Furthermore, $CO_2$–SF has distinct features, such as non-toxicity, inexpensive, readily available, does not catch fire easily, chemically inert, has a lower viscosity and surface tension, higher diffusivity, and appropriate density. It is gaseous at normal temperature and pressure, thus it removes the step of solvent evaporation once extraction is over. In addition, the degradation of extracted compounds is avoided, because $CO_2$ provides a non-oxidizing atmosphere during extraction. Thus, the product obtained is of better quality and purity and is safe for use as nutritional additives and for pharmaceutical applications.[122]

Supercritical fluid extraction (SFE) has been extensively used in countless industrial applications, such as the decaffeination of coffee, hops extraction, polymer synthesis, and purification and formation of nanoparticles. Additionally, SFE-$CO_2$ is an environmentally friendly technique for the procurement of value-added compounds from diverse matrices. Furthermore, at low temperatures, $CO_2$ is not reactive, and therefore, it is recovered easily after each extraction stage.

Supercritical extraction has principally been used for carotenoids and lipids extraction. Supercritical fluid extraction has been employed for the extraction of phytochemicals from grains in a small number of investigations. Soares et al.[179] obtained rice bran oil using a mixture of $CO_2$ and compressed LPG and found that the fatty acid profiles were similar. In this study, for the supercritical extraction, the effect of pressure and temperature on the extraction yield was assessed from 150 to 250 bars at 40–80°C. In the case of compressed LPG extractions, 5–25 bars at 20–40°C were used. The antioxidant activity and chemical composition of the extracts were assayed. The chemical profile of fatty acids was alike for both solvents. β-sitosterol, an antioxidant compound, was found in both processes. Xu and Godber[209] attempted to explore the potential of SFE-$CO_2$ for the extraction of ϒ-oryzanol from rice bran. Authors concluded that SFE under a temperature of 50°C, pressure of 69 MPa for 25 min resulted in approximately four times higher yield of ϒ-oryzanol than the solvent extraction. Also, it was observed that a high concentration of γ-oryzanol in extract (50–80%) was achieved by accumulating the extract after 15–20 min of extraction under optimized process conditions. In the near, future SFE will prove to be an innovative technology with great prospective for the extraction of phytochemicals from cereals, which can be used as food additives in the fortified cereal products.

### 9.6.2 PRESSURIZED LIQUID EXTRACTION (PLE)

Accelerated solvent extraction (ASE) is another name for pressurized liquid extraction (PLE). This approach involves a separation procedure that involves the transfer of solutes from a matrix. Liquid solvents are utilized at high temperatures and pressures. As a result, the surface tension of the solvent is reduced, allowing the solvent to penetrate the matrix pores. As a result, the matrix is disturbed, allowing for better analyte transfer from the solvent sample.

The solvents are selected based on the solubility features of the solute, which is desired. Originally, this process was developed for its usage in the laboratory analysis but was later used on a broader scale. The benefits of PLE are the enhanced analyte solubility in solvents under high temperature conditions. Higher pressure results in enhanced contact among the extracting liquid and the solid matrix,[203] whereas high temperature results in the breakage of the phenolic-matrix bonds and raises the solubility of compounds in the extraction solvent.[130]

The extraction efficiency of the PLE process depends on the extraction temperature, pressure, time, and solvents. Furthermore, in order to give enough driving force for the transfer of analyte into the extraction solvent, the extraction must be carried out using the most appropriate solvent-to-sample ratio. It has been reported that since water and ethanol are generally recognized as safe solvents, PLE using them is even more promising.[125]

Under optimal conditions, PLE is effective in the extraction of plants phenolic compounds in less time using organic solvents and water.[192] Presently, PLE has been efficaciously used in the withdrawal of thermo labile phytochemicals from several plants, for example, Brazilian cherry seeds,[138] olive leaves,[211] and blackberry.[113] It has been efficaciously utilized to extract phytochemicals from several vegetable sources.[143,207] Other than plants, this technique has also been used to extract active compounds from milk and eggs and to extract veterinary drugs from animal organs.

In case of cereals, a study demonstrating the effect of PLE on the extraction of β-glucans and phenolic compounds from waxy barley has been reported.[22] In this study, the effects of temperature, extraction time, and ethanol content on the extraction of beta-glucans and phenolics from waxy barley were evaluated. It was found that the temperature is inversely related to the extraction of both compounds; higher temperature resulted in lower β-glucan extraction yield because of fragmentation. However, a substantial increase in the phenolic recovery was seen. Extraction of β-glucans at low temperatures was favored at longer extraction durations. The content of ethanol was not statistically significant on the β-glucan extraction, but facilitated to sustain the molecular weight of the extracted β-glucan.

Povilaitis et al.[154] evaluated the antioxidant potential of rye and wheat bran using PLE with different solvents. Rye and wheat brans were ground to obtain different particle size fractions and extracted at 10.3 MPa pressure and 80°C temperature by the consecutive application of hexane, acetone, and methanol:water (80:20%). The highest extract yield was achieved from rye bran with methanol–water and also a significant effect of particle size was observed in the extraction yield in all cases.

### 9.6.3 ULTRASOUND-ASSISTED EXTRACTION (UAE)

Based on the variations in sound intensity (20 kHz–100 MHz) and frequency, ultrasound methods can be divided into two categories: (1) high-frequency ultrasound functions at frequencies of 2–20 MHz and sound intensities ranging from 0.1 to 1 W/cm$^2$. These have utility in analysis of food quality, non-destructive inspection, and medical imaging. (2) high-intensity ultrasound works at frequencies between 20 and 100 kHz and sound intensity between 10 and 1000 W/cm$^2$.[56] These are also known as power ultrasound. Using power, ultrasound is known as sonication, and liquid medium is included in most of the applications.[220]

Due to high energy level of power ultrasound waves, it apposite to increase procedures like enzyme inactivation,[161] surface decontamination,[228] oil extraction,[172] starch–protein separation,[222] bioactive compound extraction,[203] and microbial inactivation.[20] In the solvent, the ultrasonic waves produce a cavitation effect; consequently, molecules move faster and the infiltration of the solvent into the target material is enhanced. As a result of this, target compounds are released speedily.[16] Therefore, UAE is extensively used for the extraction of bioactive compounds from natural sources.[38] The competence of the UAE, as compared to conventional extraction methods, is accredited to the cavitational effect, which helps the discharge of extractable compounds and increases the mass transport by either diffusion or by disruption of cell walls of the plant.[35] For the extraction of phenolic compounds, UAE is more rapid, more effective, and less solvent utilizing technique in comparison with the conventional solvent extraction.[191] High-intensity ultrasound makes the extraction process of bioactive compounds superior and is used as an cheap, reproducible, straightforward, and effective alternate method of industrial importance.

Moreover, if a bioactive compound has to be extracted for the human consumption, using ultrasound waves cuts the use of solvents, which may be highly toxic. UAE has been used for the extraction of compounds from fruits and vegetables. It has been used for the extraction of antioxidants from pomegranate peel[145] and Zyzyphus lotus fruit,[69] extraction of carotenoids from tomato pomace,[107] extraction of phenol and anthocyanin from Jabuticaba peel,[164] extraction of aroma from garlic,[92] extraction of phenol, antioxidants, anthocyanins from grape seeds.[61] Also, UAE has been used for the extraction of oils from oleaginous seeds like soybean, sunflower, rape,[112] almond,[221] flaxseed,[223] and soybean.[99] UAE has also been used for the extraction of compounds from herbs and spices, like, antioxidants from rosemary,[80] capsaisinoids from pepper,[19] carvone, and limonene from caraway seeds,[14] etc. In case of micro-organisms, it has been utilized for the extraction of lipids from *Nannochloropsis oculata* microalgae,[5] phenolic compounds, and chlorophylls from Nannochloropsis spp microalgae,[148] carotenoids, and chlorophyll from *Dunaliella salina* microalgae Synechococcus sp. cyanobacteria.[32]

Flavonoids were extracted from corn silk using ultrasonic-assisted extraction (UAE) by Zheng et al.[227] Experiments were designed to obtain optimized process conditions of ethanol concentration, extraction time, and material to solvent ratio for maximum yield of flavonoids. A maximum extraction yield of total flavonoids 1.13% with 467.59 μmol/L FRAP value at ultrasonic power of 500 W, extraction time of 20 min, material solvent ratio of 1:20, and ethanol concentration of 30% was attained. UAE, with aqueous ethanol was utilized to extract polyphenolic compounds from red sorghum bran. The effect of ultrasonic time (min), ethanol concentration (%), and solvent to solid ratio (mL/g) on polyphenolic yield was investigated. The optimized yield of polyphenolics (49.743 mg GAE/g dw) was obtained at 21 min of ultrasonic treatment, 53% of ethanol concentration, and 52:1 ml/g of solvent to solid ratio. When compared to conventional solvent extraction, the yield was better in the case of UAE with extracts showing excellent DPPH radical scavenging activity, reducing power, and anti-proliferative activity in a level-dependent approach.[109]

Wang et al.[202] designed an experimental set up for the extraction of phenolic compounds from wheat bran that included five levels, a three changeable central composite rotatable design, and the optimum feasible combination of solvent concentration, extraction temperature, and extraction duration. The highest extraction (3.12 mg gallic acid equivalents/g) was achieved under ideal circumstances of 64% ethanol, 60°C extraction temperature, and a 25-min extraction period. Anthocyanins, total phenolic compounds, and their antioxidant properties were extracted from black glutinous rice using UAE. Using a Box–Behnken design, the effects of ethanol concentration, amplitude, and duration were explored, and extraction conditions were improved. The maximal anthocyanin content, TPC, and antioxidant activity were 50.33 mg cyn-3-glucosidase/gram fresh weight ml, 136.17 mg GAE/g, 80.14% for DPPH, and 239.94 mg AAE/g, respectively.[146]

### 9.6.4 MICROWAVE-ASSISTED EXTRACTION (MAE)

In this technique, electromagnetic radiation ranging in frequency from 0.3 to 300 GHz is applied. Due to these radiations, rapid and localized heating of both sample and solvent occurs. Due to the localized heating, pressure builds-up inside the sample cells, consequently speedy transfer of chemical of interest present in the cells to the extraction solvent occurs.[68] By diffusion, the solvent enters in the solid matrix. As MAE reduces both extraction time and solvent consumption, it is widely used as a promising green-extraction.[225] The efficiency of this process depends on the composition and type of solvent, solid–liquid ratio, extraction time, and extraction temperature.[181] For studying the influence of the diverse variables and their interactions, and to find out the ideal conditions, different experimental designs are frequently studied. In MAE, the heat generation is rapid and this distinguishing feature leads to reduced extraction time and high quality of extracts is obtained with improved target compound retrieval.[101]

Chemical substances differ in their microwaves absorption capacity, making MAE an effective method for extraction. Notably due to differences in the absorption capacity, selective extraction of the target compounds from intricate food matrices is possible.[74]

MAE has been extensively used for the extraction of phenolics and anthocyanins from a diverse range of vegetable matrices, like tomatoes,[100] grapes,[101] and blackberries.[206] In myrtle leaves, MAE was used for the extraction of polyphenols, tannins, and flavonoids.[43] The MAE seems to be one of the finest approaches for the extraction of phenolic compounds due to the distinct microwave/matter interactions and speedy extraction time.[98] In another study, the total phenolic content of cereal brans was studied using MAE and conventional extraction.[45] MAE was shown to considerably enhance the total phenolic component concentration in solvents when tested for 3.5 min at 2450 MHZ. However, at different temperatures and time intervals, the total phenolic content of traditional and microwave-assisted extractions differed considerably (p0.05). The mean TPC of brans extracted using traditional methods varied from 1.24 to 2.87 mg GAE $g^{-1}$ compared to 2.20–4.09 mg GAE $g^{-1}$ using MAE.

Aboelmaaty et al.[4] optimized the microwave-assisted extraction of total phenolics content from *Triticum aestivum* L. The MAE was compared with Soxhlet, stirring extraction and maceration to extract total phenolic content (TPC). Authors concluded that MAE provided higher extraction efficacy and time savings. In an investigation, conventional solvent extraction, ultra-sonic- and microwave-assisted extraction methods were compared with the aim of getting high yield of γ-oryzanol from rice bran.[96] Extraction yield was evaluated forgiving high crude oil yield, TPC, and γ-oryzanol content. The maximum yield of γ-oryzanol was detected in methanolic extracts of microwave treatment (85.0 ppm) followed by ultrasonication (82.0 ppm) and conventional extraction method (73.5 ppm). A combination of 80% methanolic concentration and 55 min digestion time of microwave treatment depicted the best extraction method for TPC and γ-oryzanol (105 ppm). Concentration of γ-oryzanol present in the extract was directly proportional to the total phenolic content.

### 9.6.5 PULSED ELECTRIC FIELD (PEF)

Pulse electric field (PEF) is an effective technique that enhances liquid food shelf-life by making the microorganisms inactive. The product's shelf-life is increased without having any effect on the nutritional content. Moreover, this technique does not pose any environmental risks and no case of toxicity has been yet been reported. The only main hurdle to use this technology on extensive scale is higher starting expanses.

The principle of PEF extraction is to increase the extraction yield by inducing the electroporation of the cell membrane. On passage of electric potential through the cell membrane, molecules are separated on the basis of their charge. Thus, the cell exposed to an external electric field is electroporated, making membrane permeable to water. As a result of this distension, rupture and consequently lysis of cell happen.

For application in food industry, liquid, or semi-solid food samples are treated with the short pulses of electricity. This treatment has been extensively applied for microbial inactivation mostly in dairy products, juices, and different types of liquid foods. It has been used for the preservation of fruit juices, and milk.[76] Electroporation and electro-permeabilization of microorganism's caused by PEF treatment have been attributed as key factors for bacterium death in samples.[104] It has been reported that PEF is very effective in inactivating the molds, yeast, and vegetative bacteria.

It was reported that the yield of sugar beet juice was increased from 30 to 80% on PEF treatment.[81] PEF-assisted pressure extraction augmented proteins and polysaccharides content in mushroom extracts.[149] The concentrations of falcarindiol and falcarindiol-3-acetate in PEF-treated samples were 3 times higher than the untreated carrot purees.[8] In case of Arroniz olive, the extraction yield of oil production was enhanced by 13.3% after PEF treatment.[156] The literature on application of PEF in the area of extraction of phytonutrients from cereal sources is negligible.

### 9.6.6 HIGH-VOLTAGE ELECTRICAL DISCHARGES (HVED)

In high-voltage electrical discharges (HVED) technology, energy is directly introduced into an aqueous solution through a plasma channel formed by current of HVED between two submerged electrodes.[25,26] Both electrical and mechanical effects occur in the material due to the shock waves. These waves lead to the disruption of cellular walls, particle size is reduced, and mass transfer of the intracellular content is enhanced.[29] It is essential that parameters of extraction for each product are optimized. It has been reported that the efficiency of the treatment is significantly affected by the nature of the raw materials.

In the past few years, food industry has seen the increasing interest toward using this technique. The benefits of this technique are heat and mass transfer can be increased, by changing the voltage applied, fast and effective control of heat and mass transfer is possible. Also, design involved in this technique is simpler, it is conducive for diverse environments, consumption of energy is very less and it can be used for both single phase and multiphase flows. In food processing industry, it is used for enhancing shelf-life of both fruits and vegetables and for drying and thawing. This treatment has been effectively used for total phenolic compounds recovery from different winery by-products, like, skins,[24] seeds,[28] vine shoots,[158] and grape pomace.[27]

Parniakov et al.[149] compared the extraction efficiency of conventional techniques, pulsed electric field, and high voltage discharge. Authors concluded that for the recovery of high value compounds, HVED has superior extraction efficiency. However, the production of chemical electrolysis and free radical associated with this technique reduces the benefits of using this technique.

### *9.6.7 HIGH HYDROSTATIC PRESSURE (HHP)*

Shouqin et al.[175] first reported the high hydrostatic pressure extraction (HHPE) for extracting the active compounds. In this technology, high pressures ranging from 100 to 1000 MPa are applied and it has been developed as an alternative to thermal procedures. Traditional methods of thermal treatment cause damage to nutritional and sensory qualities of food. However, the HHP process can warrant high quality food and reduces the damage to delicate bioactive compounds. This is because HHP make use of low temperature and thus there are negligible alterations in thermo-labile and volatile food components.

In hydrothermal extraction, active substances are destroyed, but HHP increases extraction efficiency without causing any damage to the bioactive compounds. In HHP, disruption of cell membranes occurs, cellular compounds are easily eluted into the extracellular spaces. One of the most prominent features of HHPE is that the extraction solvents in use vary as per the polarity of the compound in question. As a result, it can be used for the extraction of highly polar, weakly polar, and non-polar compounds by making use of diverse solvents.[224]

HHPE has been effective in the extraction of anthocyanins and flavonols from fruits and juices. As this technique increases the mass transfer and cell membrane destruction enhances permeability, it had been utilized for the extraction of carotenoid extraction,[167] anthocyanins,[39] catechins,[84] anti-obesity substance,[83] bean protein extraction,[11] and anthocyanins from apple.[60]

Marti et al.[117] investigated the possibility of high hydrostatic pressure processing to promote changes in fiber structural characteristics and the interaction of fiber with water in the instance of cereal grains. The coarse and fine bran from wheat milling were used. It was reported that HHP processing modifies the physical and structural features of by-products of milling and consequently, their capability to interact with water. Morphological changes were larger in fine bran and caused extensive modifications in the bran hydration properties.

Enzymatic hydrolysis combined with high hydrostatic pressure (HPP) resulted in the greatest extraction of tricin from rice hull in another investigation. Prior to HPP (500 MPa) treatment, enzymatic hydrolysis with Celluclast (0.5%, w/w) gave the highest tricin level (32.9 mg/kg rice hull). Rice husk extract produced by HPP or a combination of enzymatic hydrolysis and HPP treatments had much higher effectiveness than extract obtained through standard solvent extraction. Traditional solvent extraction yielded 14.7 mg/kg rice hull and HPP treatment yielded 19.7 mg/kg rice hull, respectively. Trichin-containing rice husk extract has antioxidant, anti-inflammatory, and anti-adipogenic properties.[147]

## 9.7 APPLICATIONS OF BIOACTIVE RICH FRACTIONS AND THEIR EXTRACTS IN FUNCTIONAL FOODS

Phytochemicals can be used as natural food additives owing to their functional benefits on human health. The main applications of phytonutrients in processed food products include antioxidants, source of dietary fibers, and natural colorants. The phytochemical components and pharmacological profile of cereal brans are anticipated to give comprehension to their probable application in elevation of health. Several researchers have employed the phytonutrient rich bran fraction of whole grains in the preparation of numerous functional foods, such as breads, biscuits, muffins, extruded products, pasta and noodles, meat products, and dairy products. Selected examples are illustrated in this section.

The effects of supplementation of cereal brans on physiological, rheological, and sensory characteristics of both cereal and non-cereal foods have been extensively studied. The addition of aleurone-rich flour increased the dietary fiber content from 1.70 g/100 to 7.76 g/100 g in cooked pasta.[18]

Kong et al.[93] used black rice bran as an ingredient in preparation of noodles. Noodles with black rice bran were evaluated for content of polyphenolics, flavonoids, and anthocyanins and were found to be considerably superior over conventional wheat noodles. Moreover, the antioxidant activity of supplemented noodles was significantly higher than the control sample. Also, pasta supplemented with purple bran fractions showed high anthocyanin content and greater antioxidant activities with fair cooking quality.[217]

Pick et al.[151] reported a reduction in total and LDL cholesterol concentrations in subjects with type 2 diabetes after the ingestion of bread, buns, and muffins with oat bran concentrate for 12 weeks as compared to the conventional products made out of refined wheat flour. Barley β-glucan when supplemented at low levels in bread, performed similar functions as gum additives that are used as gluten alternatives.

Lee et al.[97] reported that 1% of barley β-glucan (85% purity) improved bread grain and texture while retaining the original loaf volume. Orange juice supplemented with oat bran lowered serum total cholesterol concentrations by 3.8% and LDL-cholesterol concentrations by 6.7% compared to the control.[87]

Barley grain has found a role in meat products as fat replacers, meat extenders, and also improves the storage stability. Shand[171] reported better purge control in pork bolognas with 4% waxy barley flour than wheat or regular barley. Low-fat sausages containing barley β-glucan extract were liked equivalent to the high-fat sausage and were valued more satisfactory than those containing carboxymethyl cellulose.[95] Likewise, chicken patties prepared by incorporating 15% wheat bran showed less cholesterol content than the patties prepared without bran.[189]

Rice bran oil (RBO) has been used for stabilizing effects on fats and oils due to its potential antioxidant properties since years. Incorporation of RBO to tuna oil was studied during storage; and effects on fatty acid composition, oxygen consumption, peroxide and p-anisidine values, residual tocopherol, phenolic, and ϒ-oryzanol contents were examined in comparison to BHT treatments.[37] The addition of rice bran powder at 0.1% reduced the oxidation milk powder.[134] Beef patties with added oryzanol showed higher oxidative stability in comparison to the samples with other antioxidants.[88]

In dairy sector, preparation of ice-cream using extracted anthocyanins from black rice bran was attempted by Sethi et al.,[170] who suggested that addition of anthocyanin extract amplified the total phenol content, DPPH radical scavenging activity, ABTS activity, ferric reducing ability (FRAP), and total monomeric anthocyanins in a significant manner. Wheat bran served as prebiotic and a source of antioxidants when added in yogurt at the rate of 4%, it promotes the growth of probiotic bacteria *Lactobacillus acidophilus* and *Bifidobacterium lactis.*[7]

Another important application of phytonutrients is as natural colorants. Awareness for healthy lifestyle has increased the interest in natural extracts rich in pigments, such as anthocyanins and carotenoids. At present, there are several patents, which describe extractions and processes of application for their use as colorants. Correspondingly, anthocyanins were exploited as natural food colorants in alcoholic beverages.[215]

Rice bran oil has also found its industrial importance in non-food applications, such as production of biodiesel and probable high value-added products. It is also successfully used in cosmeceuticals due to antioxidant activity, radical scavenging power, and ability to absorb UV radiations. A plethora of work has been done in last few decades for application of rice bran oil or powder in cosmetic products that are directly applied on the skin.[77]

## 9.8 SUMMARY

Non-conventional technologies for the extraction of phytochemicals reduce extraction time and yield quality products with improved stability using minimal or negligible amounts of hazardous solvents. Due to their environmental friendly operations, these techniques are often referred to as "green technologies." However, the extraction of cereal bioactives using advanced extraction techniques requires concerted efforts through scientific investigations. Further improvements in the extraction yields and quality of extracted bioactives can be achieved by combining two of more technologies. The non-thermal/eco-friendly processes are ultrasound-assisted enzyme extraction, enzyme-assisted supercritical extraction, microwave-assisted enzymatic extraction, and high pressure-assisted enzymatic extraction. The scientific investigations involving combination techniques are in infancy stage. Therefore, extensive research exertions are obligatory to employ these techniques on commercial scale.

## KEYWORDS


- **bioactive compounds**
- **cereal**
- **extraction technology**
- **health benefits**
- **phytochemicals**
- **phytonutrients**
- **pigmented grains**
- **therapeutic benefits**


## REFERENCES


1. Abdel-Aal, E. S. M.; Choo, T. M.; Dhillon, S.; Rabalski, I. Free and Bound Phenolic Acids and Total Phenolics in Black, Blue, and Yellow Barley and their Contribution to Free Radical Scavenging Capacity. *Cereal Chem.* **2012,** *89*, 198–204.
2. Abdel-Aal, E. S. M.; Hucl, P. Rapid Method for Quantify Total Anthocyanins in Blue Aleurone and Purple Pericarp Wheats. *Cereal Chem.* **1999,** *76*, 350–354.
3. Abderrahim, F.; Huanatico, E.; Segura, R. Physical Features, Phenolic Compounds, Betalains and Total Antioxidant Capacity of Colored Quinoa Seeds (*Chenopodium quinoa* Wild.) from Peruvian Altiplano. *Food Chem.* **2015,** *183*, 83–90.
4. Aboelmaaty, W. S.; Gedara, S. R.; Afify, S. A.; Zaglol, M. G. Optimization of Microwave Assisted Extraction of Total Phenolic Contents from *Triticum aestivum* L. *Whole Grain Cereal* **2017,** *3*, 110–118.
5. Adam, F.; Abert-Vian, M.; Peltier, G.; Chemat, F. "Solvent-free" Ultrasound-assisted Extraction of Lipids from Fresh Microalgae Cells: Green, Clean and Scalable Process. *Bioresour. Technol.* **2012,** *114*, 457–465.
6. Adom, K.; Liu, R. Antioxidant Activity of Grains. *J. Agric. Food Chem.* **2002,** *50*, 6182–6187.
7. Agil, R.; Hosseinian, F. Dual Functionality of Triticale as a Novel Dietary Source of Prebiotics with Antioxidant Activity in Fermented Dairy Products. *Plant Foods Human Nutr.* **2012,** *67*, 88–93.
8. Aguiló-Aguayo, I.; Hossain, M.; Brunton, N.; Lyng, J.; Valverde, J.; Rai, D. Pulsed Electric Fields Pre-treatment of Carrot Purees to Enhance Their Polyacetylene and Sugar Contents. *Innov. Food Sci. Emergi. Technol.* **2014,** *23*, 79–86.
9. Andersson, A. A.; Lampi, A. M.; Nystrom, L. Phytochemical and Dietary Fiber Components in Barley Varieties in the Health Grain Diversity Screen. *J. Agric. Food Chem.* **2008,** *56*, 9767–9776.
10. Andreasen, M. F.; Kroon, P. A. Intestinal Release and Uptake of Phenolic Antioxidant Diferulic Acids. *Free Radical Biol. Med.* **2001,** *31*, 304–314.

11. Andrés, V.; Mateo-Vivaracho, L.; Guillamón, E.; Villanueva, M. J.; Tenorio, M. D. High Hydrostatic Pressure Treatment and Storage of Soy-Smoothies: Color, Bioactive Compounds and Antioxidant Capacity. *LWT-Food Sci. Technol.* **2016,** *69*, 123–130.
12. Annison, G.; Topping, D. P. Nutritional Role of Resistant Starch: Chemical Structure and Physiological Function. *Annu. Rev. Nutr.* **1994,** *14*, 297–320.
13. Ardiansyah, Shirakawa. H.; Koseki, T.; Ohinata, K.; Hazhizume, K.; Komai, M. Rice Bran Fractions Improve Blood Pressure, Lipid Profile, and Glucose Metabolism in Stroke-Prone Spontaneously Hypertensive Rats. *J. Agric. Food Chem.* **2006,** *54*, 1914–1920.
14. Assami, K.; Pingret, D.; Chemat, S.; Meklati, B. Y.; Chemat, F. Ultrasound Induced Intensification and Selective Extraction of Essential Oil from *Carumcarvi* L. Seeds. *Chem. Eng. Process.* **2012,** *62*, 99–105.
15. Avato, P.; Bianchi, G.; Murelli, C. Aliphatic and Cyclic Lipid Components of Sorghum Plant Organs. *Phytochemistry* **1990,** *29*, 1073–1078.
16. Avhad, D. N.; Rathod, V. K. Ultrasound Assisted Production of a Fibrinolytic Enzyme in a Bioreactor. *Ultrason. Sonochem.* **2015,** *22*, 257–264.
17. Awika, J. M.; Rooney, L. W.; Wu, X.; Prior, R. L.; Cisneros-Zevallos, L. Screening Methods to Measure Antioxidant Activity of Sorghum (*Sorghum bicolor*) and Sorghum Products. *J. Agric. Food Chem.* **2003,** *51*, 6657–6662.
18. Bagdi, A.; Szabo, F.; Gere, A.; Kokaib, Z.; Siposb, L.; Tomoskozi, S. Effect of Aleurone-Rich Flour on Composition, Cooking, Textural, and Sensory Properties of Pasta. *LWT-Food Sci. Technol.* **2014,** *59*, 996–1002.
19. Barbero, G. F.; Liazid, A.; Palma, M.; Barroso, C. G. Ultrasound-Assisted Extraction of Capsaicinoids from Peppers. *Talanta* **2008,** *75*, 1332–1337.
20. Baumann, A.; Martin, S. E.; Feng, H. Removal of *Listeria monocytogenes* Biofilms from Stainless Steel using Ultrasound and Ozone. *J. Food Protect.* **2009,** *72*, 1306–1309.
21. Becker, F. Tocopherols in Wheat and Rye: Review of Research within the Subject. M.Sc. Thesis; Department of Food Science, University of Helsinki; **2013**; p 215.
22. Benito-Roman, O.; Alvarez, V. H.; Alonso, E. Pressurized Aqueous Ethanol Extraction of Β-Glucans and Phenolic Compounds from Waxy Barley. *Food Res. Int.* **2015,** *75*, 252–259.
23. Bingham, S. Definitions and Intakes of Dietary Fiber. *Am. J. Clin. Nutr.* **1987,** *45*, 1226–1231.
24. Boussetta, N.; Lanoisellé, J. L.; Bedel-Cloutour, C.; Vorobiev, E. Extraction of Soluble Matter from Grape Pomace by High Voltage Electrical Discharges for Polyphenol Recovery: Effect of Sulphur Dioxide and Thermal Treatments. *J. Food Eng.* **2009,** *95*, 192–198.
25. Boussetta, N.; Lesaint, O.; Vorobiev, E. A. Study of Mechanisms Involved during the Extraction of Polyphenols from Grape Seeds by Pulsed Electrical Discharges. *Innov. Food Sci. Emergi. Technol.* **2013,** *19*, 124–132.
26. Boussetta, N.; Vorobiev, E. Extraction of Valuable Biocompounds Assisted by High Voltage Electrical Discharges: A Review. *Comptes Rendus Chimie* (Chemistry Reports) **2014,** *17*, 197–203.
27. Boussetta, N.; Vorobiev, E.; Deloison, V.; Pochez, F. Valorization of Grape Pomace by the Extraction of Phenolic Antioxidants: Application of High Voltage Electrical Discharges. *Food Chem.* **2011,** *128*, 364–370.
28. Boussetta, N.; Vorobiev, E.; Le, L. H.; Cordin, A. Application of Electrical Treatments in Alcoholic Solvent for Polyphenols Extraction from Grape Seeds. *LWT-Food Sci. Technol.* **2012,** *46*, 127–134.

29. Brianceau, S.; Turk, M.; Vitrac, X.; Vorobiev, E. High Voltage Electric Discharges Assisted Extraction of Phenolic Compounds from Grape Stems: Effect of Processing Parameters on Flavan-3-Ols, Flavonols and Stilbenes Recovery. *Innov. Food Sci. Emergi. Technol.* **2016,** *35*, 67–74.
30. Bunzel, M.; Ralph, J.; Marita, J. M.; Hatfield, R. D.; Steinhart, H. Diferulates as Structural Components in Soluble and Insoluble Cereal Dietary Fiber. *J. Sci. Food Agric.* **2001,** *81*,653–660.
31. Burton, G. W.; Ingold, K. U. Vitamin E as an In vitro and In vivo Antioxidants. *Ann. NY Acad. Sci.* **1989,** *570*, 7–22.
32. Cardoso, L. C.; Serrano, C. M.; Rodríguez, M. R. Extraction of Carotenoids and Fatty Acids from Microalgae Using Supercritical Technology. *Am. J. Analy. Chem.* **2012,** *3*, 877–883.
33. Carson, G. R.; Edwards, N. M. Criteria of Wheat and Flour Quality. In *Wheat Chemistry and Technology*; Khan, K., Shewry, P., Eds., 4th ed.; American Association of Cereal Chemists: St Paul, MN, 2009; pp 97–118.
34. Chang, H.; Yu, B.; Yu, X. P. Anti-cancer Activities of an Anthocyanin-Rich Extract from Black Rice against Breast Cancer Cells In Vitro and In Vivo. *Nutr. Cancer* **2010,** *62*, 1128–1136.
35. Chemat, F.; Zill-e-Huma, Khan, M. K. Applications of Ultrasound in Food Technology: Processing, Preservation and Extraction. *Ultrason. Sonochem.* **2011,** *18*, 813–835.
36. Chethan, S.; Sreerama, Y. N.; Malleshi, N. G. Mode of Inhibition of Finger Millet Malt Amylases by the Millet Phenolics. *Food Chem.* **2008,** *111*, 187–191.
37. Chotimarkorn, C.; Benjakul, S.; Silalai, N. Antioxidative Effects of Rice Bran Extracts on Refined Tuna Oil during Storage. *Food Res. Int.* **2008,** *41*, 616–622.
38. Corbin, C.; Fidel, T.; Leclerc, E. A.; Barakzoy, E.; Sagot, N. Development and Validation of an Efficient Ultrasound Assisted Extraction of Phenolic Compounds from Flax (*Linum usitatissimum* L.) Seeds. *Ultrason. Sonochem.* **2015,** *26*, 176–185.
39. Corrales, M.; García, A. F.; Butz, P.; Tauscher, B. Extraction of Anthocyanin from Grape Skins Assisted by High Hydrostatic Pressure. *J. Food Eng.* **2009,** *90*, 415–421.
40. Costabile, A.; Klinder, A.; Fava, F.; Napolitano, A. Wholegrain Wheat Breakfast Cereal has s Prebiotic Effect on the Human Gut Microbiota. *Br. J. Nutr.* **2008,** *99*, 110–120.
41. Coulibaly, A.; Kouakou, B.; Chen, J. Phytic Acid in Cereal Grains: Structure, Healthy or Harmful Ways to Reduce Phytic Acid in Cereal Grains and their Effects on Nutritional Quality. *Am. J. Plant Nutr. Fertiliz. Technol.* **2011,** *1*, 1–22.
42. Czerwiński, J.; Bartnikowska, E.; Leontowicz, H. Oat (*Avena sativa L.)* and Amaranth (*Amaranthus hypochondriacus*) Meals Positively Affect Plasma Lipid Profile in Rats Fed on Cholesterol-Containing Diets. *J. Nutr. Biochem.* **2004,** *15*, 622–629.
43. Dahmoune, F.; Nayak, B.; Moussi, K.; Remini, H.; Madani, K. Optimization of Microwave-Assisted Extraction of Polyphenols from *Myrtuscommunis* L. Leaves. *Food Chem.* **2015,** *166*, 585–595.
44. Danielski, L.; Zetzl, C.; Hense, H.; Brunner, G. Process Line for the Production of Raffinated Rice Oil from Rice Bran. *J. Supercrit. Fluids* **2005,** *34*, 133–141.
45. Dar, B. N.; Sharma, S. Total Phenolic Content of Cereal Brans using Conventional and Microwave Assisted Extraction. *Am. J. Food Technol.* **2011,** *6*, 1045–1053.
46. De Munter, J. S.; Hu, F.; Spiegelman, D.; Franz, M.; Van Dam, R. M. Wholegrain, Bran and Germ Intake and Risk of Type 2 Diabetes: a Prospective Cohort Study and Systematic Review. *PLoS Medicine,* **2007,** *4*, 1389–1395.

47. Dewanto, V.; Wu, X. Z.; Liu, R. H. Processed Sweet Corn Has Higher Antioxidant Activity. *J. Agric. Food Chem.* **2002,** *50*, 4959–4964.
48. Dlamini, N. R.; Taylor, J. R. N.; Rooney, L. W. The Effect of Sorghum Type and Processing on the Antioxidant Properties of African Sorghum-based Foods. *Food Chem.* **2007,** *105*, 1412–1419.
49. Donkor, O. N.; Stojanovska, L.; Ginn, P.; Ashton, J.; Vasiljevic, T. Germinated Grains—Sources of Bioactive Compounds. *Food Chem.* **2012,** *135*, 950–959.
50. Dykes, L.; Rooney, L. M. Sorghum and Millet Phenols and Antioxidants. *J. Cereal Sci.* **2006,** *44*, 236–251.
51. Emmanuel, Idehen; Yao, T. Bioactive Phytochemicals in Barley. *J. Food Drug Analy.* **2017,** *25,* 148–161.
52. Esatbeyoglu, T.; Wagner, A. E.; Schini-Kerth, V. B.; Rimbach, G. Betanin: A food Colorant with Biological Activity. *Mol. Nutr. Food Res.* **2015,** *59,* 36–47.
53. Fang, K.; Kang, Z.; Wong, C. Vitamin E Tocotrienols Improve Insulin Sensitivity through Activating Peroxisome Proliferator-Activated Receptors. *Mol. Nutr. Food Res.* **2010,** *54*, 345–352.
54. Fardet, A. New hypotheses for the Health Protective Mechanisms of Wholegrain Cereals: What Is Beyond Fiber? *Nutr. Res. Rev.* **2010,** *23*, 65–134.
55. Felix-Valenzyela, L.; Higuera-Ciapara, I.; Goycoolea-Valencia, F. Supercritical $CO_2$/Ethanol of Astaxanthin from Blue Crab (*Callinectes sapidus*) Shell Waste. *J. Food Process Eng.* **2001,** *24*, 101–112.
56. Feng, H.; Yang, W. *Power Ultrasound.* In *Handbook of Food Science, Technology and Engineering*; Hui, Y. H., Ed.; CRC Press: Boca Raton; Food Science & Technology Series, 2005; pp 121–127.
57. Fogarasi, A. L.; Kun, S.; Tanko, G. Comparative Assessment of Antioxidant Properties, Total Phenolic Content of Einkorn, Wheat, Barley and Their Malts. *Food Chem.* **2015,** *167*, 1–6.
58. Food and Drug Administration (FDA). FDA Allows Barley Products to Claim Reduction in Risk of Coronary Heart Disease. FDA News Release (Dec. 23, 2005). http://origin.www.fda.gov/bbs/topics/news/2005/NEW01287.html (accessed on July 05, 2021).
59. Gamel, T.; Abdel-Aal, E. S. M. Phenolic Acids and Antioxidant Properties of Barley Wholegrain and Pearling Fractions. *Agric. Food Sci.* **2012,** *21*, 118–131.
60. George, J. M.; Selvan, T. S.; Rastogi, N. K. High-Pressure-Assisted Infusion of Bioactive Compounds in Apple Slices. *Innov. Food Sci. Emergi. Technol.* **2016,** *33*, 100–107.
61. Ghafoor, K.; Choi, Y. H.; Jeon, J. Y.; Jo, I. H. Optimization of Ultrasound-Assisted Extraction of Phenolic Compounds, Antioxidants, and Anthocyanins from Grape (*Vitis vinifera*) Seeds. *J. Agric. Food Chem.* **2009,** *57*, 4988–4994.
62. Gibson, G. R.; Roberfroid, M. B. Dietary Modulation of the Human Colonic Microbiota: Introducing the Concept of Prebiotics. *J. Nutr.* **1995,** *125*, 1401–1412.
63. Gopalan, C. Carbohydrates in Diabetic Diet. *India: Bull. Nutr. Foundation* **1981,** *1981*, 1–3.
64. Gorinstein, S.; Medina Vargas, O.; Jaramillo, N.; Salas, I. The Total Polyphenols and the Antioxidant Potentials of Some Selected Cereals and Pseudo Cereals. *European Food Res. Technol.* **2007,** *225*, 321–28.
65. Gous, F. Tannins and Phenols in Black Sorghum. PhD dissertation; College Station, TX: Texas A&M University, 1989; p 215.
66. Grams, G. W.; Blessin, C. W.; Inglett, G. E. Distribution of Tocopherols within the Corn Kernel. *Journal of the American Oil Chemists' Society,* **1970,** *47*, 337–339.

67. Grist, D. H. *Rice*. 5th edition; London, UK: Longman; **1985**; pp 218.
68. Haddadi-Guemghar, H.; Janel, N.; Dairou, J.; Remini, H.; Madani, K. Optimization of Microwave-Assisted Extraction of Prune (*Prunus domestica*) Antioxidants by Response Surface Methodology. *Int. J. Food Sci. Technol.* **2014,** *49*, 2158–2166.
69. Hammi, K. M.; Jdey, A.; Abdelly, C.; Majdoub, H.; Ksouri, R. Optimization of Ultrasound-Assisted Extraction of Antioxidant Compounds from *Tunisian Zizyphus* Lotus Fruits using Response Surface Methodology. *Food Chem.* **2015,** *184,* 80–89.
70. Harborne, J. B. Classes and Functions of Secondary Products from Plants. In: *Chemicals from Plants-Perspectives on Plant Secondary Products*; Walton, J. N., Brown, D. E., Eds.; Imperial College Press: London, 1999; pp 1–25.
71. Harrabi, S.; Amand, A.; Sakouhi, F.; Sebei, K.; Kallel, H. Phytostanols and Phytosterols Distributions in Corn Kernel. *Food Chem.* **2008,** *111*, 115–120.
72. Hegde, P. S.; Chandra, T. S. ESR Spectroscopic Study Reveals Higher Free Radical Quenching Potential in Kodo Millet (*Paspalum scrobiculatum*) Compared to Other Millets. *Food Chem.* **2005,** *92*, 177–182.
73. Hemery, Y. M.; Anson, N. M.; Havenaar, R. Dry-Fractionation of Wheat Bran Increases the Bioaccessibility of Phenolic Acids in Breads made from Processed Bran Fractions. *Food Res. Int.* **2010,** *43*, 1429–1438.
74. Hemwimon, S.; Pavasant, P.; Shotipruk, A. Microwave-assisted Extraction of Antioxidative Anthraquinones from Roots of *Morinda citrifolia*. *Sep. Purif. Technol.* **2007,** *54*, 44–50.
75. Hentschel, V.; Kranl, K.; Hollmann, J.; Lindhauer, M. G. Spectrophotometric Determination of Yellow Pigment Content and Evaluation of Carotenoids by High Performance Liquid Chromatography in Durum Wheat Grain. *J. Agric. Food Chem.* **2002,** *50*, 6663–6668.
76. Hong, J.; Chen, R.; Zeng, X. A.; Han, Z. Effect of Pulsed Electric Fields Assisted Acetylation on Morphological, Structural and Functional Characteristics of Potato Starch. *Food Chem.* **2016,** *192*, 15–24.
77. Indira, T. N.; Narayan, A. V.; Barhate, R. S. *Production of Oryzanol-Enriched Fraction Useful in the Treatment of Melanin Related Disorders and for Minimizing Wrinkles Involves Saponification; Dehydration and Leaching of RBO Soap-Stock.* Patents US2004192948-A1, US6896911-B2 and JP2004300034-A; Council Sci. Indian Res., Assignee (September 30, 2004; May 24, 2005; October 28, 2004).
78. Iqbal, S.; Bhanger, M. I.; Anwar, F. Antioxidant Properties and Components of Some Commercially Available Varieties of Rice Bran in Pakistan. *Food Chem.* **2005,** *93*, 265–272.
79. Izydorczyk, M. S.; Dexter, J. E. Barley B-Glucans and Arabinoxylans: Molecular Structure, Physicochemical Properties and Uses in Food Products—A Review. *Food Res. Int.* **2008,** *41*, 850–868.
80. Jacotet-Navarro, M.; Rombaut, N.; Fabiano-Tixier, A. S. Ultrasound versus Microwave as Green Processes for Extraction of Rosmarinic, Carnosic and Ursolic Acids from Rosemary. *Ultrason. Sonochem.* **2015,** *27,* 102–109.
81. Jemai, A.; Vorobiev, E. Pulsed Electric Field Assisted Pressing of Sugar Beet Slices: Towards a Novel Process of Cold Juice Extraction. *Biosyst. Eng.* **2006,** *93*, 57–68.
82. Jones, P. J.; Raeini-Sarjaz, M.; Ntanios, F. Y.; Vanstone, C. A. Modulation of Plasma Lipid Levels and Cholesterol Kinetics by Phytosterol versus Phytostanol Esters. *J. Lipid Res.* **2000,** *41*, 697–705.

83. Joo, H.; Kim, C. T.; Kim, I. H.; Kim, Y. Anti-Obesity Effects of Hot Water Extract and High Hydrostatic Pressure Extract of Garlic in Rats Fed A High-Fat Diet. *Food Chem. Toxicol.* **2013,** *55*, 100–105.
84. Jun, X.; Shuo, Z.; Bingbing, L.; Rui, Z. Separation of Major Catechins from Green Tea by Ultra High-Pressure Extraction. *Int. J. Pharm.* **2010,** *386*, 229–231.
85. Jung, E. H.; Kim, S. R.; Hwang, I. K.; Ha, T. Y. Hypoglycemic Effects of a Phenolic Acid Fraction of Rice Bran and Ferulic Acid in C57BL/KsJ-db/db mice. *J. Agric. Food Chem.* **2007,** *5*, 9800–9804.
86. Kavitha, M. S.; Prema, L. Post Prandial Blood Glucose Response to Meals Containing Different CHO in Diabetics. *Ind. J. Nutr. Dietetics* **1995,** *32*, 123–126.
87. Kerckhoffs, D. A. J. M.; Hornstra, G.; Mensink, R. P. Cholesterol-Lowering Effect of β-Glucan from Oat Bran in Mildly Hypercholesterolemic Subjects can Decrease, when β-Glucan is Incorporated into Bread and Cookies. *Am. J. Clin. Nutr.* **2003,** *78*, 221–225.
88. Kim, J. S.; Suh, M. H.; Yang, C. B.; Lee, H. G.; Effect of γ-Oryzanol on the Flavor and Oxidative Stability of Refrigerated Cooked Beef. *J. Food Sci.* **2003,** *68*, 2423–2429.
89. Kim, M. J.; Hyun, J. N.; Kim, J. A.; Park, J. C. Relationship Between Phenolic Compounds, Anthocyanins Content and Antioxidant Activity in Colored Barley Germplasm. *J. Agric. Food Chem.* **2007,** *55*, 4802–4809.
90. Kim, S. L.; Kim, S. K.; Park, C. H. Comparisons of Lipid, Fatty Acids and Tocopherols of Different Buckwheat Species. *Food Sci. Biotechnol.* **2002,** *11*, 332–336.
91. Kim, S. P.; Kang, M. Y.; Nam, S. H.; Friedman, M. Dietary Rice Bran Component *γ-Oryzanol* Inhibits Tumor Growth in Tumor Bearing Mice. *Mol. Nutr. Food Res.* **2012,** *56*, 935–944.
92. Kimbaris, A. C.; Siatis, N. G.; Daferera, D. J. Comparison of Distillation and Ultrasound-Assisted Extraction Methods for the Isolation of Sensitive Aroma Compounds from Garlic (*Allium sativum*). *Ultrason. Sonochem.* **2006,** *13*, 54–60.
93. Kong, S.; Kim, D. J.; Oh, S. K.; Choi, I. S.; Jeong, H. S.; Lee. J. Black Rice Bran as an Ingredient in Noodles: Chemical and Functional Evaluation. *J. Food Sci.* **2012,** *77*, 3–10.
94. Krueger, C. G.; Vestling, M. A.; Reed, J. D. Matrix-Assisted Laser Desorption/ Ionization Time-of-Flight Mass Spectrometry of Heteropolyflavan-3-Ols and Glucosylated Hetero Poly-Flavans in Sorghum (*Sorghum bicolor*). *J. Agric. Food Chem.* **2003,** *51*, 538–543.
95. Kumar, M.; Sharma, B. D. Quality and Storage Stability of Low-Fat Pork Patties Containing Barley Flour as Fat Substitute. *J. Food Sci. Technol.* **2004,** *41*, 496–501.
96. Kumar, P.; Yadav, D.; Kumar, P.; Panesar, P. S.; Bunkar, D. S. Comparative Study on Conventional, Ultrasonication and Microwave Assisted Extraction of γ-oryzanol from Rice Bran. *J. Food Sci. Technol.* **2016,** *53* (4), 2047–2053.
97. Lee, Y. T.; Schwarz, P. B.; Appolonia, B. L. Effects of β-(1–3),(1–4)-D-glucans from Hull-Less Barley on the Properties of Wheat, Starch, Flour and Bread. *Barley Newslett.* **1995,** *39*, 60–63.
98. Li, H.; Deng, Z.; Wu, T.; Liu, R.; Loewen, S.; Tsao, R. Microwave-Assisted Extraction of Phenolics with Maximal Antioxidant Activities in Tomatoes. *Food Chem.* **2012,** *130*, 928–936.
99. Li, H.; Pordesimo, L.; Weiss, J. High Intensity Ultrasound-Assisted Extraction of Oil from Soybeans. *Food Res. Int.* **2004,** *37*, 731–738.
100. Li, Y.; Skouroumounis, G. K.; Elsey, G. M.; Taylor, D. K. Microwave Assistance Provides Very Rapid and Efficient Extraction of Grape Seed Polyphenols. *Food Chem.* **2011,** *129*, 570–576.

101. Liazid, A.; Guerrero, R. F.; Cantos, E.; Palma, M.; Barroso, C. G. Microwave Assisted Extraction of Anthocyanins from Grape Skins. *Food Chem.* **2011,** *124*, 1238–1243.
102. Liu, R. H. Whole Grain Phytochemicals and Health, *J. Cereal Sci.* **2007,** *46*, 207–219.
103. Liu, R. H. Potential Synergy of Phytochemicals in Cancer Prevention Mechanism of Action. *J. Nutr.* **2004,** *134*, 3479–3485.
104. Liu, Z. W.; Zeng, X. A.; Sun, D. W.; Han, Z.; Aadil, R. M. Synergistic Effect of Thermal and Pulsed Electric Field (PEF) treatment on the Permeability of Soya PC and DPPC Vesicles. *J. Food Eng.* **2015,** *153*, 124–131.
105. Liyana-Pathirana, C. M.; Shahidi, F. The Antioxidant Potential of Milling Fractions from Bread, Wheat and Durum. *J. Cereal Sci.* **2007,** *45*, 238–247.
106. Lopez, H. W.; Levrat, M. A.; Guy, C. Effects of Soluble Corn Bran Arabinoxylans on Cecal Digestion, Lipid Metabolism, and Mineral Balance (Ca, Mg) in Rats. *J. Nutr. Biochem.* **1999,** *10*, 500–509.
107. Luengo, E.; Condón-Abanto, S.; Condón, S.; Álvarez, I.; Raso, J. Improving the Extraction of Carotenoids from Tomato Waste by Application of Ultrasound under Pressure. *Sep. Purif. Technol.* **2014,** *136*, 130–136.
108. Luna-Vital, D.; Weiss, M.; Gonzalez de Mejia, E.; Anthocyanins from Purple Corn Ameliorated Tumor Necrosis Factor-Alpha-Induced Inflammation and Insulin Resistance in 3T3-L1 Adipocytes via Activation of Insulin Signaling and Enhanced GLUT4 Translocation. *Mol. Nutr. Food Res.* **2017,** *61*, 1–10.
109. Luoa, X.; Cuia, J.; Zhanga, H.; Duana, Y.; Zhanga, D.; Caia, M.; Chen, G. Ultrasound Assisted Extraction of Polyphenolic Compounds from Red Sorghum (*Sorghum bicolor* L.) Bran and their Biological Activities and Polyphenolic Compositions. *Ind. Crops Product.* **2018,** *112*, 296–304.
110. Lupton, J. R.; Turner, N. D. Potential Protective Mechanisms of Wheat Bran Fiber. *Am. J. Med.* **1999,** *106*, 24–27.
111. Luque de Castro, M. D.; Valcarcel, M.; Tena, M. T., Eds. *Analytical Supercritical Fluid Extraction*, 1st ed.; Springer-Verlag: Germany; 1994; pp 318.
112. Luque-Garcıa, J. L.; Luque de Castro, M. D. Ultrasound-Assisted Soxhlet Extraction: An Expeditive Approach for Solid Sample Treatment: Application to the Extraction of Total Fat from Oleaginous Seeds. *J. Chromatogr. A* **2004,** *1034*, 237–242.
113. Machado, A. P. D. F.; Pasquel-Reátegui, J. L.; Barbero, G. F.; Martínez, J. Pressurized Liquid Extraction of Bioactive Compounds from Blackberry (*Rubus fruticosus* L.) Residues: A Comparison with Conventional Methods. *Food Res. Int.* **2015,** *77*, 675–683.
114. Maes, C.; Delcour, J. A. Structural Characterization of Water Extractable and Water Unextractable Arabinoxylans in Wheat Bran. *J. Cereal Sci.* **2002,** *35*, 315–326.
115. Marathe, S. J.; Jadhav, S. B.; Bankar, S. B.; Dubey, K. K.; Singhal, R S. Improvements in the Extraction of Bioactive Compounds by Enzymes. *Curr. Opin. Food Sci.* **2019,** *25*, 62–72.
116. Marcone, M.; Kakuda, Y.; Yada, R. Amaranth as a Rich Dietary Source of β-sitosterol and other Phytosterols. *Plant Foods Human Nutr.* **2003,** *58*, 207–11.
117. Marti, A.; Barbiroli, A.; Bonomi, F.; Brutti, A.; Iametti, S. Effect of High-Pressure Processing on the Features of Wheat Milling By-Products. *Cereal Chem.* **2014,** *91*, 318–320.
118. Mateo Anson, N.; Van den Berg, R.; Havenaar, R. Bioavailability of Ferulic Acid is Determined by Its Bioaccessibility. *J. Cereal Sci.* **2009,** *49*, 296–300.
119. Matsubara, Y.; Kumamoto, H.; Iizuka, Y.; Murakami, T. Structure and Hypotensive Effect of Flavonoid Glycosides in Citrus Peelings. *Agric. Biol. Chem.* **1985,** *49*, 909–914.

120. Melanson, K. J.; Angelopoulos, T. J.; Nguyen, V. T. Consumption of Whole-Grain Cereals during Weight Loss: Effects on Dietary Quality, Dietary Fiber, Magnesium, Vitamin B-6, and Obesity. *J. Am. Dietetic Assoc.* **2006,** *106*, 1380–1388.
121. Mellen, B. P.; Walsh, T. F.; Herrington, D. M. Wholegrains Intake and Cardiovascular Disease: A Meta-Analysis. *Nutr. Metabol. Cardiovasc. Dis.* **2009,** *18*, 283–290.
122. Mendes, R. L.; Nobre, B. P.; Cardoso, M. T.; Pereira, A. P.; Palavra, A. F. Supercritical Carbon Dioxide Extraction of Compounds with Pharmaceutical Importance from Microalgae. *Inorg. Chim. Acta*, **2003,** *356*, 328–334.
123. Metayer, S.; Seiliez, I.; Collin, A.; Duchene, S.; Mercier, Y. Mechanisms Through Which Sulphur Amino Acids Control Metabolism and Oxidative Status. *J. Nutr. Biochem.* **2008,** *19,* 207–221.
124. Moldenhauer, K. A., Champagne, E. T., McCaskill, D. R., Guraya, H. *Functional Products from Rice*. In *Functional Foods*; Mazza, G., Ed.; Technomic Publishing Co., Inc: Lancaster, Basel, Switzerland, 2003; pp 71–89.
125. Monrad, J. K.; Howard, L. R.; King, J. W.; Srinivas, K. Subcritical Solvent Extraction of Anthocyanins from Dried Red Grape Pomace. *J. Agric. Food Chem.* **2010,** *58*, 2862–2868.
126. Moreau, R. A., Whitaker, B. D., Hicks, K. B. Phytosterols, Phytostanols and their Conjugates in Foods: Structural Diversity, Quantitative Analysis, and Health-Promoting Uses. *Progress Lipid Res.* **2002,** *41*, 457–500.
127. Moreau, R. A.; Flores, R. A.; Hicks, K. B. Composition of Functional Lipids in Hulled Barley in Fractions Obtained by Scarification and in Barley Oil. *Cereal Chem.* **2007,** *84*, 1–5.
128. Moreno, Y. S.; Sanchez, G. S.; Hernandez, D. R.; Lobato, N. R. Characterization of Anthocyanin Extracts from Maize Kernels. *J. Chromatogr. Sci.* **2005,** *43*, 483–487.
129. Mortensen, A.; Skibsted, L. H. Importance of Carotenoid Structure in Radical-Scavenging Reactions. *J. Agric. Food Chem.* **1997,** *45*, 2970–2977.
130. Mota, F. L.; Queimada, A. J.; Pinho, S. P.; Macedo, E. A. Aqueous Solubility of Some Natural Phenolic Compounds. *Ind. Eng. Chem. Res.* **2008,** *47*, 5182–5189.
131. Mourouti, N.; Kontogianni, M. D.; Papavagelis, C.; Psaltopoulou, T. Whole Grain Consumption and Breast Cancer: A Case-Control Study in Women. *J. Am. College Nutr.* **2016,** *35*, 143–149.
132. Murphy, M. M.; Douglass, J. S.; Birkett, A. Resistant Starch Intakes in the United States. *J. Am. Dietetic Assoc.* **2008,** *108*, 67–78.
133. Nam, S. H.; Choi, S. P.; Kang, M. Y.; Koh, J. H. Antioxidative Activities of Bran Extracts From Twenty One Pigmented Rice Cultivars. *Food Chem.* **2006,** *94*, 613–620.
134. Nanua, J. N.; McGregor, J. U.; Godber, J. S. Influence of High Oryzanol RBO on the Oxidative Stability of Whole Milk Powder. *J. Dairy Sci.* **2000,** *83*, 2426–2431.
135. Nawawi, H. M. Tocotrienols and Atherosclerosis Potential in Cardio-Protection. In *Tocotrienols: Vitamin E Beyond Tocopherols*; Tan, B., Watson, R. R., Preedy. V. R., Eds.; CRC Press: Boca Raton, 2012; pp 150–163.
136. Nissinen, M.; Gylling, H.; Vuoristo, M.; Miettinen, T. A. Micellar Distribution of Cholesterol and Phytosterols After Duodenal Plant Stanol Ester Infusion. *Am. J. Physiol.—Gastrointestinal Liver Physiol.* **2002,** *282*, 1009–1015.
137. Oh, S. H. Stimulation of γ-Aminobutyric Acid Synthesis Activity in Brown Rice by a Chitosan/Glutamic Acid Germination Solution and Calcium/Calmodulin. *J. Biochem. Mol. Biol.* **2003,** *36*, 319–325.

138. Oliveira, A. L.; Destandau, E.; Fougere, L.; Lafosse, M. Isolation by Pressurized Fluid Extraction (PFE) and Identification Using CPC and HPLC/ESI/MS of Phenolic Compounds from Brazilian Cherry Seeds (*Eugenia uniflora L.*). *Food Chem.* **2014,** *145*, 522–529.
139. Omary, M. B.; Fong, C.; Rothschild, J.; Finney, P. Effects of Germination on the Nutritional Profile of Gluten-Free Cereals and Pseudo Cereals: A Review. *Cereal Chem. J.* **2012,** *89*, 1–14.
140. Onyango, C.; Noetzold, H.; Bley, T.; Henle, T. Proximate Composition and Digestibility of Fermented and Extruded Uji from Maize-Finger Millet Blend. *LWT Food Sci. Technol.* **2004,** *37*, 827–832.
141. Oomah, B.; Mazza, G. Flavonoids and Antioxidative Activities in Buckwheat. *J. Agric. Food Chem.* **1996,** *44*, 1746–1750.
142. Osawa,T. Protective Role of Rice Polyphenols in Oxidative Stress. *Anticancer Res.* **1999,** *19*, 3645–3650.
143. Paes, J.; Dotta, R.; Barbero, G. F.; Martínez, J. Extraction of Phenolic Compounds and Anthocyanins from Blueberry (*Vaccinium myrtillus L.*) Residues Using Supercritical $CO_2$ and Pressurized Liquids. *J. Supercrit. Fluids* **2014,** *95*, 8–16.
144. Paiva, F. F.; Vanier, N. L.; Berrios, J. D. J.; Pan, J. Physicochemical and Nutritional Properties of Pigmented Rice Subjected to Different Degrees of Milling. *J. Food Composition Analy.* **2014,** *35*, 10–17.
145. Pan, Z.; Qu, W.; Ma, H.; Atungulu, G. G.; McHugh, T. H. Continuous and Pulsed Ultrasound-Assisted Extractions of Antioxidants from Pomegranate Peel. *Ultrason. Sonochem.* **2012,** *19*, 365–372.
146. Pariyarath, A. M.; Anal, A. K.; Sadiq, M. B. Optimization of Ultrasonic Assisted Extraction (UAE) Of Anthocyanins from Black Glutinous Rice and Evaluation of Their Antioxidant Properties. *Int. J. Food Eng.* **2018,** *4*, 4–10.
147. Park, C. Y.; Kim, S.; Lee, D.; Park, D. J.; Imm, J. Y. Enzyme and High Pressure Assisted Extraction of Tricin from Rice Hull and Biological Activities of Rice Hull Extract. *Food Sci. Biotechnol.* **2016,** *25* (1), 159–164.
148. Parniakov, O.; Apicella, E.; Koubaa, M.; Barba, F. J. Ultrasound-Assisted Green Solvent Extraction of High-Added Value Compounds from Microalgae *Nannochloropsis spp. Bioresour. Technol.* **2015,** *198*, 262–267.
149. Parniakov, O.; Barba, F. J.; Grimi, N.; Lebovka, N.; Vorobiev, E. Impact of Pulsed Electric Fields and High Voltage Electrical Discharges on Extraction of High-Added Value Compounds from Papaya Peels. *Food Res. Int.* **2014,** 65, 337–343.
150. Peanparkdee, M.; Patrawart, J.; Iwamoto, S. Effect of Extraction Conditions on Phenolic Content, Anthocyanin Content and Antioxidant Activity of Bran Extracts from Thai Rice. *J. Cereal Sci.* **2019,** *86*, 86–91.
151. Pick, M. E.; Hawrysh, Z. J..; Gee, M. I.; Toth, E.; Garg, M. L.; Hardin, R. T. Oat Bran Concentrate Bread Products Improve Long-Term Control of Diabetes: A Pilot Study. *J. Am. Dietetic Assoc.* **1996,** *96*, 1250–1254.
152. Piironen, V.; Lindsay, D. G.; Miettinen, T. A.; Toivo, J.; Lampi, A. M. Plant Sterols: Biosynthesis, Biological Function and their Importance to Human Nutrition. *J. Sci. Food Agric.* **2000,** *80*, 939–966.
153. Pinzino, C.; Nanni, B.; Zandomeneghi, M. Aging, Free Radicals, and Antioxidants in Wheat Seeds. *J. Agric. Food Chem.* **1999,** *47*, 1333–1339.

154. Povilaitis, D.; Sulniute, V.; Venskutonis, P. R.; Kraujaliene, V. Antioxidant Properties of Wheat and Rye Bran Extracts Obtained by Pressurized Liquid Extraction with Different Solvents. *J. Cereal Sci.* **2015,** *62*, 117–123.
155. Price, M. L.; Van Scoyoc, S.; Butler, L. G. A Critical Evaluation of Vanillin Reaction as an Assay for Tannin in Sorghum. *J. Agric. Food Chem.* **1978,** *26*, 1214–1218.
156. Puértolas, E.; De Marañón, I. M. Olive Oil Pilot-Production Assisted by Pulsed Electric Field: Impact on Extraction Yield, Chemical Parameters and Sensory Properties. *Food Chem.* **2015,** *167*, 497–502.
157. Rajasekaran, N. S.; Nithya, M.; Rose, C.; Chandra, T. S. The Effect of Finger Millet Feeding on the Early Responses During the Process of Wound Healing in Diabetic Rats. *Biochim. Biophys. Acta* **2004,** *1689*, 190–201.
158. Rajha, H. N.; Boussetta, N.; Louka, N.; Maroun, R. G.; Vorobiev, E. Comparative Study of Physical Pretreatments for the Extraction of Polyphenols and Proteins from Vine Shoots. *Food Res. Int.* **2014,** *65*, 462–468.
159. Ramanathan, N.; Tan, E.; Loh, L. J.; Soh, B. S.; Yap, W. N. Tocotrienol is a Cardio-Protective Agent Against Ageing-Associated Cardiovascular Disease and its Associated Morbidities. *Nutr. Metabol.* **2018,** *15*, 6–10.
160. Rao, M. V.; Muralikrishna, G. Evaluation of the Antioxidant Properties of Free and Bound Phenolic Acids from Native and Malted Finger Millet (ragi: *Eleusinecoracana Indaf-15*). *J. Agric. Food Chem.* **2002,** *50*, 889–892.
161. Raviyan, P.; Zhang, Z.; Feng, H. Ultrasonication for Tomato Pectin Methyl Esterase Inactivation: Effect of Cavitation Intensity and Temperature on Inactivation. *J. Food Eng.* **2005,** *70*, 189–196.
162. Ribas-Agustí, A.; Martín-Belloso, O.; Soliva-Fortuny R.; Elez-Martínez, P. Food Processing Strategies to Enhance Phenolic Compounds Bioaccessibility and Bioavailability in Plant-Based Foods. *Crit. Rev. Food Sci. Nutr.* **2018,** *58*, 10–15.
163. Roberts, J. E.; Dennison, J. The Photobiology of *Lutein* and *Zeaxanthin* in the Eye. *J. Ophthalmol.* **2015,** *2015*, E-article ID: 687173.
164. Rodrigues, S.; Fernández, F. A. N.; De Brito, E. S.; Sousa, A. D.; Narain, N. Ultrasound Extraction of Phenolics and Anthocyanins from Jabuticaba Peel. *Ind. Crops Product.* **2015,** *69*, 400–407.
165. Ryu, S. N.; Park, S. Z.; Ho, C. T. High Performance Liquid Chromatographic Determination of Anthocyanin Pigments in Some Varieties of Black Rice. *J. Food Drug Analy.* **1998,** *6*, 729–736.
166. Salar, R. K.; Certik, M.; Brezova, V. Modulation of Phenolic Content and Antioxidant Activity of Maize By Solid State Fermentation with *Thamnidium elegans* CCF 1456. *Biotechnol. Bioprocess Eng.* **2012,** *17*, 109–116.
167. Sánchez-Moreno, C.; Plaza, L.; Ancos, B.; Cano, M. P. Effect of Combined Treatments of High-Pressure and Natural Additives on Carotenoid Extractability and Antioxidant Activity of Tomato Puree. *Eur. Food Res. Technol.* **2004,** *219*, 151–160.
168. Schley, P. D.; Field, C. J. The Immune Enhancing Effects of Dietary Fibers and Prebiotics. *Br. J. Nutr.* **2002,** *97*, S221–S230.
169. Seetharamaiah, G. S.; Chandrasekhara, N. Studies on Hypocholesterolemic Activity of Rice Bran Oil. *Atherosclerosis* **1989,** *78*, 219–223.
170. Sethi, S.; Nanda, S. K.; and Bala, M. *Pigmented Rice Varieties of India: Processing and Value Addition*. Technical Bulletin No.: ICAR-CIPHET/Pub./2018-19/04; ICAR-Central Institute of Post- Harvest Engineering and Technology, Ludhiana—141004; 2019; p 25.

171. Shand, P. J. Textural, Water Holding, and Sensory Properties of Low-Fat Pork Bologna with Normal or Waxy Starch Hull-Less Barley. *J. Food Sci.* **2000,** *65*, 97–101.
172. Sharma, A.; Gupta, M. N. Oil Extraction from Almond, Apricot and Rice by Three-Phase Partitioning After Ultrasonication. *Eur. J. Lipid Sci. Technol.* **2004,** *106*, 183–186.
173. Shobana, S.; Malleshi, N. G. Preparation and Functional Properties of Decorticated Finger Millet (*Eleusine coracana*). *J. Food Eng.* **2007,** *79*, 529–538.
174. Shobana, S.; Sreerama, Y. N.; Malleshi, N. G. Composition and Enzyme Inhibitory Properties of Finger Millet (*Eleusine coracana L.)* Seed Coat Phenolics: Mode of Inhibition of α-Glucosidase and Pancreatic Amylase. *Food Chem.* **2009,** *115*, 1268–1273.
175. Shouqin, Z.; Junjic, Z.; Changzhen, W. Novel High Pressure Extraction Technology. *Int. J. Pharm.* **2004,** *278*, 471–474.
176. Singh, A.; Rehal, J.; Kaur, A.; Jyot, G. Enhancement of Attributes of Cereals by Germination and Fermentation: A Review. *Crit. Rev. Food Sci. Nutr.* **2015,** *55*, 1575–1589.
177. Singh, V.; Moreau, R. A.; Hicks, K. B. Yield and Phytosterol Composition of Oil Extracted From Grain Sorghum and its Wet-Milled Fractions. *Cereal Chem.* **2003,** *80*, 126–129.
178. Smeds, A. I.; Eklund, P. C.; Sjoholm, R. E. Quantification of a Broad Spectrum of Lignans in Cereals, Oilseeds, and Nuts. *J. Agric. Food Chem.* **2007,** *55*, 1337–1346.
179. Soares, J. F.; Dal Prá, V.; De Souza, M.; Lunelli, F. C. Extraction of Rice Bran Oil Using Supercritical $CO_2$ and Compressed Liquefied Petroleum Gas. *J. Food Eng.* **2015,** *170*, 58–63.
180. Sompong, R.; Siebenhandl, E. S.; Linsberger, M. G.; Berghofer, E. Physicochemical and Antioxidative Properties of Red and Black Rice Varieties from Thailand. China and Sri Lanka. *Food Chem.* **2011,** *124,* 132–140.
181. Song, J.; Li, D.; Liu, C.; Zhang, Y. Optimized Microwave-Assisted Extraction of Total Phenolics from *Ipomoea Batatas* Leaves and its Antioxidant Activity. *Innov. Food Sci. Emergi. Technol.* **2011,** *12*, 282–287.
182. Soquetta, M. B.; Terra, L. M.; Bastos, C. P. Green Technologies for the Extraction of Bioactive Compounds in Fruits and Vegetables. *CyTA—J. Food* **2018,** *16* (1), 400–412.
183. Sourki, A. H. *Optimization of Beta-Glucan Extraction from Hull-Less Barley Cultivar Lut by Conventional and Ultrasound—Assisted Methods and Investigation of It's Physicochemical, Rheological and Molecular Properties*; PhD Thesis; Ferdowsi University, Mashhad, Iran, 2016; p 218.
184. Stafford, H. A. Flavonoids and Related Phenolic Compounds Produced in the First Internode of Sorghum Vulgare Preserved in Darkness and in Light. *Plant Physiol.* **1965,** *40*, 130–139.
185. Stevenson, L.; Phillips, F.; O'sullivan, K.; Walton, J. Wheat Bran its Composition and Benefits to Health: A European Perspective. *Int. J. Food Sci. Nutr.* **2012,** *63*, 1001–1013.
186. Sullivan, P.; Arendt, E.; Gallagher, E. The Increasing Use of Barley and Barley By-Products in the Production of Healthier Baked Goods. *Trends Food Sci. Technol.* **2013,** *29*, 124–134.
187. Suman, S.; Datta, K.; Chakraborty, K.; Kulkarni, S. S. Gamma Tocotrienol, a Potent Radioprotector, Preferentially Upregulates Expression of Anti-Apoptotic Genes to Promote Intestinal Cell Survival. *Food Chem. Toxicol.* **2013,** *60*, 488–496.
188. Sweeny, J. G.; Iacobucci, G. A. Synthesis of Anthocyanidins, Part III: Total Synthesis of Apigeninidin and Luteolinidin Chlorides. *Tetrahedron* **1981,** *37*, 1481–1483.
189. Talukder, S.; Sharma, D. P. Development of Dietary Fiber Rich Chicken Meat Patties Using Wheat and Oat Bran. *J. Food Sci. Technol.* **2010,** *47*, 224–229.

190. Tang, Y.; Li, X.; Chen, P. X.; Zhang, B.; Liu, R. Assessing The Fatty Acid, Carotenoid, and Tocopherol Compositions of Amaranth and Quinoa Seeds Grown in Ontario and their Overall Contribution to Nutritional Quality. *J. Agric. Food Chem.* **2016,** *64* (5), 1103–1110.
191. Tao, Y.; Wu, D.; Zhang, Q. A.; Sun, D. W. Ultrasound-Assisted Extraction of Phenolics from Wine Lees: Modeling, Optimization and Stability of Extracts during Storage. *Ultrason. Sonochem.* **2014,** *21,* 706–715.
192. Teixeira, A.; Baenas, N.; Dominguez-Perles, R. Natural Bioactive Compounds from Winery By-Products as Health Promoters: A Review. *International Journal of Molecular Sciences*, **2014,** *15*, 15638–15678.
193. Tharanathan, R. N.; Mahadevamma, S. Grain Legumes: A Boon to Human Nutrition. *Trends Food Sci. Technol.* **2003,** *14*, 507–518.
194. Thebaudin, J. Y.; Lefebvre, A. C.; Harrington, M. Bourgeois, C. M. Dietary Fiber: Nutritional and Technological Interest. *Trends Food Sci. Technol.* **1997,** *8*, 41–48.
195. Tian, S.; Sun, Y.; Chen, Z.; Yang, Y.; Wang Y. Functional Properties of Polyphenols in Grains and Effects of Physicochemical Processing on Polyphenols. *J. Food Qual.* **2019,** *2019* Article ID 2793973, 1–8.
196. Tighe,P.; Duthie, G.; Vaughan, N.; Brittenden, J. Effect of Increased Consumption of Wholegrain Foods on Blood Pressure and Other Cardiovascular Risk Markers in Healthy Middle-Aged Persons: A Randomized Controlled Trial. *Am. J. Clin. Nutr.* **2010,** *92*, 733–740.
197. Topping, D. L.; Clifton, P. M. Short Chain Fatty Acids and Human Colonic Function: Roles of Resistant Starch and Non-Starch Polysaccharides. *Physiol. Rev.* **2001,** *81*, 1031–1064.
198. Tovey, F. I. Diet and Duodenal Ulcer. *J. Gastroenterol. Hepatol.* **1994,** *9*, 177–185.
199. Upadhyay, J.; Misra, K. Towards the Interaction Mechanism of Tocopherols and Tocotrienols (Vitamin E) with Selected Metabolizing Enzymes. *Bioinformation* **2009,** *3*, 326–331.
200. USA Rice Federation; www.usarice.com; Accessed on Aug 13, 2019.
201. Wang, H.; Chen, C.; Chang, C. Carbon Dioxide Extraction of Ginseng Root Hair Oil and Ginsenosides. *Food Chem.* **2001,** *72*, 505–509.
202. Wang, J.; Sun, B.; Cao, Y.; Tian, Y.; Li, X. Optimization of Ultrasound-Assisted Extraction of Phenolic Compounds from Wheat Bran. *Food Chem.* **2008,** *106*, 804–810.
203. Wang, L.; Weller, C. L. Recent Advances in Extraction of Nutraceuticals from Plants. *Trends Food Sci. Technol.* **2006,** *17*, 300–312.
204. Wang,J.; Sun, B.; Cao, Y.; Tian, Y. Protection of Wheat Bran Feruloyl Oligosaccharides Against Free Radical Induced Oxidative Damage in Normal Human Erythrocytes. *Food Chem. Technol.* **2009,** *47*, 1591–1599.
205. Waniska, R. D.; Poe, J. H.; Bandyopadhyay, R. Effects of Growth Conditions on Grain Molding and Phenols in Sorghum Caryopsis. *J. Cereal Sci.* **1989,** *10*, 217–225.
206. Wen, Y.; Chen, H.; Zhou, X.; Deng, Q.; Zhao, Y.; Zhao, C.; Gong, X. Optimization of the Microwave-Assisted Extraction and Antioxidant Activities of Anthocyanins from Blackberry using a Response Surface Methodology. *RSC Adv.* **2015,** *5*, 19686–19695.
207. Wijngaard, H. H.; Ballay, M.; Brunton, N. The Optimization of Extraction of Antioxidants from Potato Peel by Pressurized Liquids. *Food Chem.* **2012,** *133*, 1123–1130.
208. Wu, T.; Guo, X. Q.; Zhang, M.; Yang, L.; Liu, R.; Yin, J. J. Anthocyanins in Black Rice, Soybean and Purple Corn Increase Fecal Butyric Acid and Prevent Liver Inflammation in High Fat Diet-Induced Obese Mice. *Food Funct.* **2017,** *8*, 3178–3186.

209. Xu, Z.; Godber, J. S. Comparison of Supercritical Fluid and Solvent Extraction Methods in Extracting γ-oryzanol from Rice Bran. *J. Am. Oil Chem. Soc.* **2000,** *77* (5), 547–551.
210. Xu, Z.; Hua, N.; Godber, J. S. Antioxidant Activity of Tocopherols, Toco- Trienols, and γ-Oryzanol Components from Rice Bran Against Cholesterol Oxidation Accelerated by 2,2-Azobis (2-Methyl Propionamidine) Dihydrochloride. *J. Agric. Food Chem.* **2001,** *49*, 2077–2081.
211. Xynos, N.; Papaefstathiou, G.; Gikas, E.; Argyropoulou, A. Design Optimization Study of the Extraction of Olive Leaves Performed with Pressurized Liquid Extraction Using Response Surface Methodology. *Sep. Purif. Technol.* **2014,** *122*, 323–330.
212. Yasukawa, K.; Akihisa, T.; Kimura, Y.; Tamura, T.; Takido, M. Inhibitory Effect of Cycloartenol Ferulate, a Component of Rice Bran, On Tumor Promotion in Two-Stage Carcinogenesis in Mouse Skin. *Biol. Pharma. Bull.* **1998,** *21*, 1072–1076.
213. Yawadio, R.; Tanimori, S.; Morita, N. Identification of Phenolic Compounds Isolated from Pigmented Rices and their Aldose Reductase Inhibitory Activities. *Food Chem.* **2007,** *101*, 1616–1625.
214. Ye, E. Q.; Chacko, S. A.; Chou, E. L.; Kugizaki, M.; Liu, S. Greater Wholegrain Intake is Associated with Lower Risk of Type 2 Diabetes, Cardio Vascular Disease, and Weight Gain. *J. Nutr.* **2012,** *142*, 1304–1313.
215. Yoshinaga, K.; Yakahashi, K.; Yoshizawa, K. Liquor with Pigments of Red Rice. *J. Brewing Soc. Jpn.* **1986,** *81*, 337–342.
216. Yu, L.; Zhou, K.; Parry, J. W. 2005. Inhibitory Effect of Wheat Bran Extracts on Human LDL Oxidation and Free Radicals. *LWT Food Sci. Technol.* **2005,** *38*, 463–470.
217. Zanoletti, M.; Parizad, P. A.; Lavelli, V.; Cecchini, C. Debranning of Purple Wheat: Recovery of Anthocyanin-Rich Fractions and their Use in Pasta Production. *LWT Food Sci. Technol.* **2017,** *75*, 663–669.
218. Zawistowski, J.; Kopec, A.; Kitts, D. D. Effects of a Black Rice Extract (*Oryza sativa L. Indica*) on Cholesterol Levels and Plasma Lipid Parameters in Wistar Kyoto Rats. *J. Funct. Foods* **2009,** *1*, 50–56.
219. Zhai, F. H.; Wang, Q.; Han, J. R. Nutritional Components and Antioxidant Properties of Seven Kinds of Cereals Fermented by the Basidiomycete *Agaricus blazei*. *J. Cereal Sci.* **2015,** *65*, 202–208.
220. Zhang, H. Q.; Barbosa-Canovas, G. V., Eds., *Non-thermal Processing Technologies for Food*; International Food Technology Press: London, 2011; p 314.
221. Zhang, Q. A.; Zhang, Z. Q.; Yue, X. F.; Fan, X. H.; Li, T.; Chen, S. F. Response Surface Optimization of Ultrasound-Assisted Oil Extraction from Autoclaved Almond Powder. *Food Chem.* **2009,** *116*, 513–518.
222. Zhang, S.; Xi, J.; Wang, C. High Hydrostatic Pressure Extraction of Flavonoids from Propolis. *J. Chem. Technol. Biotechnol.* **2005,** *80*, 50–54.
223. Zhang, Z. S.; Wang, L. J.; Li, D.; Jiao, S. S.; Chen, X. D.; Mao, Z. H.; Ultrasound Assisted Extraction of Oil from Flaxseed. *Sep. Purif. Technol.* **2008,** *62*, 192– 198.
224. Zhang, Z.; Feng, H.; Niu, Y.; Eckhoff, S. R. Starch Recovery from Degermed Corn Flour and Hominy Feed Using Power Ultrasound. *Cereal Chem.* **2005,** *82*, 447–449.
225. Zhao, C. N.; Zhang, J.; Li, Y.; Meng, X.; Li, H. Microwave Assisted Extraction of Phenolic Compounds from *Melastoma Sanguineum* Fruit: Optimization and Identification. *Molecules* **2018,** *23*, 2490–2498.
226. Zhao, C.; Giusti, M. M.; Malik, M.; Moyer, M. P.; Magnuson, B. A. Effects of Commercial Anthocyanin-Rich Extracts on Colonic Cancer and Non-Tumorigenic Colonic Cell Growth. *J. Agric. Food Chem.* **2004,** *52*, 6122 6128.

227. Zheng, L.; Wen, G.; Yuan, M. Y.; Gao, F. Ultrasound-Assisted Extraction of Total Flavonoids from Corn Silk and their Antioxidant Activity. *J. Chem.* **2016,** 1–5.
228. Zhou, B.; Feng, H.; Lou, Y. Ultrasound Enhanced Sanitizer Efficacy in Reduction of *Escherichia Coli* O157: H7 Population on Spinach Leaves. *J. Food Sci.* **2009,** *74*, 308–313.
229. Zhou, K.; Su, L.; Yu, L. Phytochemicals and Antioxidant Properties in Wheat Bran. *J. Agric. Food Chem.* **2004,** *52*, 6108–6114.
230. Zieliński, H.; Kozłowska, H. Antioxidant Activity and Total Phenolics in Selected Cereal Grains and Their Different Morphological Fractions. *J. Agric. Food Chem.* **2000,** *48*, 2008–2016.

## CHAPTER 10

# HEALTH BENEFITS OF PHYTOCHEMICALS IN HOT PEPPER (*CAPSICUM ANNUUM* L.)

ANUP KUMAR SARKAR and SANJOY SADHUKHAN


## ABSTRACT

*Capsicum annuum* L. (hot pepper) fruits are being used as a spice, condiment, nutraceutical supplement, culinary supplement, vegetable, and medicine. It has organoleptic and culinary properties, with nutritional, nutraceutical, dietetic, and health benefits. Several health characteristics are antioxidant, cardiovascular protective, anticancer, anti-inflammatory, immunomodulatory, antidiabetic, antiobesity, and antibacterial characteristics. Phytochemicals present in chili modulate several cellular and biochemical processes of humans. In this chapter, we have enlightened the nutritional values, physicochemical composition, volatile chemicals, and health advantages of *Capsicum annum*, as well as its physicochemical makeup. Now it is predominantly grown throughout the globe and appreciated for its therapeutic and bioactive properties in treating various diseases and disorders.


## 10.1 INTRODUCTION

*Capsicum annuum* L. is a herbaceous plant of the family Solanaceae, which is recognized as one of the cash crops in international trading for its fruits. Globally these fruits are consumed in many forms and their popularity has progressively increased due to their taste, flavor, and health benefits. Chili is a ubiquitous spice that is regularly used in a wide range of cuisines throughout

Plant-Based Bioactive Compounds and Food Ingredients: Encapsulation, Functional, and Safety Aspects.
Junaid Ahmad Malik, Megh R. Goyal, Preeti Birwal, & Ritesh B. Watharkar (Eds.)

the globe and is also frequently used as a flavoring agent, colorant, to impart tang and flavor to otherwise bland foods. Chili fruits are consumed either fresh or processed in form or as spices.

Functional foods are dietary items that have nutritional components and perform one or more critical goal activities in the body, such as lowering or minimizing the risk of developing certain illnesses and improving overall health.[24] In addition to its nutritional advantages, it also has therapeutic value and pharmacological applications. Chili is now considered a functional food. Incorporating hot pepper into our daily diet may help to treat nutritional insufficiency in people.[96] The hot pepper is an excellent source of bioactive compounds like proteins, fats, oils, phenolic substances, minerals, ascorbic acid, and other vitamins.

For a long time, different varieties of *C. annuum* have been utilized in Indian ethnomedicine, native American indigenous medicine, and Chinese traditional medicine to cure a number of human diseases including gastrointestinal problems, arthritis and rheumatoid disorders, and skin diseases[105,147]. In terms of vegetables, chili has the advantage of being inexpensive and containing an array of minerals and vitamins including vitamin B6, vitamin B12, vitamin C, vitamin E, and provitamin A, which may result in considerably improved nutrition. Medical journals highlighted some evidence indicating persons who consume spicy meals; particularly fresh chili fruits are at a lower risk of dying of cancer or diabetes.[117] Since ancient times, the species has been regarded as one of the most essential spices with incredible therapeutic properties, yet it appears to be under-appreciated as a herbal medicine in the modern pharmaceutical drug industry.

This chapter explores the potential and adaptability of hot chili (*C. annuum* L.) for pharmaceutical and medicinal uses, nutritional value, and therapeutic properties.

## 10.2 AGRONOMIC CHARACTERIZATION OF CHILI

Among 35 species of *Capsicum*, 5 species can be cultivated or domesticated, namely, *C. annuum* L., *Capsicum baccatum* L., *Capsicum chinense* Jacq., *Capsicum frutescens* L., and *Capsicum pubescens* L.[52,110] Throughout the globe amongst the five domesticated species, *C. annuum* is the most extensively cultivated because of its high yield and low cost of production.[112]

*C. annuum* thrives in tropical climates since it requires a warm, humid environment to exist. It might be difficult to distinguish this species from another cultivated species *Capsicum Chinese* due to certain similarities in botanical and agronomic characteristics.[49] Additionally, they have a common ancestor gene pool designated as the "annuum-Chinese-frutescens complex."[145] *C. annuum* is mostly grown in tropical and subtropical countries. Both pungent and nonpungent cultivars are found in this species with herb or sub-shrubs habit.[5,129] Cultivars of this species show a great variation in size, shape, and color of mature fruit.[13,136] Most of the cultivars of this species need 4–5 months to complete the life cycle.

There are several varieties of *C. annuum*, but no appropriate categorization of the variations has been developed for this species, however, based on fruit morphology provided the classification of the varieties, using the term Abbreviatum, Acuminatum, Cerasiforme, Conoides, Fasciculatum, Grossum, and Longum.[103,145] Hot chili usage is widely recognized around the globe for its high nutritional value, which minimizes human micronutrient shortages. Most of the chili-growing regions are the Indian Subcontinent, China, Japan, the United States of America, Mexico, Turkey, and African countries. Chili is considered a cold-season crop, yet it may be grown all year if suitable temperatures and optimum moisture conditions are maintained. It needs 25–30°C day temperature and 18–20°C night temperature along with 50–60% relative humidity for optimal.[117] The crop is susceptible to temperature fluctuation, and if the temperature goes beyond 35°C or falls below 12°C, it will suffer negative consequences.[31,71]

## 10.3 FRUIT CHARACTERISTICS AND DEMAND OF CHILI IN THE INTERNATIONAL MARKET

The shape, size, and color of chili fruits are highly variable.[108] Usually, it is oblong, conical, 10–20 mm long, and 4–7 mm broad in general but may be of various shapes and sizes which depend upon the commercial variety. The glabrous pericarp is attached to a cup-shaped five-toothed calyx and a straight pedicel and is divided into two halves by the membranous dissepiments. Each half contains around 5–10 tiny, flattened, disc-shaped white golden seeds. Pungent cultivars of chili are used to flavor a variety of foods and also as a spice in a variety of salt pickling, marinades, and preserves preparations; however, sweet cultivars are used directly in food as a stuffed chili or salad. Chili has medicinal potential due to its active component like capsaicinoids; more specifically, capsaicin found mostly in tissues of the placenta and pericarp.[28] Capsaicinoids, which are found in chili fruit, bind to the pain receptors throughout the mouth, tongue, and throat which are normally associated with heat-sensing. Then receptors provide a signal to the brain that the individual has consumed something hot. Finally, the brain responds to the burning sensation by increasing heart rate, increasing sweat, and endorphin creation.

## 10.4 NUTRITIONAL VALUE OF CHILI AND ITS PHYTOCONSTITUENTS

Hot chili peppers are appreciated as a beneficial component of the diet.[26] The hot chili comprises protein, carbohydrates, vitamins minerals, and other measurable substances that are all present in traceable amounts.[92] Its vibrant color is a result of a mixture of capsorubin, capsanthin, cryptoxanthin, zeaxanthin, and other carotenoid esters.[69] When ingested, chili has a pungent flavor due to the presence of capsaicin (8-methyl-*N*-vanillyl-6-nonenamide) and a large number of other substances categorized as capsaicinoids, which are chemical variants of benzylamin.[44,110]

Major capsaicinoids of chili include nordihydrocapsaicin, capsaicin, dihydrocapsaicin, homocapsaicin, homodihydrocapsaicin, and nonivamide.[12,118] Among the capsaicinoids of chili, capsaicin accounts for 69%, dihydrocapsacinoids represent 22%, nordihydrocapsacinoids represent 7%, and homocapsaicin and homohydrocapsaicin collectively represent 1%.[110] High levels of volatile compounds have been reported in the crude and alcoholic extract of chili fruit and some of them are 4-methoxyphenol, ethylhexadecanoate, hexanal, or isopulegol.[56]

According to a study, the primary antioxidant components found in red hot chili are sinapoyl and feruloyl glycosides, while the primary antioxidant compound found in green chili is quercetin-3-*O*-*L*-rhamnoside, along with capsaicin and dihydrocapsaicin.[90] The benefits of chili and its phytoconstituents have since been suggested by various *in vitro*, *in vivo*, and experimental models. The phytoconstituents significantly reduce the chances of cardiovascular diseases (CVDs) like diabetes, hypertension, hyperlipoproteinemia, and hypercholesterolemia, problems of digestive system, liver cirrhosis, neurogenic bladder, and different types of cancers. In addition, it also exhibits antimicrobial activity, antioxidant activity, anti-inflammatory effect, antilithogenic effect, antianalgesic, and many more.

## 10.5 THERAPEUTIC PROPERTIES OF CHILI

Besides its culinary use, *C. annuum* has been the focus of therapeutic applications owing to the occurrence of several chemical constituents having possible medicinal characteristics. Due to the obvious existence of medically essential active ingredients, it must be utilized in numerous forms of folk medicine to treat a variety of diseases and/or ailments.[89] The consumption of hot red chili pepper is reported to be associated with reduced mortality.[24,140] The processes by which chili peppers may reduce the chance of death remain unknown. Chili has medicinal potential due to its unpleasant potent component capsaicin, a significant capsaicinoid found mostly in placental and pericarp tissues.[28] Although innumerable studies have been published to support the medicinal value of chili in food, it is worth noting that the chemical composition and extent of various bioactivities vary significantly depending on the cultivar studied, as well as the extraction conditions used in the experiments.

### 10.5.1 ANTIOXIDANT ACTIVITY AND ANTIMICROBIAL ACTIVITY

Several studies have shown that *C. annuum* has antioxidant effects. It is a worldwide famous vegetable crop that is high in antioxidants and is a good origin of provitamins A (carotene) and C (ascorbic acid).[82,119] It is also appreciated as the good origin of vitamin B complex, notably vitamin B5.[44] Green fruits provide vitamin A, C, as well as minerals, whereas dry chilies are an outstanding origin of vitamin A as well as vitamin D.[64] Alcoholic extracts of *C. annuum* were shown to exhibit considerable antioxidant and antibacterial properties (723.795 mg/L GAE, comparable to 3.315 mg/L Trolox, and 327.394 mg/L QE).[32]

Isolated bioactive compounds of chili like lycopene, vitamin C, *p*-coumaryl alcohol, ethoxyquin, and capsaicinoids protect lipids from Fe(II)-induced peroxidation.[99] Park et al.[106] conducted an experiment on human leucocytes and revealed methanolic extracts of *C. annuum* have been shown to prevent 4-hydroxy-2-nonenal-induced as well as $H_2O_2$-induced DNA degradation. Most of the compounds have promising properties to neutralize free radicals associated with the oxidation of lipids, proteins, and cellular DNA damage, as well as chronic human diseases (Figure 10.1).

Chili contains vital antimicrobial action for vital in-public health. Antimicrobial substances have always been important in the medical field, particularly since those prevalent disease-causing microorganisms are becoming increasingly resistant to antibiotics. The antibacterial effects of *C. annuum* fruit extract could be attributed to its phenolic, flavonoid, as well as tannin levels.[40] Extracts of chili and isolated peptides were also found to be beneficial in the case of diverse fungus.[36] Antimicrobial capabilities were discovered in several types of chili peppers, having potency values comparable to certain contemporary antibiotics. Chili is generally regarded for its antibacterial capabilities since its components are capable of killing or inhibiting up to 70% of bacteria.[101] Extracts of fruit in water, methanol, and ether were proven to significantly limit microbe multiplication and expansion.[39] Aqueous as well as methanolic extracts of *C. annuum* are efficient toward *Bacillus cereus*, *Bacillus subtilis*, *Clostridium sporogenes*, *Clostridium tetani*, *Escherichia coli*, *Enterococcus faecalis*, *Proteus mirabilis*, *Pseudomonas aeruginosa*, *Salmonella typhimurium*, *Staphylococcus aureus*, *Vibrio cholerae*, and *Streptococcus mutans*.[1,23,32,67,70,72,122] The occurrence of capsicinoids as well as other active ingredients in chilies could explain its antimicrobial and antioxidant capabilities.[39]

On an annual basis, microorganisms are found to deteriorate food and cause a variety of foodborne diseases. The disease can vary between stomachaches through unplanned abortions in pregnant women, and in extreme situations, they also can lead to death. Hafiz et al.[47] demonstrated that unrefined extract of *C. annuum* has antiviral and cytotoxic potential toward herpes simplex virus-1 as well as herpes simplex virus-2. Researchers discovered that the existence of several phytochemical compounds, such as ascorbic acid and beta-carotene, was responsible for this activity. Recent research indicated that the methanolic extract of *C. annuum* fruit had antibacterial and therapeutic effects when applied to an infected lesion in a rat model.[40] Oleoresins of some cultivars have a bactericidal property toward *E. coli* and *Listeria innocua*.[116]

## 10.5.2 NEUROPROTECTIVE EFFECT

Many recent studies have indicated that chili and its bioactive compounds have significant importance on nervous organs. Capsaicin-induced neuroprotection was related to a decrease in the production of proinflammatory cytokines (signaling molecules) in addition to ROS produced by stimulated microglia-derived NADPH oxidase.[9] Growing data reveal that ROS-induced oxidative stress is the primary pathogenic component in neurodegenerative disorders such as Alzheimer's disease, Huntington's disease, and Parkinson's disease (PD).[22] Chili intake has been suggested in numerous researches as a possible treatment for a variety of neurodegenerative illnesses.

Byun et al.[23] showed that *C. annuum* var. *abbreviatum* has a neuroprotective effect toward $H_2O_2$-induced oxidative stress in HT22 hippocampal cells. Further studies confirmed that this chili variety possesses a strong neuroprotective activity by restraining mitochondrial-related apoptosis signaling, including the modulation of bcl-2 families, inhibition of caspase cascades, and activation of the Nrf2/HO-1 signaling pathway. Oboh and Rocha[99] conducted in-vitro studies that showed that ripe and unripe fruit of *C. annuum* extracted with water protect the brain against oxidative stress caused by sodium nitroferricyanide(III) dihydrate and quinolinic acid. It is reported that capsaicin stimulates nigrostriatal dopamine neurons, enhances striatal function, and improves behavioral recovery in mice.[9] Capsaicin of chili and its analogs could be useful remedial medicines for PD and other neurological illnesses linked with neuroinflammation and oxidative damage caused by glial activation (Figure 10.1).[29]

### 10.5.3 OVERWEIGHT CONTROL AND ANTIOBESITY EFFECT

Obesity-related issues have developed into serious public health concerns on a global scale since obesity is a significant threat issue for a variety of syndromes, involving hypertension, dyslipidemia, inflammatory, type II diabetes, and atherosclerosis.[124] Consumption of chili-containing meals has been demonstrated to boost calorie expenses as well as fat oxidation, which may aid during the reduction of obesity and associated illnesses.[2]

Numerous preclinical and clinical investigations have revealed that capsaicinoids derived from C. *annuum* may have significant health advantages in obese individuals.[59] Many in-vitro and in-vivo investigations show that having capsaicin, the component responsible for the strong aroma of hot chili peppers may be a potential antiobesity drug-induced through oxidative stress and adipogenesis of cells.[50,78] In a study, chili consumption was proven to inhibit obesity by decreasing adipocyte formation despite causing deadly side effects such as lower insulin responsiveness.[11]

It is a fact that total adipocyte suppression is detrimental to human health since adipocytes are important for the physiological function of mammals, fatty acid metabolism, energy balance maintenance, as well as adipocytokines production (Figure 10.1). Diet-induced thermogenesis is increased by about 50% when chili along with medium-chain triglycerides are added to food that may also promote reduced weight, also prevent obesity gradually.[35] According to research on humans, consumption of chili pepper aids weight loss by activating several receptors as well as improving insulin regulation.[132]

Capsaicin has been found to have antiobesity as well as cholesterol-lowering properties.[62] Capsaicin aids in the production of heat inside the body and the consumption of oxygen, both of which aid in the burning of calories in the body and the reduction of body weight.[91] Tsui et al.[130] reported that component P, a neuropeptide produced from capsaicin, was already found to treat diabetes among rodents. Chili peppers seem to have weight-loss effects, as supported through inverse relationships with the prevalence of overweight/obesity together in a major epidemiological study.[121]

Oral intake of red chili ethanol extract demonstrated antiobesity effects, which might have been attributed to enhanced metabolism via two putative mechanisms: hypolipidemic impact and initiation of serum thyroid-stimulating hormone.[4] Mechanism investigations utilizing cell prototypes concluded that capsaicin demonstrated an antiobesity activity via enhancing the overexpression of genes associated with heat production, lipid oxidation, as well as catabolic activity, like UCP1, PGC1α, PRDM16, PPARs, and C/EBPs.[80] According to the latest investigation, chili capsaicinoids have been shown to affect plasma lipids, aorta capability including the development of atherosclerotic plaque, cholesterol uptake indicator, fecal sterol excretion, and gene expression of significant receptors, enzymes, as well as transporters associated with cholesterol metabolism in hamsters.[81]

### 10.5.4 EFFECT ON THE GASTROINTESTINAL SYSTEM

Due to the presence of bioactive chemicals, chili has been recognized to heal a variety of degenerative human illnesses. Chili and its bioactive compounds have been implicated in a variety of activities in the gastrointestinal system. Chili seems to have both inhibiting and stimulating effects on the acid secretion of the stomach. Chili peppers, particularly in powder form, are high in antioxidants and other components that may help to alleviate various digestive problems, such as upset stomachs, intestinal gas, diarrhea, and cramping.[88]

Ericson et al.[41] demonstrated that capsaicin promotes the release of gastrin by isolated human antral glands as well as verified that a chili-rich food inhibits its release. There are two probable mechanisms by which chili or its active ingredient capsaicin might stimulate digestion: (a) hepatic activation to produce extra bile richer in bile acids and (b) activation of enzyme actions that contribute to digestion in the pancreas and the intestines (Figure 10.1).[126] Consumption of sufficient red chili in combination with a standard/healthy diet or as an ingredient in deep-fried fast food has a beneficial effect on lipid profiles, lowering cholesterol, triglycerides, and LDL while increasing HDL, as well as liver function, lowering glucose, glutamic-oxaloacetic transaminase, and alanine aminotransferase.[21]

Capsaicin also reduces acid release, boosts alkali as well as mucus production and activates blood flow to the stomach mucosa, which aids in the prevention and repair of gastric ulcers.[126] Fruit crude extract has been found to reduce intestinal glucose absorption that might help to lower blood sugar levels.

For millennia, *C. annuum* has been commonly used as a spicy food condiment to enhance hunger.[138] In persons with IBS-D, chronic ingestion of chili significantly reduced belly burning generated by conventional spiced food, as well as rectal feeling in response to balloon distention (irritable bowel syndrome).[8] In a meta-analysis, it is found that consuming a high dose of chili or capsaicin increased the overall incidence of gastric cancer by 1.51 times, while substantial variances and even opposite trends were seen in certain locations in the subgroup analysis.[86] Capsaicin also inhibits the viability and proliferation of pancreatic neuroendocrine tumor cells and promotes apoptosis. It also lowers the potential of the mitochondrial membrane, lowering the amount of ATP, ROS, and mitochondrial Bcl-2 protein production, and increasing cytochrome *c* levels.[125]

### 10.5.5 CARDIOVASCULAR PROTECTION

Chili pepper intake has been linked to a reduced incidence of overall cardiovascular illnesses.[19] Chili peppers may aid in the creation of new blood cells in addition to supporting heart function as it is high in copper and iron, which contributes to this.[88] The metabolic and vascular threat for CVD involves hypercholesterolemia, hyperglycemia, hyperinsulinemia, obesity, hypertension, and arterial stiffness.[2]

Chili promotes vasodilation, potentially via increasing the release of a peptide linked to the calcitonin gene and/or by increasing the creation of nitric oxide.[43] Capsaicin (0.015%) combined with high cholesterol diet provided not just a hypolipidemic impact and rectified the enhanced osmotic instability in erythrocytes by counteracting the changed lipid panel of erythrocyte membranes.[61] Chili consumption has been related to a reduced incidence of cardiovascular mortality, but not from cerebrovascular disease.[100] In an animal model, capsaicin stimulation of ion channel transient receptor potential vanilloid subtype 1 (TRPV1) inhibited vascular oxidative stress.[127] Additionally, it has been shown that activating TRPV1 inhibits myocardial hypertrophy and fibrosis (Figure 10.1).[73]

### 10.5.6 EFFECT ON RENAL AND URINARY DISORDER

Current research reveals that consuming an aqueous extract of the *C. annum* fruit protects kidney cells against ethanol-mediated structural and functional damage.[37] Additionally, the research demonstrated that administering this extract may effectively inhibit the generation of excess free radicals caused by ethanol and modification of antioxidant status. Another research discovered that a methanolic extract of *C. annuum* fruits considerably decreased toxicant-induced elevations in urea, uric acid, and creatinine values, as well as a rise in the protein level of blood and urine.[113] Furthermore, *C. annuum* fruits extracted with methanol have been shown to attenuate aberrant alterations in renal microscopic anatomy caused with rifampicin, showing nephroprotection. Capsaicin also increases urinary sodium excretion.[127]

Capsaicin has been found to improve kidney function in animal models by acting as a diuretic in normal and diabetic mice and lowering urine, epidermal growth factor an indicator of renal disorders.[115] Many studies have shown that activating the capsaicin receptor TRPV1 prevents salt-induced kidney injury in mice and enhances glomerular filtration rate as well as renal excretion via increasing natriuresis along with dieresis.[70,143,146]

Aitken et al.[3] concluded that topical capsaicin 8% (Qutenza®) treatment is safe and efficacious to manage neuropathic pain of severe ischemic patients having the end-stage renal disease, which is the final phase of renal failure. Capsaicin is also being studied as a possible treatment for neurogenic bladder diseases such as neurogenic detrusor overactive as well as detrusor hyperreflexia.[42]

### *10.5.7 EFFECT ON DERMATOLOGICAL SYSTEM*

Externally, chili is used as a soothing agent for rheumatism, lumbago, and neuralgia. Capsaicin is shown to be effective in the treatment of a variety of skin conditions, including shingles, muscle strain, arthritis, diabetic peripheral retinopathy, as well as pain problems when used in a tropical setting.[139] Capsaicin changes profilaggrin and CDSN proteolysis and dysregulates epidermal proteases including matriptase and KLKs, which might explain the reported skin manifestations.[14]

### *10.5.8 ANTI-INFLAMMATORY ACTIVITY*

Besides having a burning flavor, chili is also an excellent source of anti-inflammatory chemicals, which makes it a popular medicine. Inflammation is a tissue response to infection or irritants and is a necessary component of the host defensive system. For many decades, inflammation was seen as a basic allergic response. However, it is now believed to underpin the pathophysiology of a considerably larger range of disorders. Redness, heat, swelling, and pain are the four cardinal indicators of inflammatory disorders. Chili contains phenolic and flavonoid compounds that are shown to have anti-inflammatory properties.[18]

Chili had the best ability to reduce inflammation in a model where lipopolysaccharide-stimulated macrophages were exposed to it.[97] Using chili could be an effective treatment method for managing allergic airway illness and inflammation. Chili effectively reduces primary symptoms of allergic asthma, involving airway hyper-responsiveness, airway inflammation, and high intensity of Th2 cytokines in BAL fluid and lung tissues.[57] Capsaicin protects human umbilical vein endothelial cells from oxLDL-induced malfunction by causing mitochondrial membrane potential instability, obstructing oxLDL-induced ROS generation, and inhibiting caspase-3 activation.[53] Capsaicin effectively prevents the formation of foam cells in macrophage RAW 264.7 cells.[16,109]

### 10.5.9 PAIN RELIEF AND ANALGESIC ACTIVITY

In the mouse model, intake of both chili and capsaicin reduced pain sensation, enhanced anxiety symptoms, and hampered memory and learning while improving social interactions and motor regulation.[102] Natural capsaicinoids found in chili have garnered significant interest for external pain treatments, and this exclusive feature of capsaicinoids is explored using creams, ointments, and patches to cure a variety of forms of pain, most notably neuropathic pain (Figure 10.1).[38,87] Capsaicinoids interact with sensory nerves in the oral cavity, causing heat to be felt shortly after consumption. The brain responds to the sensation by raising perspiration production, increasing heartbeat, as well as releasing endorphins.[44]

Capsaicin stimulates main afferent nerve neurons via interacting with vanilloid receptor-1, a member of the transient receptor potential (TRP) channel that works as a structural mediator of unpleasant chemical and physical stimuli, such as noxious temperature and acidic pH.[15,33] Capsaicin therapy is often widely accepted, indicating it can be administered initially in osteoarthritis management, particularly in exterior joints of the hand and knee.[74] The rear latencies of the chili and capsaicin-fed mice were way longer than the control mice in the immersion test.[102] Chili or capsaicin feed enhanced the mice's tolerance for pain, resulting in decreased pain intensity.[98]

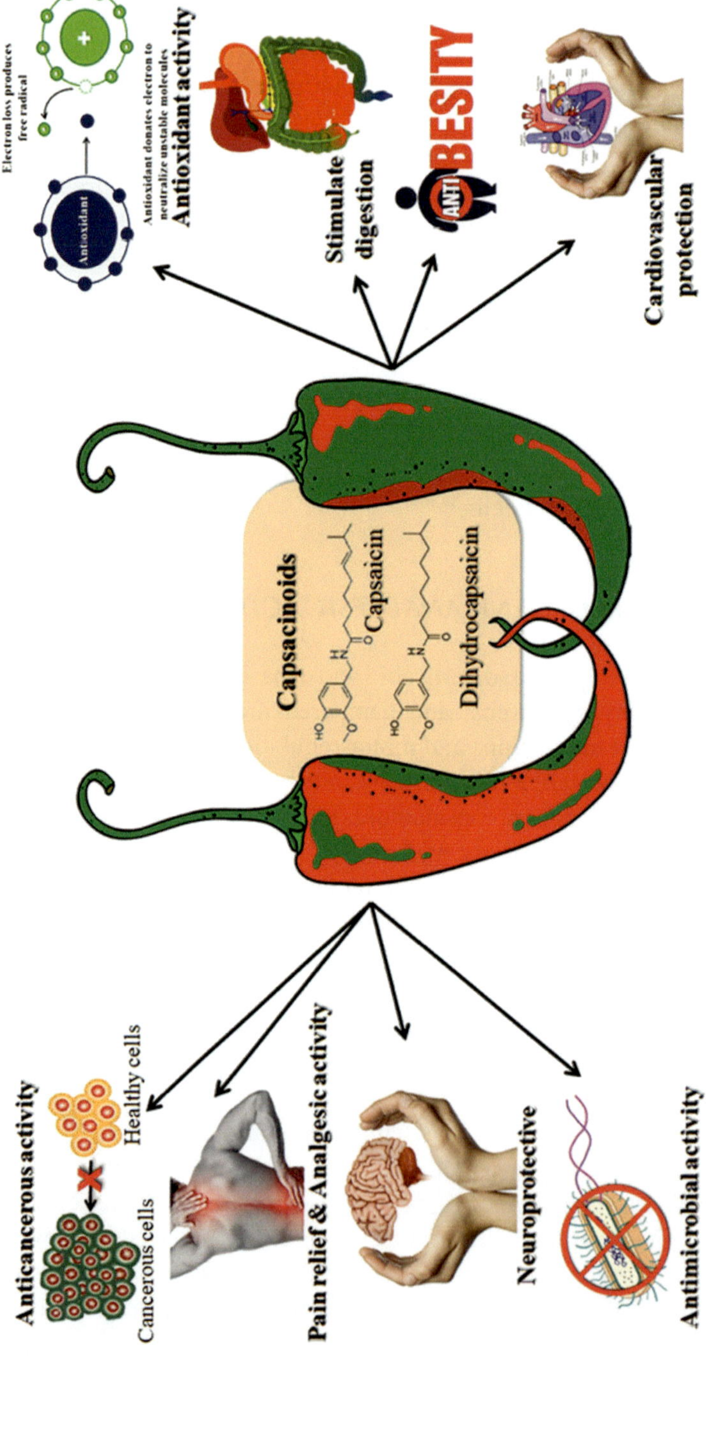


FIGURE 10.1 Diverse therapeutic properties of chili.

### 10.5.10 EFFECT ON EYE DISORDER

*C. annuum* L. is recognized to contain a wide range of nutrients that were proved to improve human health and protect against age-linked macular degeneration (AMD) and cataracts.[65] AMD is a deteriorating eye illness that results from the degradation of photoreceptor as well as retinal pigment epithelial cells close to the core area of the macula.[131] Kim et al.[65] reported that fermented *C. annuum* can safeguard macrophages and retinal impairment from oxidative stress. According to them due to its nonapoptotic action with modulation of PI3K/AKT as well as MAPK signaling, it protects against oxidative stress-mediated injury.

### 10.5.11 EFFECT ON REPRODUCTIVE AND SEXUAL DISORDERS

*C. annum* leaves and fruits are used by certain tribal communities in India for abortion and to treat menstrual disorders.[129]

### 10.5.12 ANTICANCER ACTIVITY

Many researchers have conducted many in-vitro and in-vivo researches to prove the efficiency of chili and its bioactive compounds against several cancer lines. Capsaicin present in *C. annuum* inhibited the development of various cancerous cell lines through induction of autophagy, cycle arrest, and apoptosis and by suppressing metabolic activity at the cellular level.[10] Capsaicin, a major alkaloid in chili, has been studied a fair amount because it can cause apoptosis in a variety of cancer cell lines, like pancreatic, colonic, prostatic, liver, esophageal, bladder, skin, leukemia, lung, and endothelium cells but healthy cells are unaffected.[34]

Capsaicin was shown to inhibit the growth and multiplication of prostate cancer cells both in vitro and in vivo.[95] Capsaicin induces autophagy in basal carcinoma cells through disrupting redox status, affecting the adenosine diphosphate/adenosine diphosphate balance, thus generating mitochondrial depolarization, as well as stimulating the activated protein kinase cascade.[54]

Capsaicin was recently discovered to target several signaling pathways, oncogenes (genes involved in malignant alterations in cells), as well as tumor suppressor genes in a variety of cancers. According to research, capsaicin may suppress the transcription of COX-2 and B-catenin mRNA, boosting apoptosis by activating caspase 3 and limiting cell growth.[142] According to Xu et al.,[138] capsaicin suppresses numerous processes of metastasis of papillary thyroid cancer B-CPAP cells, involving cell migration, invasion, and adhesion. Ito et al.[55] reported that capsaicin inhibits the formation of leukemic cells straight away.

Lee et al.[76] showed capsaicin-induced apoptosis of human lung cancer cells as well. According to Kim et al.,[66] capsaicin causes apoptosis to cancer cells of the stomach (SNU-1), probably for the increased expression of the p53 gene and tumor suppressor. Capsaicin inhibits proliferation due to VEGF (cancer angiogenesis), synthesis of DNA, expansion, and constitution of endothelial cells in vitro and in vivo, stopping the cellular cycle at G1 phases.[93]

TRPV1 is mainly accountable for capsaicin's proapoptotic effects in diverse cancers[7,144] and TRPV6.[27,75] *C. annuum* also suppressed 12-*O*-tetradecanoylphorbol 13-acetate–induced activation of NF-κB in mouse skin,[48] in human promyelocytic leukemia HL-60 cells[45] and human myeloid ML-1a cells.[104] It is also found that capsaicin interacts synergistically with other cancer-preventative medicines, allowing for the possibility of using capsaicin in cancer treatment in conjunction with other chemotherapeutic drugs.[34] TRPV1 is primarily responsible for capsaicin's proapoptotic effects in different cancers[7,144] as well as TRPV6.[27,75]

## 10.6 HEALTH RISK OF OVERCONSUMPTION OF *CAPSICUM ANNUUM*

Many researchers have confirmed that the safe consumption of *C. annuum* is beneficial to health but the high dose is responsible for gastric erosion. Al-Jumayi et al.[4] have shown that excessive consumption of red chili induces liver and kidney damage. It should not be taken during pregnancy or breastfeeding, in persons who are allergic to it, or in children.[5] Fruits or any other parts of the plant must not be applied directly to open injuries or bruises or close to the eyes.[84] Many responses and conflicts have been recorded when chili was taken concurrently with aspirin and salicylic chemicals.[5] Additionally, it inhibited the effect of α-adrenergic blockers such as clonidine and methyldopa.[68,84]

In a study on rats, it was shown that inhalation of capsaicinoids found in pepper sprays harmed bronchial, tracheal, nasal, and alveolar cells in the rat, resulting in acute inflammations.[30] Capsaicin also reduces spermatogonia viability in vitro, through triggering apoptosis within these cells, and TRPV-1, a capsaicin sensor expressed in mitotic germ cells, may be involved

**TABLE 10.1** Health Perspectives of Bioactive Compounds in *Capsicum annum*.

| Nature of bioactivity | Mode of use/ Active compounds | Activity | Ref. |
|---|---|---|---|
| Antimicrobial activity | Ethanolic extract (100 mg/mL) | Antimicrobial activity against *Salmonella typhimurium* and *Pseudomonas aeruginosa* | 120 |
| | Butanol extract | High antimicrobial activity against *Micrococcus* sp., *Bacillus*, *E. coli*, *Pseudomonas* sp., and *Citrobacter* sp. | 5 |
| | Methanolic extract | Antimicrobial properties against *Vibrio cholerae* strains that are multidrug-resistant | 141 |
| | Methanolic extract | Antiviral effects against herpes simplex virus-1 (HSV-1) as well as herpes simplex virus-2 (HSV-2) | 47 |
| | Cis-capsaicin (civamide) | Block viral replication cycle | 63 |
| | Capsaicin | Inhibit spreading and presence of HSV | 20 |
| | Extracts | Inhibition of growth of *Aspergillus niger* | 123 |
| | Peptides | Antifungal activity against *Candida cerevisiae*, *Kluyveromyces marxiannus*, *Pichia membranifaciens*, *Saccharomyces cerevisiae*, *Candida tropicalis*, *Candida parapsilosis*, and *Candida albicans* | 114 |
| | Oleoresin extracted from *C. annuum* var. Cacho de cabra | Bacteriostatic effect against *Escherichia coli* and *Listeria innocua* | 116 |
| Antiobesity and antidiabetic | Alcoholic extract | Suppressing adipocyte development while causing life-threatening consequences like lower insulin sensitivity | 11 |
| | Capsicum oleoresin | A rat model fed an increased fat diet, lowers blood cholesterol, and triglycerides | 46 |
| | Capsaicin | Reduced Bcl-2 protein synthesis in mitochondria; elevated cytochrome *c* levels | 125 |

**TABLE 10.1** *(Continued)*

| Nature of bioactivity | Mode of use/ Active compounds | Activity | Ref. |
|---|---|---|---|
| | Capsaicin | Lower triglycerides and fasting glucose levels; improved expression of the adiponectin gene/protein as well as its receptor (AdipoR2); activated hepatic AMP-induced activation of protein kinase | 60 |
| Cardioprotective effect | Capsaicin | Inhibits myocardial hypertrophy and fibrosis | 73 |
| | Methanolic extracts | Hinder DNA damage caused by 4-hydroxy-2-nonenal and $H_2O_2$ among leucocytes | 106 |
| | Capsaicin | Triacylglycerols, total cholesterol, non-high-density lipoprotein cholesterol, and triacylglycerols in plasma were all decreased. The stimulation of cholesterol 7-hydroxylase production had a deleterious impact on hepatic X receptor alpha activity | 25 |
| Wound healing | Methanolic extract | Chili extract contains active ingredients that prevent biofilm formation, ATPases/$H^+$ proton pumps, and dehydrogenase activity and also alters the bacterial cell membrane by nucleic acid leakage and sugar reduction, including total protein reduction | 40 |
| Neuroprotective effect | *C. annuum* var. *abbreviatum* extract | In HT22 hippocampal cells, there is a protective effect against hydrogen peroxide-induced oxidative damage | 23 |
| Anti-inflammatory | Hydroalcoholic extract | Anti-inflammatory potential in the carageenan-induced edema model | 134 |
| | Methanolic extract | A mouse model of asthma suppresses ovalbumin-mediated airway inflammation and oxidative stress | 57 |
| | Capsaicin | In macrophage RAW 264.7 cells, it prevents the formation of foam cells | 16,109 |
| | Capsaicin | The mitochondrial membrane potential collapses, preventing oxLDL-stimulated ROS production and caspase-3 activation | 54 |
| | Capsaicin | EC-LPS-mediated MAPK cascade stimulation was suppressed | 135 |

**TABLE 10.1** *(Continued)*

| Nature of bioactivity | Mode of use/ Active compounds | Activity | Ref. |
|---|---|---|---|
| Antiaging | Capsaicin | Reduced apoptosis by decreasing the KA-facilitated increase in the cytokines IL-1 and TNF | 77 |
| Pain relief | Capsaicin | Reduces osteoarthritic pain in superficial joints like hand and knee joint | 74 |
| | Capsaicinoids | Topical application results in desensitization of the local neuropathic pain | 107 |
| | Fresh fruit; capsaicin | Decreased pain perception | 102 |
| Renoprotective activity | Aqueous extract | Reduce the vulnerability of structural and functional integrity of renal cells caused by ethanol; prevent the formation of surplus free radicals | 37 |
| | Methanolic extract | Attenuate aberrant alterations in kidney histology caused by rifampicin; decreased toxicant-induced elevations in urea, uric acid, and creatinine | 113 |
| | Capsaicin | Increase urinary sodium excretion | 127 |
| Anticancerous activity | Capsaicin | Numerous malignant cell lines were slowed down in their proliferation | 10 |
| | Capsaicin | Represses the development of leukemic cells directly | 55 |
| | Capsaicin | Apoptosis of human lung cancer cells occurs quickly | 76 |
| | Capsaicin | Cell cycle arrest in the G0/G1 phase was triggered. Increased FOXO3a production | 111,133 |
| | Capsaicin | Modified histone acetylation | 137 |
| | Capsaicin | Reduced the synthesis and release of IL-1, TNF-, IL-10, IFN-, and IL-1ra; suppressed colon cancer cell propagation | 17 |

**TABLE 10.1** *(Continued)*

| Nature of bioactivity | Mode of use/ Active compounds | Activity | Ref. |
|---|---|---|---|
| | Capsaicin | IFN-γ secretion was enhanced, apoptosis was triggered, and tNOX expression was negatively regulated | 58 |
| | Capsaicin | ERK activation was inhibited, while paxillin and FAK phosphorylation was lowered | 83 |
| | Capsaicin | Dhh/Ptch2/Zeb2 components of the Hedgehog signaling pathway are better regulated; CD24, VEGFA, and TIMP1 are improved, but CD44 and ALCAM mRNA expression is decreased | 6 |
| | | Renal cancer cell growth was inhibited; capsazepine (CPZ), a typical antagonist of attenuated transient receptor potential vanilloid type 1 (TRPV1); P-38 and JNK MAPK pathways are stimulated | 85 |

in this impact.[94] Reports of experiments on animal models also revealed that consumption of chili has been linked to a carcinogenic or cocarcinogenic effect.[128] According to Hwang et al.,[51] capsaicin may be associated with skin cancer.

## 10.7 SUMMARY

Chili pepper is widely used as a spice across the globe, and many of its health benefits are linked to its phytochemical constitution. In terms of both commercial and medicinal value, chili is among the most significant solanaceous vegetable crops in the world. More than 200 active components have been found in chili pepper, and many of these active constituents perform numerous functions in the body. Chili and/or its active ingredient capsaicin has indeed been linked to a variety of therapeutic advantages. The chemical composition including capsaicin of chili varieties enables them to be applied for different purposes. As a resource of therapeutic chemical agents and medicines, this is a potential model plant. However, further research is essential to have a detailed knowledge of the modes of action of fruits and their bioactive substances in various illnesses or disorders, as well as their health consequences.

## KEYWORDS

- **antimicrobial activity**
- **antioxidant activity**
- **ascorbic acid**
- **bioactive compound**
- **capsaicin**
- ***Capsicum annuum***
- **therapeutic properties**

## REFERENCES

1. Adamu, H. M.; Abayeh, O.; Agho, M.; Abdullahi, A.; Uba, A.; Dukku, H.; Wufem, B. Ethnobotanical Survey of Bauchi State Herbal Plants and Their Antimicrobial Activity. *J. Ethnopharmacolo.* **2005,** *99* (1), 1–4.

2. Ahuja, K.; Robertson, I.; Geraghty, D.; Ball, M. The Effect of 4-Week Chilli Supplementation on Metabolic and Arterial Function in Humans. *Eur. J. Clin. Nutr.* **2007,** *61* (3), 326–333.
3. Aitken, E.; McColl, G.; Kingsmore, D. The Role of Qutenza® (Topical Capsaicin 8%) in Treating Neuropathic Pain from Critical Ischemia in Patients with End-Stage Renal Disease: An Observational Cohort Study. *Pain Med.* **2017,** *18* (2), 330–340.
4. Al-Jumayi, H. A.; Elhendy, H. A.; Darwish, A. M. Biological Effects of Red Chili Pepper (*Capsicum annuum*) Consumption on High Fat Diet Female Albino Rats. *Pak. J. Biol. Sci.* **2020,** *23* (2), 150–158.
5. Al-Snafi, A. The Pharmacological Importance of Capsicum Species (*Capsicum annuum* and *Capsicum frutescens*) Grown in Iraq. *J. Pharma. Biol.* **2015,** *5* (3), 124–142.
6. Amantini, C.; Morelli, M. B.; Nabissi, M.; Cardinali, C.; Santoni, M.; Gismondi, A.; Santoni, G. Capsaicin Triggers Autophagic Cell Survival Which Drives Epithelial Mesenchymal Transition and Chemoresistance in Bladder Cancer Cells in an Hedgehog-Dependent Manner. *Oncotarget* **2016,** *7* (31), E-article 50180.
7. Amantini, C.; Mosca, M.; Nabissi, M.; Lucciarini, R.; Caprodossi, S.; Arcella, A.; Giangaspero, F.; Santoni, G. Capsaicin-Induced Apoptosis of Glioma Cells Is Mediated By TRPV1 Vanilloid Receptor and Requires p38 MAPK Activation. *J. Neurochem.* **2007,** *102* (3), 977–990.
8. Aniwan, S.; Gonlachanvit, S. Effects of Chili Treatment on Gastrointestinal and Rectal Sensation in Diarrhea-Predominant Irritable Bowel Syndrome: A Randomized, Double-Blinded, Crossover Study. *J. Neurogastroenterol. Motility* **2014,** *20* (3), 400–406.
9. Antonious, G. F. Capsaicinoids and Vitamins in Hot Pepper and Their Role in Disease Therapy. In *Capsaicin and Its Human Therapeutic Development*; Mozsik, G. Ed. Chapter 2; *IntechOpen*, **2018**; pp 13–40. https://doi.org/10.5772/intechopen.78243
10. Arora, R.; Gill, N.; Chauhan, G.; Rana, A. An Overview About Versatile Molecule Capsaicin. *Int. J. Pharma. Sci. Drug Res.* **2011,** *3* (4), 280–286.
11. Arumugam, M.; Vijayan, P.; Raghu, C.; Ashok, G.; Dhanaraj, S.; Kumarappan, C. Anti-Adipogenic Activity of *Capsicum annum* (Solanaceae) in 3T3 L1. *J. Complement. Integr. Med.* **2008,** *5* (1), 28–32.
12. Barbero, G. F.; Ruiz, A. G.; Liazid, A.; Palma, M.; Vera, J. C.; Barroso, C. G. Evolution of Total and Individual Capsaicinoids in Peppers During Ripening of the Cayenne Pepper Plant (*Capsicum annuum* L.). *Food Chem.* **2014,** *153*, 200–206.
13. Bargavi, R.; Elumalai, S. Chilli (*Capsicum annuum*) Cultivation, Diseases, Breeding, Advanced Techniques in Biotechnology-General Review. *Biosci. Biotechnol. Res. Asia* **2016,** *7* (1), 171–178.
14. Basharat, S.; Gilani, S. A.; Iftikhar, F.; Murtaza, M. A.; Basharat, A.; Sattar, A.; Qamar, M. M.; Ali, M. Capsaicin: Plants of the Genus *Capsicum* and Positive Effect of Oriental Spice on Skin Health. *Skin Pharmacol. Physiol.* **2020,** *33* (6), 331–341.
15. Benham, C. D.; Gunthorpe, M. J.; Davis, J. B. TRPV Channels as Temperature Sensors. *Cell Calcium* **2003,** *33* (5–6), 479–487.
16. Bert, J.; Mahowald, M. L.; Frizelle, S.; Dorman, C. W.; Funkenbusch, S. C.; Krug, H. E. The Effect of Treatment with Resiniferatoxin and Capsaicin on Dynamic Weight Bearing Measures and Evoked Pain Responses in a Chronic Inflammatory Arthritis Murine Model. *Intern. Med. Rev.* **2016,** *2016* (6), E-article 27774522.

17. Bessler, H.; Djaldetti, M. Capsaicin Modulates the Immune Cross Talk Between Human Mononuclears and Cells from Two Colon Carcinoma Lines. *Nutr. Cancer* **2017,** *69* (1), 14–20.
18. Bhattacharya, A.; Chattopadhyay, A.; Mazumdar, D.; Chakravarty, A.; Pal, S. Antioxidant Constituents and Enzyme Activities in Chilli Peppers. *Int. J. Veg. Sci.* **2010,** *16* (3), 201–211.
19. Bonaccio, M.; Di Castelnuovo, A.; Costanzo, S.; Ruggiero, E.; De Curtis, A.; Persichillo, M.; Tabolacci, C.; Facchiano, F.; Cerletti, C.; Donati, M. B. Chili Pepper Consumption and Mortality in Italian Adults. *J. Am. College Cardiol.* **2019,** *74* (25), 3139–3149.
20. Bourne, N.; Bernstein, D.; Stanberry, L. Civamide (cis-capsaicin) for Treatment of Primary or Recurrent Experimental Genital Herpes. *Antimicrob. Agents Chemother.* **1999,** *43* (11), 2685–2688.
21. Bukhari, H. M.; Header, E. A.; Shaikhomar, O. A.; Alharbi, S. A. Effect of Spicy Fast Foods and Red Chili Powder on Liver Status of Hypercholesterolemic Rats. *Br. J. Med. Health Res.* **2020,** *7* (9), 1–10.
22. Butterfield, D. A.; Howard, B.; Yatin, S.; Koppal, T.; Drake, J.; Hensley, K.; Aksenov, M.; Aksenova, M.; Subramaniam, R.; Varadarajan, S. Elevated Oxidative Stress in Models of Normal Brain Aging and Alzheimer's Disease. *Life Sci.* **1999,** *65* (18–19), 1883–1892.
23. Byun, E.-B.; Park, W.-Y.; Kim, W. S.; Song, H.-Y.; Sung, N.-Y.; Byun, E.-H. Neuroprotective Effect of *Capsicum annuum* Against Hydrogen Peroxide-Induced Oxidative Stress in HT22 Hippocampus Cells. *Biosci. Biotechnol. Biochem.* **2018,** *82* (12), 2149–2157.
24. Chakrabarty, S.; Islam, A. M.; Islam, A. A. Nutritional Benefits and Pharmaceutical Potentialities of Chili: A Review. *Fundamental Appl. Agric.* **2017,** *2* (2), 227–232.
25. Chen, J.; Li, L.; Li, Y.; Liang, X.; Sun, Q.; Yu, H.; Zhong, J.; Ni, Y.; Chen, J.; Zhao, Z. Activation of TRPV1 Channel by Dietary Capsaicin Improves Visceral Fat Remodeling Through Connexin43-Mediated $Ca^{2+}$ Influx. *Cardiovasc. Diabetol.* **2015,** *14* (1), 1–14.
26. Chopan, M.; Littenberg, B. The Association of Hot Red Chili Pepper Consumption and Mortality: A Large Population-Based Cohort Study. *PloS One* **2017,** *12* (1), E-article 0169876. https://doi.org/10.1371/journal.pone.0169876
27. Chow, J.; Norng, M.; Zhang, J.; Chai, J. TRPV6 Mediates Capsaicin-Induced Apoptosis in Gastric Cancer Cells: Mechanisms Behind a Possible New "Hot" Cancer Treatment. *Biochim. Biophys. Acta (BBA)-Mol. Cell Res.* **2007,** *1773* (4), 565–576.
28. Chuichulcherm, S.; Prommakort, S.; Srinophakun, P.; Thanapimmetha, A. Optimization of Capsaicin Purification from *Capsicum frutescens* Linn. with Column Chromatography Using Taguchi Design. *Ind. Crops Product.* **2013,** *44*, 473–479.
29. Chung, Y. C.; Baek, J. Y.; Kim, S. R.; Ko, H. W.; Bok, E.; Shin, W.-H.; Won, S.-Y.; Jin, B. K. Capsaicin Prevents Degeneration of Dopamine Neurons by Inhibiting Glial Activation and Oxidative Stress in the MPTP Model of Parkinson's Disease. *Exp. Mol. Med.* **2017,** *49* (3), e298–e298.
30. Cil, H.; Atilgan, Z.; Islamoglu, Y.; Tekbas, E.; Dostbil, Z. Is the Pepper Spray a Triggering Factor in Myocardial Infarction? A Case Report. *Eur. Rev. Med. Pharmacol. Sci.* **2012,** *16* (Suppl 1), 73–74.
31. Cimen, I.; Pirinc, V.; Sagır, A.; Akpinar, C.; Guzel, S. Effects of Solarization and Vesicular Arbuscularmycorrhizal Fungus (VAM) on Phytopthora Blight (*Phytophthoracapsici* leonian) and Yield in Pepper. *Afr. J. Biotechnol.* **2009,** *8* (19), 4884–4894.

32. Ciulu-Costinescu, F.; Chifiriuc, M. C.; Popa, M.; Bleotu, C.; Neamtu, J.; Averis, L. M. E.; Bubulica, M. V.; Simionescu, A.; Aldea, I. M.; Belu, I. Screening of Polyphenol Content and In Vitro Studies of Antioxidant, Antibacterial and Cytotoxic Activities of *Capsicum annuum* Extracts. *Rev. Chim.* **2015,** *66*, 1261–1266.
33. Clapham, D. E. TRP Channels as Cellular Sensors. *Nature* **2003,** *426* (6966), 517–524.
34. Clark, R.; Lee, S.-H. Anticancer Properties of Capsaicin Against Human Cancer. *Anticancer Res.* **2016,** *36* (3), 837–843.
35. Clegg, M. E.; Golsorkhi, M.; Henry, C. J. Combined Medium-Chain Triglyceride and Chilli Feeding Increases Diet-Induced Thermogenesis in Normal-Weight Humans. *Eur. J. Nutr.* **2013,** *52* (6), 1579–1585.
36. Cruz, L. P.; Ribeiro, S. F.; Carvalho, A. O.; Vasconcelos, I. M.; Rodrigues, R.; Cunha, M. D.; Gomes, V. M. Isolation and Partial Characterization of a Novel Lipid Transfer Protein (LTP) and Antifungal Activity of Peptides from Chilli Pepper Seeds. *Protein Peptide Lett.* **2010,** *17* (3), 311–318.
37. Das, M.; Basu, S.; Banerjee, B.; Jana, K.; Sen, A.; Datta, G. Renoprotective Effect of *Capsicum annum* Against Ethanol-Induced Oxidative Stress and Renal Apoptosis. *J. Food Biochem.* **2021,** *45* (3), E-article 13325. https://doi.org/10.1111/jfbc.13325
38. Derry, S.; Rice, A. S.; Cole, P.; Tan, T.; Moore, R. A. Topical Capsaicin (High Concentration) for Chronic Neuropathic Pain in Adults. *Cochrane Database System. Rev.* **2017,** *1* (1), E-article CD007393. https://doi.org/10.1002/14651858.CD007393.pub4
39. Dogan, A. N. C.; Çelik, E.; Kılıçle, P. A.; Atalay, E.; Saglam, A. G.; Dogan, A.; Otlu, S. Antibacterial Effect of Hot Peppers (*Capsicum annuum*, *Capsicum annuum*varglobriusculum, *Capsicum frutescens*) on Some *Arcobacter*, *Campylobacter* and *Helicobacter* Species. *Pak. Vet. J.* **2018,** *38* (3), 266–270.
40. Ekom, S. E.; Tamokou, J.-D.-D.; Kuete, V. Antibacterial and Therapeutic Potentials of the *Capsicum annuum* Extract Against Infected Wound in a Rat Model with Its Mechanisms of Antibacterial Action. *BioMed Res. Int.* **2021,** E-article 4303902. https://doi.org/10.1155/2021/4303902
41. Ericson, A.; Nur, E. M.; Petersson, F.; Kechagias, S. The Effects of Capsaicin on Gastrin Secretion in Isolated Human Antral Glands: Before and After Ingestion of Red Chilli. *Digest. Dis. Sci.* **2009,** *54* (3), 491–498.
42. Foster, H. E.; Lake, A. G. Use of Vanilloids in Urologic Disorders. *Progress Drug Res.* **2014,** *68*, 307–317.
43. Gaete, P. S.; Lillo, M. A.; Puebla, M.; Poblete, I.; Figueroa, X. F. CGRP Signaling Inhibits NO Production Through Pannexin-1 Channel Activation in Endothelial Cells. *Sci. Rep.* **2019,** *9* (1), 1–17.
44. Ganguly, S.; Praveen, K.; Para, P. A.; Sharma, V. Medicinal Properties of Chilli Pepper in Human Diet. *J. Public Health Commun. Med.* **2017,** *2* (1), 6–7.
45. Gueders, M. M.; Paulissen, G.; Crahay, C.; Quesada-Calvo, F.; Hacha, J.; Van Hove, C.; Tournoy, K.; Louis, R.; Foidart, J.-M.; Noel, A. Mouse Models of Asthma: A Comparison Between C57BL/6 and BALB/c Strains Regarding Bronchial Responsiveness, Inflammation, and Cytokine Production. *Inflamm. Res.* **2009,** *58* (12), 845–854.
46. Gupta, R.; Dixit, V.; Dobhal, M. Hypocholesterolaemic Effect of the Oleoresin of *Capsicum annum* L. in Gerbils (*Merioneshurrianae*Jerdon). *Phytother. Res.* **2002,** *16* (3), 273–275.
47. Hafiz, T. A.; Mubaraki, M.; Dkhil, M.; Al-Quraishy, S. Antiviral Activities of *Capsicum annuum*methanolic Extract Against Herpes Simplex Virus 1 and 2. *Pak. J. Zool.* **2017,** *49* (1), 251–259. http://dx.doi.org/10.17582/journal.pjz/2017.49.1.267.272

48. Han, S.-S.; Keum, Y.-S.; Chun, K.-S.; Surh, Y.-J. Suppression of Phorbol Ester-Induced NF-κB Activation by Capsaicin in Cultured Human Promyelocytic Leukemia Cells. *Arch. Pharma. Res.* **2002,** *25* (4), 475–479.
49. Hernandez-Perez, T.; Gomez-Garcia, M. d. R.; Valverde, M. E.; Paredes-Lopez, O. *Capsicumannuum* (Hot Pepper): An Ancient Latin-American Crop with Outstanding Bioactive Compounds and Nutraceutical Potential: A Review. *Compreh. Rev. Food Sci. Food Saf.* **2020,** *19* (6), 2972–2993.
50. Huei, C. S.; Azlan, A.; Ismail, A.; Shafie, N. H.; Sultana, S. Antioxidant and Anti-Obesity Properties of Local Chilies Varieties in Malaysia. *J. Food Sci. Technol.* **2020,** *57* (10), 3677–3687.
51. Hwang, M. K.; Bode, A. M.; Byun, S.; Song, N. R.; Lee, H. J.; Lee, K. W.; Dong, Z. Cocarcinogenic Effect of Capsaicin Involves Activation of EGFR Signaling But Not TRPV1. *Cancer Res.* **2010,** *70* (17), 6859–6869.
52. Ibarra-Torres, P.; Valadez-Moctezuma, E.; Perez-Grajales, M.; Rodriguez-Campos, J.; Jaramillo-Flores, M. E. Inter-and Intraspecific Differentiation of *Capsicum annuum* and *Capsicum pubescens* Using ISSR and SSR Markers. *ScientiaHorticulturae* **2015,** *181*, 137–146.
53. Imran, M.; Butt, M. S.; Suleria, H. A. R. *Capsicum annuum* Bioactive Compounds: Health Promotion Perspectives. In *Bioactive Molecules in Food*; Merillon, J. M., Ramawat, K., Eds., Chapter 1; Springer: Cham, Switzerland, 2018; pp 1–22.
54. Imran, M.; Rauf, A.; Abu-Izneid, T.; Nadeem, M.; Shariati, M. A.; Khan, I. A.; Imran, A.; Orhan, I. E.; Rizwan, M.; Atif, M. Luteolin, a Flavonoid, as an Anticancer Agent: A Review. *Biomed. Pharmacother.* **2019,** *112*, E-article 108612. https://doi.org/10.1016/j.biopha.2019.108612
55. Ito, K.; Nakazato, T.; Yamato, K.; Miyakawa, Y.; Yamada, T.; Hozumi, N.; Segawa, K.; Ikeda, Y.; Kizaki, M. Induction of Apoptosis in Leukemic Cells by Homovanillic Acid Derivative, Capsaicin, Through Oxidative Stress: Implication of Phosphorylation of p53 at Ser-15 Residue by Reactive Oxygen Species. *Cancer Res.* **2004,** *64* (3), 1071–1078.
56. Jang, H. W.; Ka, M. H.; Lee, K. G. Antioxidant Activity and Characterization of Volatile Extracts of *Capsicum annuum* L. and *Allium* spp. *Flavor Fragrance J.* **2008,** *23* (3), 178–184.
57. Jang, H.-Y.; Kim, S.-M.; Yuk, J.-E.; Kwon, O.-K.; Oh, S.-R.; Lee, H.-K.; Jeong, H.; Ahn, K.-S. *Capsicum annuum* L. Methanolic Extract Inhibits Ovalbumin-Induced Airway Inflammation and Oxidative Stress in a Mouse Model of Asthma. *J. Med. Food* **2011,** *14* (10), 1144–1151.
58. Jin, T.; Wu, H.; Wang, Y.; Peng, H. Capsaicin Induces Immunogenic Cell Death in Human Osteosarcoma Cells. *Exp. Therap. Med.* **2016,** *12* (2), 765–770.
59. Joseph, A.; Johannah, N.; Kumar, S.; Maliakel, B.; Krishnakumar, I. Safety Assessment of a Fenugreek Dietary Fiber-Based Formulation of Capsaicinoids-Rich Red Chili (*Capsicum annum*) Extract (Capsifen®): Acute and Sub-Chronic Studies. *Toxicol. Rep.* **2020,** *7*, 602–609.
60. Kang, J. H.; Tsuyoshi, G.; Le Ngoc, H.; Kim, H. M.; Tu, T. H.; Noh, H. J.; Kim, C. S.; Choe, S. Y.; Kawada, T.; Yoo, H. Dietary Capsaicin Attenuates Metabolic Dysregulation in Genetically Obese Diabetic Mice. *J. Med. Food* **2011,** *14* (3), 310–315.
61. Kempaiah, R. K.; Srinivasan, K. Beneficial Influence of Dietary Curcumin, Capsaicin and Garlic on Erythrocyte Integrity in High-Fat Fed Rats. *J. Nutr. Biochem.* **2006,** *17* (7), 471–478.

62. Kempaiah, R.; Manjunatha, H.; Srinivasan, K. Protective Effect of Dietary Capsaicin on Induced Oxidation of Low-Density Lipoprotein in Rats. *Mol. Cell. Biochem.* **2005,** *275* (1), 7–13.
63. Khan, F. A.; Mahmood, T.; Ali, M.; Saeed, A.; Maalik, A. Pharmacological Importance of an Ethnobotanical Plant: *Capsicum annuum* L. *Nat. Product Res.* **2014,** *28* (16), 1267–1274.
64. Khan, M. T. I.; Qamar, A.; Muhammad, A.; Muhammad, W. Economic Analysis of Open Field Chilli (*Capsicum annuum* L.) Production in Punjab. *Pak. J. Exp. Biol. Agric. Sci.* **2017,** *5* (1), 120–125.
65. Kim, H. R.; Kim, S.; Lee, S. W.; Sin, H. S.; Kim, S. Y. Protective Effects of Fermented Paprika (*Capsicum annuum* L.) on Sodium Iodate-Induced Retinal Damage. *Nutrients* **2021,** *13* (1), 25–30.
66. Kim, J. D.; Kim, J. M.; Pyo, J. O.; Kim, S. Y.; Kim, B. S.; Yu, R.; Han, I. S. Capsaicin Can Alter the Expression of Tumor Forming-Related Genes Which Might be Followed by Induction of Apoptosis of a Korean Stomach Cancer Cell Line, SNU-1. *Cancer Lett.* **1997,** *120* (2), 235–241.
67. Koffi-Nevry, R.; Kouassi, K. C.; Nanga, Z. Y.; Koussemon, M.; Loukou, G. Y. Antibacterial Activity of Two Bell Pepper Extracts: *Capsicum annuum* L. and *Capsicum frutescens*. *Int. J. Food Prop.* **2012,** *15* (5), 961–971.
68. Koh, H. L.; Tan, C. H.; Chua, T. K. Guide to Medicinal Plants. *Illustrated Sci. Med. Approach* **2009,** *2009,* 1277–1286.
69. Kothari, S.; Joshi, A.; Kachhwaha, S.; Ochoa-Alejo, N. Chilli Peppers—A Review on Tissue Culture and Transgenesis. *Biotechnol. Adv.* **2010,** *28* (1), 35–48.
70. Kuda, T.; Iwai, A.; Yano, T. Effect of Red Pepper *Capsicum annuum* var. Conoides and Garlic *Allium sativum* on Plasma Lipid Levels and Cecalmicroflora in Mice Fed Beef Tallow. *Food Chem. Toxicol.* **2004,** *42* (10), 1695–1700.
71. Kurt, S.; Emir, B. Effect of Soil Solarization, Chicken Litter and Viscera on Populations of Soilborne Fungal Pathogens and Pepper Growth. *Plant Pathol. J.* **2004,** *3* (2), 118–124.
72. Lampe, J. W. Spicing Up a Vegetarian Diet: Chemopreventive Effects of Phytochemicals. *Am. J. Clin. Nutr.* **2003,** *78* (3), 579S–583S.
73. Lang, H.; Li, Q.; Yu, H.; Li, P.; Lu, Z.; Xiong, S.; Yang, T.; Zhao, Y.; Huang, X.; Gao, P. Activation of TRPV1 Attenuates High Salt-Induced Cardiac Hypertrophy Through Improvement of Mitochondrial Function. *Br. J. Pharmacol.* **2015,** *172* (23), 5548–5558.
74. Laslett, L. L.; Jones, G. Capsaicin for Osteoarthritis Pain. *Prog Drug Res.* **2014,***68*, 277–291.
75. Lau, J. K.; Brown, K. C.; Dom, A. M.; Witte, T. R.; Thornhill, B. A.; Crabtree, C. M.; Perry, H. E.; Brown, J. M.; Ball, J. G.; Creel, R. G. Capsaicin Induces Apoptosis in Human Small Cell Lung Cancer via the TRPV6 Receptor and the Calpain Pathway. *Apoptosis* **2014,** *19* (8), 1190–1201.
76. Lee, S. H.; Krisanapun, C.; Baek, S. J. NSAID-Activated Gene-1 as a Molecular Target for Capsaicin-Induced Apoptosis Through a Novel Molecular Mechanism Involving GSK3β, C/EBPβ and ATF3. *Carcinogenesis* **2010,** *31* (4), 719–728.
77. Lee, T. H.; Lee, J. G.; Yon, J. M.; Oh, K. W.; Baek, I. J.; Nahm, S. S.; Lee, B. J.; Yun, Y. W.; Nam, S. Y. Capsaicin Prevents Kainic Acid-Induced Epileptogenesis in Mice. *Neurochem. Int.* **2011,** *58* (6), 634–640.
78. Leung, F. W. Capsaicin-Sensitive Intestinal Mucosal Afferent Mechanism and Body Fat Distribution. *Life Sci.* **2008,** *83* (1–2), 1–5.

79. Li, J.; Wang, D. H. Increased GFR and Renal Excretory Function by Activation of TRPV1 in the Isolated Perfused Kidney. *Pharmacol. Res.* **2008,** *57* (3), 239–246.
80. Li, R.; Lan, Y.; Chen, C.; Cao, Y.; Huang, Q.; Ho, C. T.; Lu, M. Anti-Obesity Effects of Capsaicin and the Underlying Mechanisms: A Review. *Food Funct.* **2020,** *11* (9), 7356–7370.
81. Liang, Y. T.; Tian, X.-Y.; Chen, J. N.; Peng, C.; Ma, K. Y.; Zuo, Y.; Jiao, R.; Lu, Y.; Huang, Y.; Chen, Z.-Y. Capsaicinoids Lower Plasma Cholesterol and Improve Endothelial Function in Hamsters. *Eur. J. Nutr.* **2013,** *52* (1), 379–388.
82. Lidikova, J.; Ceryova, N.; Snirc, M.; Vollmannova, A.; Musilova, J.; Tóthova, M.; Hegedusova, A. Determination of Bioactive Components in Selected Varieties of Pepper (*Capsicum* L.). *Int. J. Food Prop.* **2021,** *24* (1), 1148–1163.
83. Lin, M. H.; Lee, Y. H.; Cheng, H. L.; Chen, H. Y.; Jhuang, F. H.; Chueh, P. J. Capsaicin Inhibits Multiple Bladder Cancer Cell Phenotypes by Inhibiting Tumor-Associated NADH Oxidase (tNOX) and Sirtuin1 (SIRT1). *Molecules* **2016,** *21* (7), 849–853.
84. Lininger, S. W. Ed. *A-Z Guide to Drug-Herb-Vitamin Interactions: How to Improve Your Health and Avoid Problems When Using Common Medications and Natural Supplements Together*; Three Rivers Press: New York, 1999; pp. 1–436.
85. Liu, T.; Wang, G.; Tao, H.; Yang, Z.; Wang, Y.; Meng, Z.; Cao, R.; Xiao, Y.; Wang, X.; Zhou, J. Capsaicin Mediates Caspases Activation and Induces Apoptosis Through P38 and JNK MAPK Pathways in Human Renal Carcinoma. *BMC Cancer* **2016,** *16* (1), 1–13.
86. Luo, L.; Yan, J.; Wang, X.; Sun, Z. The Correlation Between Chili Pepper Consumption and Gastric Cancer Risk: A Meta-Analysis. *Asia Pac. J. Clin. Nutr.* **2021,** *30* (1), 130–139.
87. Luo, X. J.; Peng, J.; Li, Y.-J. Recent Advances in the Study on Capsaicinoids and Capsinoids. *Eur. J. Pharm.* **2011,** *650* (1), 1–7. https://doi.org/10.1016/j.ejphar.2010.09.074
88. Madala, N.; Nutakki, M. K. Hot Pepper: History, Health and Dietary Benefits and Production. *Int. J. Curr. Microbiol. App. Sci.* **2020,** *9* (4), 2532–2538.
89. Maji, A. K.; Banerji, P. Phytochemistry and Gastrointestinal Benefits of the Medicinal Spice, *Capsicum annuum* L. (Chilli): A Review. *J. Complement. Integr. Med.* **2016,** *13* (2), 97–122.
90. Materska, M.; Perucka, I. Antioxidant Activity of the Main Phenolic Compounds Isolated from Hot Pepper Fruit (*Capsicum annuum* L.). *J. Agric. Food Chem.* **2005,** *53* (5), 1750–1756.
91. Mercy, E.; David, U. Potential Health Benefits of Conventional Nutrients and Phytochemicals of Capsicum Peppers. *Pharm. Pharmacol. Int. J.* **2018,** *6* (1), 62–69.
92. Mgbeje, B. Quality and Safety of Medicinal Plants. In *Phytomedicine: Panacea for Primary Health Care Delivery in Nigeria*; Ebong, P. E., Ed.; University of Calabar Press: Calabar, 2015, pp 293–310.
93. Min, J. K.; Han, K. Y.; Kim, E. C.; Kim, Y. M.; Lee, S. W.; Kim, O. H.; Kim, K. W.; Gho, Y. S.; Kwon, Y. G. Capsaicin Inhibits In Vitro and In Vivo Angiogenesis. *Cancer Res.* **2004,** *64* (2), 644–651.
94. Mizrak, S. C.; Gadella, B. M.; Erdost, H.; Ozer, A.; van Pelt, A. M.; van Dissel-Emiliani, F. M. Spermatogonial Stem Cell Sensitivity to Capsaicin: An In Vitro Study. *Reproduct. Biol. Endocrinol.* **2008,** *6* (1), 1–9.
95. Mori, A.; Lehmann, S.; O'Kelly, J.; Kumagai, T.; Desmond, J. C.; Pervan, M.; McBride, W. H.; Kizaki, M.; Koeffler, H. P. Capsaicin, a Component of Red Peppers, Inhibits the Growth of Androgen-Independent, p53 Mutant Prostate Cancer Cells. *Cancer Res.* **2006,** *66* (6), 3222–3229.

96. Motukuri, S. K.; Jaswanthi, N. Hot Pepper (*Capsicum annuum* L.): An Alternative Food to Reduce Micronutrient Deficiencies in Human. In *Capsicum*; Dekebo, A., Ed., Chapter 6; IntechOpen: London, 2020. https://doi.org/10.5772/intechopen.92198
97. Mueller, M.; Hobiger, S.; Jungbauer, A. Anti-Inflammatory Activity of Extracts from Fruits, Herbs and Spices. *Food Chem.* **2010,** *122* (4), 987–996.
98. Nmaju, A. U.; Joshua, I. E.; Okon, U. E.; Nwankwo, A. A.; Osim, E. E. Long-Term Consumption of *Capsicum annum* (Chilli Pepper) and Capsaicin Diets Suppresses Pain Perception and Improves Social Behaviour of CD-1 Mice. *Nutr. Food Sci.* **2018,** *48* (6), 911–921.
99. Oboh, G.; Rocha, J. Hot Pepper (*Capsicum* spp.) Protects Brain from Sodium Nitroprusside-and Quinolinic Acid-Induced Oxidative Stress In Vitro. *J. Med. Food* **2008,** *11* (2), 349–355.
100. Ofori-Asenso, R.; Mohsenpour, M. A.; Nouri, M.; Faghih, S.; Liew, D.; Mazidi, M. Association of Spicy Chilli Food Consumption with Cardiovascular and All-Cause Mortality: A Meta-Analysis of Prospective Cohort Studies. *Angiology* **2021,** *72* (7), 625–632.
101. Omolo, M. A.; Wong, Z. Z.; Mergen, K.; Hastings, J. C.; Le, N. C.; Reil, H. A.; Case, K. A.; Baumler, D. J. Antimicrobial Properties Of Chili Peppers. *J. Infect. Dis. Ther.* **2014,** *2* (4), E-article 1000145. https://doi.org/10.4172/2332–0877.1000145
102. Osim, E. E.; Nmaju, A.; Nwankwo, A.; Okon, U.; Iwasam, J. The Effects of Consumption of Capsicum on Some Neurobehavioural Parameters. In *Capsicum*; Dekebo, A. Ed. Chapter 11; IntechOpen; London, 2020. https://doi.org/10.5772/intechopen.91744
103. Pandit, M. K.; Pandit, R.; Bairagi, S. Chili. In: *Ethnopharmacological Investigation of Indian Spices*; Mishra, N., Ed., Chapter 18; IGI Global: New Delhi, 2020; pp 253–268.
104. Park, H. S.; Kim, S. R.; Lee, Y. C. Impact of Oxidative Stress on Lung Diseases. *Respirology* **2009,** *14* (1), 27–38.
105. Park, J. S.; Chyun, J. H.; Kim, Y. K.; Line, L. L.; Chew, B. P. Astaxanthin Decreased Oxidative Stress and Inflammation and Enhanced Immune Response in Humans. *Nutr. Metabol.* **2010,** *7* (1), 1–10.
106. Park, J.-H.; Jeon, G.-I.; Kim, J.-M.; Park, E. Antioxidant Activity and Antiproliferative Action of Methanol Extracts of 4 Different Colored Bell Peppers (*Capsicum annuum* L.). *Food Sci. Biotechnol.* **2012,** *21* (2), 543–550.
107. Peppin, J. F.; Pappagallo, M. Capsaicinoids in the Treatment of Neuropathic Pain: A Review. *Therap. Adv. Neurol. Disorders* **2014,** *7* (1), 22–32.
108. Perry, L.; Dickau, R.; Zarrillo, S.; Holst, I.; Pearsall, D. M.; Piperno, D. R.; Berman, M. J.; Cooke, R. G.; Rademaker, K.; Ranere, A. J. Starch Fossils and the Domestication and Dispersal of Chili Peppers (*Capsicum* spp. L.) in the Americas. *Science* **2007,** *315* (5814), 986–988.
109. Persson, M. S.; Fu, Y.; Bhattacharya, A.; Goh, S.-L.; van Middelkoop, M.; Bierma-Zeinstra, S. M.; Walsh, D.; Doherty, M.; Zhang, W. Relative Efficacy of Topical Non-Steroidal Anti-Inflammatory Drugs and Topical Capsaicin in Osteoarthritis: Protocol for an Individual Patient Data Meta-Analysis. *System. Rev.* **2016,** *5* (1), 1–8.
110. Popelka, P.; Jevinova, P.; Smejkal, K.; Roba, P. Determination of Capsaicin Content and Pungency Level of Different Fresh and Dried Chilli Peppers. *Folia Veterinaria* **2017,** *61* (2), 11–16.
111. Qian, K.; Wang, G.; Cao, R.; Liu, T.; Qian, G.; Guan, X.; Guo, Z.; Xiao, Y.; Wang, X. Capsaicin Suppresses Cell Proliferation, Induces Cell Cycle Arrest and ROS Production

in Bladder Cancer Cells Through FOXO3a-Mediated Pathways. *Molecules* **2016,** *21* (10), 1406–1411.
112. Rahevar, P. M.; Patel, J. N.; Kumar, S.; Acharya, R. R. Morphological, Biochemical and Molecular Characterization for Genetic Variability Analysis of *Capsicum annuum*. Vegetos **2019,** *32* (2), 131–141.
113. Raju, P. C.; Satyanarayana, P. Evaluation of Nephroprotective Activity of *Capsicum annuum* in Rifampicin Induced Nephrotoxicity. *Int. J. Appl. Res.* **2017,** *3* (6), 758–763.
114. Ribeiro, S. F.; Carvalho, A. O.; Da Cunha, M.; Rodrigues, R.; Cruz, L. P.; Melo, V. M.; Vasconcelos, I. M.; Melo, E. J.; Gomes, V. M. Isolation and Characterization of Novel Peptides from Chilli Pepper Seeds: Antimicrobial Activities Against Pathogenic Yeasts. *Toxicon* **2007,** *50* (5), 600–611.
115. Ríos-Silva, M.; Santos-Álvarez, R.; Trujillo, X.; Cárdenas-María, R. Y.; López-Zamudio, M.; Bricio-Barrios, J. A.; Leal, C.; Saavedra-Molina, A.; Huerta-Trujillo, M.; Espinoza-Mejía, K. Effects of Chronic Administration of Capsaicin on Biomarkers of Kidney Injury in Male Wistar Rats with Experimental Diabetes. *Molecules* **2019,** *24* (1), 36–42.
116. Riquelme, N.; Matiacevich, S. Characterization and Evaluation of Some Properties of Oleoresin from *Capsicum annuum* var. cacho de cabra. *CyTA-J. Food* **2017,** *15* (3), 344–351.
117. Saleh, B. K.; Omer, A.; Teweldemedhin, B. Medicinal Uses and Health Benefits of Chili Pepper (*Capsicum* spp.): A Review. *MOJ Food Process. Technol.* **2018,** *6* (4), 325–328.
118. Santos, M. M. P.; Vieira-da-Motta, O.; Vieira, I. J. C.; Braz-Filho, R.; Gonçalves, P. S.; Maria, E. J.; Terra, W. S.; Rodrigues, R.; Souza, C. L. M. Antibacterial Activity of *Capsicum annuum* Extract and Synthetic Capsaicinoid Derivatives Against Streptococcus Mutans. *J. Nat. Med.* **2012,** *66* (2), 354–356.
119. Sharma, J.; Sharma, P.; Sharma, B.; Chaudhary, P. In Vitro Estimation of Antioxidant Activity in Green Chilli (*Capsicum annuum*) and Yellow Lantern Chilli (*Capsicum chinense*). *Int. J. Res. Rev.* **2017,** *4* (6), 54–61.
120. Shayan, S.; Saeidi, S. Antibacterial and Antibiofilm Activities of Extract *Capsicum annuum* L on the Growth and Biofilm Formation of Common Pathogenic Strains. *Int. J. Sci.* **2013,** *5* (4), 513–518.
121. Shi, Z.; Riley, M.; Taylor, A.; Page, A. Chilli Consumption and the Incidence of Overweight and Obesity in a Chinese Adult Population. *Int. J. Obes.* **2017,** *41* (7), 1074–1079.
122. Silva, L. R.; Azevedo, J.; Pereira, M. J.; Valentão, P.; Andrade, P. B. Chemical Assessment and Antioxidant Capacity of Pepper (*Capsicum annuum* L.) Seeds. *Food Chem. Toxicol.* **2013,** *53*, 240–248.
123. Singh, H.; Fairs, G.; Syarhabil, M. Anti-Fungal Activity of *Capsicum frutescence* and *Zingiber officinale* Against Key Post-Harvest Pathogens in Citrus. *Int. Conf. Biomed. Eng. Technol.* **2011,** *11,* 1–6.
124. Sivasinprasasn, S.; Wikan, N.; Tocharus, J.; Pantan, R.; Chaichompoo, W.; Suksamrarn, A.; Tocharus, C. Synergistic Effects of the Capsaicinoidnonivamide and Rosuvastatin on Obesity-Related Endothelial Dysfunction in Rat Fed a High-Fat Diet. *Phytother. Res.* **2019,** *33* (7), 1815–1826.
125. Skrzypski, M.; Sassek, M.; Abdelmessih, S.; Mergler, S.; Grötzinger, C.; Metzke, D.; Wojciechowicz, T.; Nowak, K. W.; Strowski, M. Capsaicin Induces Cytotoxicity in Pancreatic Neuroendocrine Tumor Cells via Mitochondrial Action. *Cell. Signal.* **2014,** *26* (1), 41–48.
126. Srinivasan, K. Biological Activities of Red Pepper (*Capsicum annuum*) and Its Pungent Capsaicin: A Review. *Crit. Rev. Food Sci. Nutr.* **2016,** *56* (9), 1488–1500.

127. Sun, F.; Xiong, S.; Zhu, Z. Dietary Capsaicin Protects Cardiometabolic Organs from Dysfunction. *Nutrients* **2016,** *8* (5), 174–180.
128. Szallasi, A.; Blumberg, P. M. Vanilloid (Capsaicin) Receptors and Mechanisms. *Pharmacol. Rev.* **1999,** *51* (2), 159–212.
129. Tripodi, P.; Kumar, S. The Capsicum Crop: An Introduction. In *The Capsicum Genome*; Ramchiary, N. Kole, C., Eds., Chapter 1; Springer: Cham, Switzerland, 2019; pp 1–8. https://doi.org/10.1007/978-3-319-97217-6_1
130. Tsui, H.; Razavi, R.; Chan, Y.; Yantha, J.; Dosch, H.-M. Sensing Autoimmunity in Type-1 Diabetes. *Trends Mol. Med.* **2007,** *13* (10), 405–413.
131. van LookerenCampagne, M.; LeCouter, J.; Yaspan, B. L.; Ye, W. Mechanisms of Age-Related Macular Degeneration and Therapeutic Opportunities. *J. Pathol.* **2014,** *232* (2), 151–164.
132. Varghese, S.; Kubatka, P.; Rodrigo, L.; Gazdikova, K.; Caprnda, M.; Fedotova, J.; Zulli, A.; Kruzliak, P.; Büsselberg, D. Chili Pepper as a Body Weight-Loss Food. *Int. J. Food Sci. Nutr.* **2017,** *68* (4), 392–401.
133. Vendrely, V.; Peuchant, E.; Buscail, E.; Moranvillier, I.; Rousseau, B.; Bedel, A.; Brillac, A.; de Verneuil, H.; Moreau-Gaudry, F.; Dabernat, S. Resveratrol and Capsaicin Used Together as Food Complements Reduce Tumor Growth and Rescue Full Efficiency of Low Dose Gemcitabine in a Pancreatic Cancer Model. *Cancer Lett.* **2017,** *390*, 91–102.
134. Vijayalakshmi, K.; Shyamala, R.; Thirumurugan, V.; Sethuraman, M.; Rajan, S.; Badami, S.; Mukherjee, P. K. Physico- and Phytochemical Investigation and Anti-Inflammatory Screening of *Capsicum annum* L. and *Hemidesmus indicus* (Linn.) R. Br. *Ancient Sci. Life* **2010,** *29* (4), 35–40.
135. Walker, J.; Ley, J. P.; Schwerzler, J.; Lieder, B.; Beltran, L.; Ziemba, P. M.; Hatt, H.; Hans, J.; Widder, S.; Krammer, G. E. Nonivamide, a Capsaicin Analogue, Exhibits Anti-Inflammatory Properties in Peripheral Blood Mononuclear Cells and U-937 Macrophages. *Mol. Nutr. Food Res.* **2017,** *61* (2), E-article 1600474. https://doi.org/10.1002/mnfr.201600474.
136. Wang, D.; Bosland, P. W. The Genes of Capsicum. *HortScience* **2006,** *41* (5), 1169–1187.
137. Wang, F.; Zhao, J.; Liu, D.; Zhao, T.; Lu, Z.; Zhu, L.; Cao, L.; Yang, J.; Jin, J.; Cai, Y. Capsaicin Reactivates hMOF in Gastric Cancer Cells and Induces Cell Growth Inhibition. *Cancer Biol. Therap.* **2016,** *17* (11), 1117–1125.
138. Xu, S.; Zhang, L.; Cheng, X.; Yu, H.; Bao, J.; Lu, R. Capsaicin Inhibits the Metastasis of Human Papillary Thyroid Carcinoma BCPAP Cells Through the Modulation of the TRPV1 Channel. *Food Funct.* **2018,** *9* (1), 344–354.
139. Yaksh, T. L.; Di Nardo, A. Complexity of Systems and Actions Underlying Neurogenic Inflammation. *Semin. Immunopathol.* **2018,** *40*, 225–228.
140. Yamani, N.; Musheer, A.; Gosain, P.; Sarfraz, S.; Qamar, H.; Waseem, M. M.; Arshad, M. S.; Almas, T.; Figueredo, V. Meta-Analysis Evaluating the Impact of Chili-Pepper Intake on All-Cause and Cardiovascular Mortality: A Systematic Review. *Ann. Med. Surg.* **2021,** *70*, E-article 102774. https://doi.org/10.1016/j.amsu.2021.102774
141. Yamasaki, S.; Asakura, M.; Neogi, S. B.; Hinenoya, A.; Iwaoka, E.; Aoki, S. Inhibition of Virulence Potential of *Vibrio cholerae* by Natural Compounds. *Ind. J. Med. Res.* **2011,** *133* (2), 232–239.
142. Yang, K.; Pyo, J.; Kim, G.-Y.; Yu, R.; Ju, S.; Kim, W.; Kim, B.-S. Capsaicin Induces Apoptosis by Generating Reactive Oxygen Species and Disrupting Mitochondrial

Transmembrane Potential in Human Colon Cancer Cell Lines. *Cell. Mol. Biol. Lett.* **2009,** *14* (3), 497–510.
143. Yu, S. Q.; Ma, S.; Wang, D. H. Activation of TRPV1 Prevents Salt-Induced Kidney Damage and Hypertension After Renal Ischemia-Reperfusion Injury in Rats. *Kidney Blood Pressure Res.* **2018,** *43* (4), 1285–1296. https://doi.org/10.1159/000492412
144. Zheng, L.; Chen, J.; Ma, Z.; Liu, W.; Yang, F.; Yang, Z.; Wang, K.; Wang, X.; He, D.; Li, L. Capsaicin Enhances Anti-Proliferation Efficacy of Pirarubicin via Activating TRPV1 and Inhibiting PCNA Nuclear Translocation in 5637 Cells. *Mol. Med. Rep.* **2016,** *13* (1), 881–887.
145. Zhigila, D. A.; AbdulRahaman, A. A.; Kolawole, O. S.; Oladele, F. A. Fruit Morphology as Taxonomic Features in Five Varieties of *Capsicum annuum* L. Solanaceae. *J. Bot.* **2014,** E-article 540868. https://doi.org/10.1155/2014/540868
146. Zhu, Y.; Wang, Y.; Wang, D. H. Diuresis and Natriuresis Caused by Activation of VR1-Positive Sensory Nerves in Renal Pelvis of Rats. *Hypertension* **2005,** *46* (4), 992–997.
147. Zimmer, A. R.; Leonardi, B.; Miron, D.; Schapoval, E.; de Oliveira, J. R.; Gosmann, G. Antioxidant and Anti-Inflammatory Properties of *Capsicum baccatum*: From Traditional Use to Scientific Approach. *J. Ethnopharmacol.* **2012,** *139* (1), 228–233.

## CHAPTER 11

# HEALTH BENEFITS OF BIOACTIVE COMPOUNDS AND NUTRIENTS IN COFFEE SILVERSKIN

WALLAF COSTA VIMERCATI, LEANDRO LEVATE MACEDO, and CINTIA DA SILVA ARAÚJO

## ABSTRACT

In recent years, large amounts of coffee coproducts have been generated. This is due to the increase in world production and consumption of coffee, especially in relation to specialty coffees due to increased market appreciation. Coffee silverskin is a thin integument that involves the raw coffee bean, being the main coproduct generated in the coffee roasting process. This coproduct has a high content of proteins, minerals, and fibers, as well as several other bioactive compounds, especially antioxidants, such as phenolic compounds, caffeine, melanoidins, and vitamin E. The direct application of CS or the extraction of its compounds has been suggested as a potential way of valuing this coproduct, since the compounds present in CS contribute to several health benefits. In addition, the low cost of CS and the lower environmental impact that will be generated further justify the appreciation of this coproduct.

## 11.1 INTRODUCTION

Coffee is produced in high volumes because of high commercial importance.[6] The characteristic attributes of coffee (such as flavor, aroma, stimulating

Plant-Based Bioactive Compounds and Food Ingredients: Encapsulation, Functional, and Safety Aspects.
Junaid Ahmad Malik, Megh R. Goyal, Preeti Birwal, & Ritesh B. Watharkar (Eds.)

effect) and its health benefits are some of the factors that contribute to this beverage being widely consumed throughout the world.[2,39,43,61]

According to the International Coffee Organization,[40] world coffee production and consumption has increased over the past 30 years. In the year 2020/2021, world production of coffee was estimated at 169.6 million bags, comprising 99.2 million bags of Arabica coffee and 70.4 million bags of Robusta coffee. Although the COVID-19 pandemic initially influenced coffee consumption trends and patterns,[40] world consumption is estimated to increase by 1.9% compared to last year, corresponding to 167.58 million bags. In addition, a continued growth in world coffee consumption is expected due to the relaxation of restrictions imposed by the COVID-19 pandemic and the trend toward economic recovery.[41]

In this world scenario, Brazil ranks as the largest producer, exporter, and second largest consumer of coffee.[1] Brazilian coffee production in the 2019/2020 harvest was estimated at 58 million bags, with 37 million bags of Arabica coffee and 21 million bags of robusta coffee (ICO, 2020). In 2020, per capita consumption of green and roasted coffee was estimated at 6.01 and 4.81 kg of coffee/inhabitant year, respectively.[1]

The increase in world production and consumption of coffee consequently generates an important portion of coproducts by the coffee industry.[10] Coffee silverskin (CS) is the main coproduct of the coffee roasting process.[42] This coproduct is highly polluting and requires a high demand for oxygen to be degraded. Studies aimed at different and profitable applications must be developed in order to minimize the impact generated on the environment.[50] CS has been the subject of extensive research as it has a rich composition, mainly in bioactive compounds, which exert beneficial effects on health.[10,53,54]

This chapter focuses on health benefits of phytochemicals in CS.

## 11.2 COMPOSITION AND SCOPE OF COFFEE SILVERSKIN

CS is a thin integument that involves the raw coffee bean, called perisperm. The CS detachment from grain occurs during the roasting process, being separated by air flow. Therefore, this is the main coproduct generated in the coffee roasting process.[4,23,42,50,53] In the roasting process, 60 kg of CS are produced for every eight tons of roasted coffee, corresponding to 0.75% (w/w).[5,54]

Most of the time, CS is spent for composting, to landfills, or is used as a furnace fuel.[4,23,50] However, in recent years, some studies have highlighted that CS has a biochemical composition of interest for application in the foods, chemical, and pharmaceutical industries and is considered as a potential natural source of several bioactive compounds. Therefore, several studies have reported the importance of valorization this coproduct.[5,53,63]

The wide variation in percentage composition (Table 11.1) and bioactive compounds of CS[11,16,19,24,55,58] is due to factors related to geographic origin, genetics, agricultural practices, climate, fruit maturation and degree of roasting, postharvest processing, and differences in analytical procedures.[12,16,42,58]

**TABLE 11.1** Percentage Composition (%) of Coffee Silverskin.[11,16,19,24,55,58]

| | **Percentage composition (%)** |
|---|---|
| Moisture | 4.76–10.3* |
| Available carbohydrates | 0.25–6.35 |
| Carbohydrates | 62.1–80.5 |
| Fats | 1.32–3.78 |
| Insoluble dietary fiber | 44.2–53.7 |
| Minerals (ash) | 5.36–11.58 |
| Proteins | 11.8–22.86 |
| Soluble dietary fiber | 3.3–10.95 |
| Total dietary fiber | 53.4–62.4 |

*Moisture is expressed on wet basis, and the other parameters are expressed on dry basis (d.b.).

CS is considered as the most stable biomaterial compared to other coffee coproducts, as it has a low moisture content (<10%).[16] In addition, the high temperatures used in the roasting process also contribute to the reduction or elimination of microorganisms potentially associated with the raw coffee beans.[60]

Carbohydrates are the main nutrients that are present in CS, followed by protein and minerals (Table 11.1). On the other hand, the lipid content is considered low.[7,20,55]

Dietary fiber is the major constituent of the carbohydrate fraction, being composed of ~85% insoluble dietary fiber and ~15% soluble dietary fiber. These dietary fiber values are higher than those found in other vegetables/fruits (such as broccoli, apples) and oat bran. Dietary fiber helps to prevent diabetes, cardiovascular diseases, and cholesterol reduction, among others. The various physiological effects of fiber depend on its physical and chemical properties and on the sources, in which the fiber is found.[19,53,55] Dietary fiber of CS is mainly composed of cellulose (17.8–23.8% d.b.), hemicellulose (13.1–16.7% d.b.), lignin (28.2–30% d.b.), and total pectic substances (0.02% d.b.).[28,53,60]

Available carbohydrates are found in smaller amounts in CS, with sucrose, glucose, fructose, mannitol, and inositol being some of the main soluble carbohydrates reported by Toschi et al.[63] In addition, it is noteworthy that cellulose and hemicellulose are formed by glucose, fructose, xylose, galactose, mannose, and arabinose.[7,11] Therefore, extraction conditions influence the actual quantification of these components, since the degree of hydrolysis of native polysaccharides will affect the final content of available carbohydrates.[28,63]

High amounts of protein can also be found in CS (Table 11.1). In addition to quantity, protein quality is also an important parameter to be considered. Therefore, amino acid composition analysis, and the determination of total, protein, and nonprotein nitrogen can be used to measure these parameters.[25,45] Machado et al.[45] found that 25% of the CS nitrogen corresponded to the nonprotein fraction. In this way, only the protein nitrogen was considered for the protein content calculation, in order not to overestimate the real value. The actual protein value was found to be 12% on wet weight basis. Regarding the amino acid profile, CS presents all essential amino acids in free form, with the exception of methionine. Regarding total amino acids, aspartic acid (10.2 mg/g) and glutamic acid (9.2 mg/g) were the main amino acids in CS, followed by valine, isoleucine, and leucine (5–8 mg/g), as well as arginine and proline (~5 mg/g).[28]

The mineral profile of CS is composed of several mineral elements (major and trace), such as potassium (2.11–5 g/100 g d.b.), calcium (0.5–0.94 g/100 g d.b.), magnesium (0.31–2 g/100 g d.b.), sulfur (0.28 g/100 g d.b.), phosphorus (0.12 g/100 g d.b.), iron (0.0418–0.084 g/100 g d.b.), and other minerals in quantities less than 0.047 g/100 g d.b., such as strontium, barium, and manganese, among others. Some of these minerals are important for human health, being considered essential, as they contribute to various functions in the human body, such as the regulation of physiological and metabolic functions.[10,24]

Toschi et al.[63] found that the extraction method and the type of solvent influenced the lipid profile of CS. According to these authors, soxhlet extraction using *n*-hexane was the most sustainable choice, with the CS lipid fraction being essentially composed of triacylglycerols (~48%), free fatty acids (~21%), esterified sterols (~15%), free sterols (~13%), and diacylglycerols (~4%). In addition, some authors have reported that the fatty acid profile of CS mostly presents saturated fatty acids (62–86%), followed by polyunsaturated (10–29%) and monounsaturated (5–10%), of which the main ones are palmitic acid (C16:0), linoleic acid (C18:2n6), arachidic acid (C20:0), behenic acid (C22:0), oleic acid (C18:1n9c), and linoleic acid (C18:2 n6c).[16,24,63]

## 11.3 BIOACTIVE COMPOUNDS IN COFFEE SILVER SKIN

The term "bioactive" is derived from words bio- and -active. Bio from Greek word "bios" refers to life and active from the Latin from "activus" refers to energetic, dynamic, full of energy, with energy or involves an activity, that is full of vigor. Therefore, bioactive compounds are substances that exert direct effects on living organism, on the activity or functioning of life.[29,62]

According to Vilas-Boas et al.,[65] there is still no consensus in the literature on a single definition of bioactive compounds. The definition commonly used is that "bioactive compounds are essential and nonessential compounds that occur in nature, are part of the food chain, and can be shown to have an effect on human health."[18] Essien et al.[32] also described that bioactive compounds "are sourced naturally from plants, algae, foods and co-products or synthetically produced and can interact with one or more components of living tissue resulting in a wide range of effects." These compounds are phytochemicals and also called "extranutritional *constituents that typically occur in small quantities in foods*."[59]

The interest in bioactive compounds is due to their numerous health benefits, such as antioxidant, hypocholesterolemic, anticancer, immunomodulatory, anti-inflammatory, antidiabetic, antihypertensive, oxidative stress and metabolic disorders reduction, inhibition of receptor activities, inhibition or induction of enzymes, induction, and inhibition of gene expression.[14,17,22,59]

Bioactive compounds include an extremely heterogeneous class of compounds, such as polyphenolic compounds, carotenoids, tocopherols, vitamins, phytosterols, organosulfur compounds, bioactive lipids, bioactive peptides, essential oils, and probiotics.[14,67] According to Vilas-Boas et al.,[65] these compounds "can enter the market as food additives to improve technological and sensory functions, as a supplement or being presented as a functional food."

Beyond their macro- and micronutrient contents, the CS is a rich natural source of several bioactive compounds (mainly antioxidants), such as phenolic compounds, caffeine, melanoidins, and vitamin E.[16,19,24,54]

Nzekoue et al.[54] quantified about 30 bioactive compounds that are present in CS extract, such as 2 alkaloids (caffeine and quinine) and 18 phenolic compounds (phenolic acids, flavonoids, and xanthones). The total content of CS bioactive compounds ranged from 1.56% to 4% w/w, according to the type of solvent used.

### 11.3.1 PHENOLIC COMPOUNDS

Phenolic compounds, also known as polyphenols, are defined as substances produced by the secondary metabolism of plants and are considered the most abundant natural antioxidants found in the plant kingdom, with around 10,000 different phenolic structures being isolated and identified in plants. Polyphenols are made up of one or more aromatic rings bonded (nonpolar part) to one or more hydroxyl groups (polar part), generating different chemical structures. Various classifications are proposed for these compounds. Generally, they can be divided into different groups: phenolic acids, flavonoids, tannins, stilbenes, and lignans, and each of these groups can be divided into other.[3,21,56]

The polyphenols present in plants provide pigmentation, astringency, and protection against UV light, parasites, and insects. In addition, they are of great interest for human health, as they have antioxidant, antimicrobial, anti-inflammatory, antiproliferative effects, being potential substances for the prevention of various chronic diseases, such as cardiovascular diseases, diabetes, and cancer, among others.[3,21,56] In the food industry, polyphenols have been used as natural food additives, such as antimicrobials, antioxidants, and flavorings.[68]

Polyphenols are the most studied of the bioactives and are present in various vegetable sources, such as fruits, vegetables, spices, grains, and beverages.[3,56] According to Albuquerque et al.,[3] food coproducts can also be considered a natural and cheap source for obtaining polyphenols, since these coproducts generally have no economic value and are wasted.

Coffee and its coproducts have relevant amounts of phenolic compounds, mainly chlorogenic acids (CGA).[20,24,57] CGAs are esters of trans-hydroxycinnamic acids and quinic acid, with over 300 CGA major, minor, and related compounds found in coffee and other vegetables. These compounds are considered the main antioxidants in the diet, due to the high consumption of coffee.[34] According to the chemical identity, number, and position of acyl residues, CGAs can be subclassified into caffeoylquinic acids (CQAs), dicaffeoylquinic acids (diCQA), feruloylquinic acids (FQAs), *p*-coumaroylquinic acids (pCoQA), and six mixed diesters of caffeoyl-feruloyl-quinic acids, where CQA, diCQA, FQA, and pCoQA are formed by three isomers.[30,33,35]

In the CS, the CGA content varies from 0.6% to 3% d.b. CQAs are the main CGAs found in this coproduct, where the highest amounts were for 5-CQAs (0.20–0.22%), 3-CQAs (0.15%), and 4-CQAs (0.08–0.10%), respectively. Therefore, this coproduct is considered a good source for the extraction of these bioactive compounds.[28] Bresciani et al.[20] found a CGA profile of CS similar to that of coffee brews. These authors identified three CQA isomers (74% of the total CGA detected), three FQA isomers (23%), and the two coumaroylquinic acids plus the two CQA lactones (3% of total hydroxycinnamates). The degree of roasting significantly influences the CGA content, since the roasting process contributes to isomerization and degradation reactions of these compounds, generating lactone.[24,33]

### *11.3.2 CAFFEINE*

Caffeine (also called 1,3,7-trimethylxanthine) is a purine derivative (methylxanthine) belonging to the alkaloid class and is widely known and consumed throughout the world. This substance has a bitter taste and is present in small amounts in the composition of various foods, such as coffee, cocoa, tea, and yerba mate.[5,13]

In coffee beans, the caffeine content can range from 0.9% to 2% d.b. (dry basis). Caffeine has been extensively researched and used in the food, pharmaceutical, and cosmetic industries due to its stimulating effect, which enhances concentration capacity and counteracts tiredness, and for other purposes.[14] In addition, caffeine has other beneficial properties, such as antioxidant potential, immune modulator; and beneficial effects on various diseases (such as Parkinson's disease, and Alzheimer's disease, among others). On the other hand, some studies describe that excessive caffeine consumption can generate undesirable effects, such as anxiety, malabsorption of some nutrients, insomnia, and dependence.[44]

Studies have highlighted that coffee coproducts can be considered a good natural source for caffeine extraction, as they present significant amounts of this substance, as is the case with CS.[14] The caffeine content found in CS ranges from 0.17% to 1.4%, being influenced by the type of solvent, extraction method, and temperature used,[5,14,16,20,52] in addition to other factors previously described.

### *11.3.3 MELANOIDINS AND VITAMIN E*

Melanoidins are the end-products generated by the Maillard reaction. They are group of nitrogen-containing polymeric compounds with high molecular weight, brown-colored, with complex structure and high antioxidant activity. Their exact composition is yet unknown. Coffee and bakery products are the main sources of these compounds in the human diet.[27,31,49] In coffee, melanoidins are formed during the roasting process and represent about 25% of the total solids of the coffee brew.[27] CS can also be considered an important source of melanoidins.[26] Borrelli et al.[19] found that the amount of water-soluble melanoidins in CS was 4.5%. On the other hand, Mesías et al.[48] found values of 17–24% of melanoidins present in the CS extract. Melanoidins are formed by several compounds, mainly polysaccharides, proteins, and CGAs. The types and quantities found in coffee and its coproducts depend on the conditions adopted during roasting and the chemical composition of the beans.[28]

Vitamin E is an essential micronutrient for the body. It is considered the most important fat-soluble antioxidant in the cell, as it protects the cell membrane against the action of free radicals.[16,24] Furthermore, it presents modulatory effects in relation to gene expression, cellular pathways, and signal transduction.[64] Vitamin E is composed of eight different compounds, divided into two groups: tocopherol (α, β, ɣ, and δ) and tocotrienol (α, β, ɣ, and δ) and tocotrienol (α, β, ɣdifferent compounds, divided into two groups: tocopherol (α, β, ɣuty.[16,24] In recent years, research has identified an interesting vitamin E profile of CS, with a total content ranging from 3.94 to 16.79 mg/100 g.[16,24]

Bessada et al.[16] evaluated the tocopherol profile of CS from different geographic origins and found three tocopherols present (α, β, and ɣ). On the other hand, Costa et al.[24] also evaluated the vitamin E profile of CS and found a more complete profile with seven different vitamers (α-tocopherol, β-tocopherol, ɣ-tocopherol, δ-tocopherol, β-tocotrienol, ɣ-tocotrienol).

## 11.4 USES OF THE COFFEE SILVERSKIN

The rich composition of CS in fiber, protein, minerals, and bioactive compounds justifies its potential use for application in different fields, such as the food, pharmaceutical, and cosmetic industries.[5,23] This coproduct has been used as a coloring agent, source of dietary fibers and antioxidants,[7,10] and application in some products to add value, such as yogurt,[15] cookies,[38] cakes,[8,9] biscuits,[37] beverages,[46] and breads.[55]

Some studies have also highlighted other properties of CS, such as a prebiotic, stimulating the selective growth of intestinal microflora,[19] a production source for fructooligosaccharides and fructofuranosidases by fermentation,[51] anticarcinogenic, anticariogenic, anti-inflammatory, and antimicrobial effects,[26,47] hyaluronidase inhibitory activity,[36] and an antiphotoaging effect in human skin cells by isolation of atractyligenin.[66] Therefore, this coproduct can be considered as a potential functional ingredient,[7] which can be used for direct incorporation into some foods or subjected to an extraction process so that its properties can be used separately.

## 11.5 GENERAL CONSIDERATIONS

The valorization of CS is very important, as this coproduct represents a rich natural source of several bioactive compounds with beneficial properties to health. In addition, the reuse of CS reduces or eliminates impacts on the environment. Due to the large volumes of coproducts generated by coffee processing, efficient alternatives must be studied for better reuse, such as direct application of CS in a product to add nutritional value or extraction of its compounds for later application.

## 11.6 SUMMARY

In recent years, large amounts of coffee coproducts have been generated. This is due to the increase in world production and consumption of coffee, especially in relation to specialty coffees due to increased market appreciation. CS is a thin integument that involves the raw coffee bean, being the main coproduct generated in the coffee roasting process. This coproduct has a high content of proteins, minerals, and fibers, as well as several other bioactive compounds, especially antioxidants, such as phenolic compounds, caffeine, melanoidins, and vitamin E. The direct application of CS or the extraction of its compounds has been suggested as a potential way of valuing this coproduct, since the compounds present in CS contribute to several health benefits. In addition, the low cost of CS and the lower environmental impact that will be generated further justify the appreciation of this coproduct.

## KEYWORDS

- **bioactive compounds**
- **coffee**
- **coffee coproducts**
- **coffee silverskin**
- **valorization**

## REFERENCES


1. ABIC (Associação Brasileira da Indústria do Café: Brazilian Association of Coffee Industry), 2021. https://www.abic.com.br/ (accessed on Aug 30, 2021).
2. Açıkalın, B.; Sanlier, N. Coffee and its Effects on the Immune System. *Trends Food Sci. Technol.* **2021,** *114*, 625–632.
3. Albuquerque, B. R.; Heleno, S. A.; Oliveira, M. B. P. P.; Barros, L.; Ferreira, I. C. F. R. Phenolic Compounds: Current Industrial Applications, Limitations and Future Challenges. *Food Funct.* **2021,** *12*, 14–29.
4. Alghooneh, A.; Mohammad Amini, A.; Behrouzian, F.; Razavi, S. M. A. Characterisation of Cellulose from Coffee Silverskin. *Int. J. Food Prop.* **2017,** *20* (11), 2830–2843.
5. Alves, R. C.; Rodrigues, F.; Antónia Nunes, M.; Vinha, A. F.; Oliveira, M. B. P. P. State of the Art in Coffee Processing By-Products. In *Handbook of Coffee Processing By-Products: Sustainable Applications*; Galanakis, C. M., Ed.; Academic Press, Elsevier: Boston, 2017; pp 1–26.
6. Ameca-Veneroso, C.; Sánchez-Arellano, L.; Ramón-Canul, L. G.; Herrera-Corredor, J. A.; Cuervo-Osorio, V. D.; Quetz-Aguirre, E. M.; Rodríguez-Miranda, J.; Cabal-Prieto, A.; Ramírez-Rivera, E. J. Modified Version of the Sensory Pivot Technique as a Possible Tool for the Analysis of Food Adulteration: A Case of Coffee. *J. Sens. Stud.* **2021,** *2021*, 1–10, E-article 12705.
7. Arya, S. S.; Venkatram, R.; More, P. R.; Vijayan, P. The Wastes of Coffee Bean Processing for Utilization in Food: A Review. *J. Food Sci. Technol.* **2021,** 1–16.
8. Ateş, G.; Elmacı, Y. Coffee Silverskin as Fat Replacer in Cake Formulations and its Effects on Physical, Chemical and Sensory Attributes of Cakes. *LWT-Food Sci. Technol.* **2018,** *90*, 519–525.
9. Ateş, G.; Elmacı, Y. Physical, Chemical and Sensory Characteristics of Fiber-Enriched Cakes Prepared with Coffee Silverskin as Wheat Flour Substitution. *J. Food Measure. Character.* **2019,** *13*, 755–763.
10. Ballesteros, L. F.; Teixeira, J. A.; Mussatto, S. I. Chemical, Functional, and Structural Properties of Spent Coffee Grounds and Coffee Silverskin. *Food Bioprocess Technol.* **2014,** *7*, 3493–3503.
11. Ballesteros, L. F.; Teixeira, J. A.; Mussatto, S. I. Selection of the Solvent and Extraction Conditions for Maximum Recovery of Antioxidant Phenolic Compounds from Coffee Silverskin. *Food Bioprocess Technol.* **2014,** *7*, 1322–1332.
12. Barbosa-Pereira, L.; Guglielmetti, A.; Zeppa, G. Pulsed Electric Field Assisted Extraction of Bioactive Compounds from Cocoa Bean Shell and Coffee Silverskin. *Food Bioprocess Technol.* **2018,** *11*, 818–835.
13. Bartella, L.; Di Donna, L.; Napoli, A.; Siciliano, C.; Sindona, G.; Mazzotti, F. Rapid Method for the Assay of Methylxanthines Alkaloids: Theobromine, Theophylline and Caffeine, in Cocoa Products and Drugs by Paper Spray Tandem Mass Spectrometry. *Food Chem.* **2019,** *278*, 261–266.
14. Belščak-Cvitanović, A.; Komes, D. Extraction and Formulation of Bioactive Compounds. In *Handbook of Coffee Processing By-Products—Sustainable Applications*; Galanakis, C. M., Ed.; Elsevier: Boston, 2017; pp 93–140.
15. Bertolino, M.; Barbosa-Pereira, L.; Ghirardello, D.; Botta, C.; Rolle, L.; Guglielmetti, A.; Borotto-Dalla, V. S.; Zeppa, G. Coffee Silverskin as Nutraceutical Ingredient in

Yogurt: Its Effect on Functional Properties and Its Bioaccessibility. *J. Sci. Food Agric.* **2019,** *99*, 4267–4275.
16. Bessada, S. M. F., Alves, R. C., Costa, A. S. G., Nunes, M. A., & Oliveira, M. B. P. P. Coffea Canephora Silverskin from Different Geographical Origins: A Comparative Study. *Sci. Total Environ.* **2018,** *45*, 1021–1028.
17. Bhushan, I.; Sharma, M.; Mehta, M.; Badyal, S.; Sharma, V.; Sharma, I.; Singh, H.; Sistla, S. Bioactive Compounds and Probiotics: A Ray of Hope in COVID-19 management. *Food Sci. Human Wellness* **2021,** *10*, 131–140.
18. Biesalski, H.-K.; Dragsted, L. O.; Elmadfa, I.; Grossklaus, R.; Müller, M.; Schrenk, D.; Walter, P.; Weber, P. Bioactive Compounds: Definition and Assessment of Activity. *Nutrition* **2009,** *25*, 1202–1205.
19. Borrelli, R. C.; Esposito, F.; Napolitano, A.; Ritieni, A.; Fogliano, V. Characterization of a New Potential Functional Ingredient: Coffee Silverskin. *J. Agric. Food Chem.* **2004,** *52*, 1338–1343.
20. Bresciani, L.; Calani, L.; Bruni, R.; Brighenti, F.; Del Rio, D. Phenolic Composition, Caffeine Content and Antioxidant Capacity of Coffee Silverskin. *Food Res. Int.* **2014,** *61*, 196–201.
21. Chiorcea-Paquim, A.; Enache, T. A.; De Souza Gil, E.; Oliveira-Brett, A. M. Natural Phenolic Antioxidants Electrochemistry: Towards a New Food Science Methodology. *Compreh. Rev. Food Sci. Food Saf.* **2020,** *19*, 1680–1726.
22. Coelho, M. S.; Fernandes, S. S.; Salas-Mellado, M. de las M. Association Between Diet, Health, and the Presence if Bioactive Compounds in Foods. In *Bioactive Compounds: Health Benefits and Potential Applications*; Campos, M. R. S. (Ed.); Woodhead Publishing: London, 2019; pp 159–183.
23. Costa, A. S. G.; Alves, R. C.; Vinha, A. F.; Barreira, S. V. P.; Nunes, M. A.; Cunha, L. M.; Oliveira, M. B. P. P. Optimization of Antioxidants Extraction from Coffee Silverskin, a Roasting By-Product, Having in View a Sustainable Process. *Ind. Crops Product.* **2014,** *53*, 350–357.
24. Costa, A. S. G.; Alves, R. C.; Vinha, A. F.; Costa, E.; Costa, C. S. G.; Nunes, M. A.; Almeida, A. A.; Santos-Silva, A.; Oliveira, M. B. P. P. Nutritional, Chemical and Antioxidant/Pro-Oxidant Profiles of Silverskin: A Coffee Roasting By-Product. *Food Chem.* **2018,** *267*, 28–35.
25. Damodaran, S.; Parkin, K. L. *Química de Alimentos de Fennema* (Fennema Food Chemistry), 5th ed.; Artmed: Porto Alegre, 2019; p 1120.
26. De La Cruz, S. T.; Iriondo-DeHond, A.; Herrera, T.; Lopez-Tofiño, Y.; Galvez-Robleño, C.; Prodanov, M.; Velazquez-Escobar, F.; Abalo, R.; Del Castillo, M. D. An Assessment of the Bioactivity of Coffee Silverskin Melanoidins. *Foods* **2019,** *8*, 1–20.
27. De Peña, M. P.; Ludwig, I. A.; Cid, C. Beverage Preparation. In *Coffee: Production, Quality and Chemistry*; Farah, A., Ed.; Royal Society of Chemistry: Cambridge, 2019; pp 272–291.
28. Del Castillo, M. D.; Fernandez-Gomez, B.; Martinez-Saez, N.; Iriondo-DeHond, A.; Mesa, M. D. Coffee By-products. In *Coffee: Production, Quality and Chemistry*; Farah, A., Ed.; Royal Society of Chemistry: Cambridge, 2019; pp 309–334.
29. Dhaval, A.; Yadav, N.; Purwar, S. Potential Applications of Food Derived Bioactive Peptides in Management of Health. *Int. J. Peptide Res. Therap.* **2016,** *22*, 377–398.
30. Dorsey, B. M.; Jones, M. A. Healthy Components of Coffee Processing By-products. In: *Handbook of Coffee Processing By-Products—Sustainable Applications*; Galanakis, C. M., Ed.; Academic Press, Elsevier: London, 2017; pp 27–62.

31. Esquivel, P.; Jiménez, V. M. Functional Properties of Coffee and Coffee By-Products. *Food Res. Int.* **2012,** *46*, 488–495.
32. Essien, S. O.; Young, B.; Baroutian, S. Recent Advances in Subcritical Water and Supercritical Carbon Dioxide Extraction of Bioactive Compounds from Plant Materials. *Trends Food Sci. Technol.* **2020,** *97*, 156–169.
33. Farah, A.; de Paulis, T.; Trugo, L. C.; Martin, P. R. Effect of Roasting on Formation of Chlorogenic Acid Lactones in Coffee. *J. Agric. Food Chem.* **2005,** *53*, 1505–1513.
34. Farah, A.; de Paula-Lima, J. Consumption of Chlorogenic Acids Through Coffee and Health Implications. *Beverages* **2019,** *5*, 11.
35. Farah, A.; Donangelo, C. M. Phenolic Compounds in Coffee. *Braz. J. Plant Physiol.* **2006,** *18,* 23–36.
36. Furusawa, M.; Narita, Y.; Iwai, K.; Fukunaga, T.; Nakagiri, O. Inhibitory Effect of a Hot Water Extract of Coffee Silverskin on Hyaluronidase. *Biosci. Biotechnol. Biochem.* **2011,** *75* (6), 1205–1207.
37. Garcia-Serna, E.; Martinez-Saez, N.; Mesias, M.; Morales, F. J.; Castillo, M. D. Del. Use of Coffee Silverskin and Stevia to Improve Formulation of Biscuits. *Polish J. Food Nutr. Sci.* **2014,** *64* (4), 243–251.
38. Gocmen, D.; Sahan, Y.; Yildiz, E.; Coskun, M.; Aroufai, İ. A. Use of Coffee Silverskin to Improve Functional Properties of Cookies. *J Food Sci. Technol.* **2019,** *56* (6), 2979–2988.
39. Gokcen, B. B.; Sanlier, N. Coffee Consumption and Disease Correlations. *Crit. Rev. Food Sci. Nutr.* **2017,** *59* (2), 336–348.
40. ICO (International Coffee Organization). *Coffee Development Report*; 2020. https://5aa6088a-da13-41c1-b8ad-b2244f737dfa.filesusr.com/ugd/38d76b_4fc7b54a15f14a548b2f4a208c2eae6d.pdf (accessed on Aug 30, 2021).
41. ICO (International Coffee Organization). Coffee Market Report; July 2021; International Coffee Organization: London. https://www.ico.org/documents/cy2020-21/cmr-0721-e.pdf (accessed on Aug 30, 2021).
42. Iriondo-DeHond, A.; Aparicio García, N.; Fernandez-Gomez, B.; Guisantes-Batan, E.; Velázquez Escobar, F.; Blanch, G. P.; San Andres, M. I.; Sanchez-Fortun, S.; del Castillo, M. D. Validation of Coffee By-Products as Novel Food Ingredients. *Innov. Food Sci. Emerg. Technol.* **2019,** *51*, 194–204.
43. Khochapong, W.; Ketnawa, S.; Ogawa, Y.; Punbusayakul, N. Effect of In Vitro Digestion on Bioactive Compounds, Antioxidant and Antimicrobial Activities of Coffee (*Coffea Arabica* L.) Pulp Aqueous Extract. *Food Chem.* **2021,** *348*, E-article 129094.
44. Lima, J. de P.; Farah, A. Potential Negative Effects of Caffeine Consumption on Health. In: *Coffee: Consumption and Health Implications*; Farah, A., Ed.; Royal Society of Chemistry: London, 2019; pp 489–508.
45. Machado, S.; Costa, A. S. G.; Pimentel, B., F.; Oliveira, M. B. P. P.; Alves, R. C. Study on the Protein Fraction of Coffee Silverskin: Protein/Non-Protein Nitrogen and Free and Total Amino Acid Profiles. *Food Chem.* **2020,** *326*, E-article 126940.
46. Martinez-Saez, N.; Ullate, M.; Martin-Cabrejas, M. A.; Martorell, P.; Genovés, S.; Ramon, D.; Del Castillo, M. D. Novel Antioxidant Beverage for Body Weight Control Based on Coffee Silverskin. *Food Chem.* **2014,** *150*, 227–234.
47. Mesías, M.; Delgado-Andrade, C. Melanoidins as a Potential Functional Food Ingredient. *Curr. Opin. Food Sci.* **2017,** *14*, 37–42.
48. Mesías, M.; Navarro, M.; Martínez-Saez, N.; Ullate, M.; del Castillo, M. D.; Morales, F. J. Antiglycative and Carbonyl Trapping Properties of the Water Soluble Fraction of Coffee Silverskin. *Food Res. Int.* **2014,** *62*, 1120–1126.

49. Moreira, A. S. P.; Nunes, F. M.; Domingues, M. R.; Coimbra, M. A. Coffee Melanoidins: Structures, Mechanisms of Formation and Potential Health Impacts. *Food Funct.* **2012,** *3*, 903–910.
50. Mussato, S. I.; Machado, E. M. S.; Martins, S.; Teixeira, J. A. Production, Composition, and Application of Coffee and its Industrial Residues. *Food Bioprocess Technol.* **2011,** *4*, 661–672.
51. Mussatto, S. I.; Teixeira, J. A. Increase in the Fructooligosaccharides Yield and Productivity by Solid-State Fermentation with *Aspergillus Japonicus* Using Agro-Industrial Residues as Support and Nutrient Source. *Biochem. Eng. J.* **2010,** *53*, 154–157.
52. Narita, Y.; Inouye, K. High Antioxidant Activity of Coffee Silverskin Extracts Obtained by the Treatment of Coffee Silverskin with Subcritical Water. *Food Chemistry*, **2012,** *135*, 943–949.
53. Narita, Y.; Inouye, K. Review on Utilization and Composition of Coffee Silverskin. *Food Res. Int.* **2014,** *61*, 16–22.
54. Nzekoue, F. K.; Angeloni, S.; Navarini, L.; Angeloni, C.; Freschi, M.; Hrelia, S.; Vitali, L. A.; Sagratini, G.; Vittori, S.; Caprioli, G. Coffee Silverskin Extracts: Quantification of 30 Bioactive Compounds by a New HPLC-MS/MS Method and Evaluation of Their Antioxidant and Antibacterial Activities. *Food Res. Int.* **2020,** *133*, E-article 109128.
55. Pourfarzad, A.; Mahdavian-Mehr, H.; Sedaghat, N. Coffee Silverskin as a Source of Dietary Fiber in Bread-Making: Optimization of Chemical Treatment Using Response Surface Methodology. *LWT-Food Sci. Technol.* **2013,** *50*, 599–606.
56. Rashmi, H. B.; Negi, P. S. Phenolic Acids from Vegetables: A Review on Processing Stability and Health Benefits. *Food Res. Int.* **2020,** *136*, E-article 109298.
57. Regazzoni, L.; Saligari, F.; Marinello, C.; Rossoni, G.; Aldini, G.; Carini, M.; Orioli, M. Coffee Silverskin as a Source of Polyphenols: High Resolution Mass Spectrometric Profiling of Components and Antioxidant Activity. *J. Funct. Foods* **2016,** *20*, 472–485.
58. Sales, A. L.; Miguel, M. A. L.; Farah, A. Potential Prebiotic Effect of Coffee. In *Coffee: Consumption and Health Implications*; Farah, A., Ed.; Royal Society of Chemistry: Cambridge, 2019; pp 286–312.
59. Santos, D. I.; Saraiva, J. M. A.; Vicente, A. A.; Moldão-Martins, M. Methods for Determining Bioavailability and Bioaccessibility of Bioactive Compounds and Nutrients. In *Innovative Thermal and Non-Thermal Processing, Bioaccessibility and Bioavailability of Nutrients and Bioactive Compounds*; Barba, F. J., Saraiva, J. M. A., Cravotto, G., Lorenzo, J. M., Eds.; Woodhead Publishing, Elsevier: London, 2019; pp 23–54.
60. Santos, É. M. dos; Macedo, L. M. de; Tundisi, L. L.; Ataide, J. A.; Camargo, G. A.; Alves, R. C.; Oliveira, M. B. P. P.; Mazzola, P. G. Coffee By-products in Topical Formulations: A Review. *Trends Food Sci. Technol.* **2021,** *111*, 280–291.
61. Scholz, M. B. dos S.; Kitzberger, C. S. G.; Durand, N.; Rakocevic, M. From the Field to Coffee Cup: Impact of Planting Design on Chlorogenic Acid Isomers and Other Compounds in Coffee Beans and Sensory Attributes of Coffee Beverage. *Eur. Food Res. Technol.* **2018,** *244*, 1793–1802.
62. Shrinet, K.; Singh, R. K.; Chaurasia, A. K.; Tripathi, A.; Kumar, A. Bioactive Compounds and Their Future Therapeutic Applications. In *Natural Bioactive Compounds: Technological Advancements*; Sinha, R. P., Häder, D. P., Eds.; Academic Press, Elsevier: London, 2021; pp 337–362.
63. Toschi, T. G.; Cardenia, V.; Bonaga, G.; Mandrioli, M.; Rodriguez-Estrada, M. T. Coffee silverskin: Characterization, Possible Uses, and Safety Aspects. *J. Agric. Food Chem.* **2014,** *62*, 10836–10844.

64. Ungurianu, A.; Zanfirescu, A.; Nițulescu, G.; Margină, D. Vitamin E Beyond Its Antioxidant Label. *Antioxidants* **2021,** *10*, 634–640.
65. Vilas-Boas, A. A.; Pintado, M.; Oliveira, A. L. S. Natural Bioactive Compounds from Food Waste: Toxicity and Safety Concerns. *Foods* **2021,** *10*, 1564–1570.
66. Xuan, S. H.; Lee, N. H.; Park, S. N. Atractyligenin, a Terpenoid Isolated from Coffee Silverskin, Inhibits Cutaneous Photoaging. *J. Photochem. Photobiol. B Biol.* **2019,** *194*, 166–173.
67. Yun, P.; Devahastin, S.; Chiewchan, N. Microstructures of Encapsulates and Their Relations with Encapsulation Efficiency and Controlled Release of Bioactive Constituents: A Review. *Comprehen. Rev. Food Sci. Food Saf.* **2021,** *20*, 1768–1799.
68. Zeb, A. Concept, Mechanism, and Applications of Phenolic Antioxidants in Foods. *J. Food Biochem.* **2020,** *44* (9), 1–22.

# CHAPTER 12

# SPICE BIOACTIVE COMPOUNDS VERSUS LIFESTYLE DISORDERS

UMA MAHESHWARI SRINIVASA and MADENENI MADHAVA NAIDU

## ABSTRACT

Spices play a significant role in food flavoring and as adjuncts in traditional medicine. Recent developments have shown that the spice extracts and their bioactive molecules play a vital part in either preventing, ameliorating, or curing different diseases or disorders. Lifestyle diseases have increased globally and are a significant public health challenge to every country. Cardiovascular diseases and cancer, the first and second leading cause of death, respectively, and diabetes and obesity, are on par through the exponential increase in the incidents. Though pharmaceuticals strive hard to be active and efficient, their synergistic effect makes them partially unreliable. This surges the need for natural therapeutic medicines to outmaneuver the existing global health burden. Hence, nutrition-based medicine is the present vogue used to target the pathogeny and their complications. Here we illustrate the most critical investigations which have been validated to be potential in combating these complications through rhizome spice (turmeric and ginger), seed spice (fenugreek), seed/leaf spice (coriander), and their bioactive components. Drug safety is also the primary concern of medical therapy; hence the usage is recommended in the safety range without surpassing the toxicity levels. Nutraceuticals and functional foods are a promising source of natural medicines for controlling various health complications used in proper concentrations and methods. In this chapter, selective spices and their effects and the mechanisms of action taken by them to circumvent the significant lifestyle diseases have been discussed.

Plant-Based Bioactive Compounds and Food Ingredients: Encapsulation, Functional, and Safety Aspects.
Junaid Ahmad Malik, Megh R. Goyal, Preeti Birwal, & Ritesh B. Watharkar (Eds.)

## 12.1 INTRODUCTION

Nutraceutical and functional foods are the new trends that provide health benefits and serve as an alternative to modern medicine. They constitute dietary supplements, nutrients, herbals, and bioactive compounds and are instrumental in balancing health and work against various diseases and disorders, improving the quality of life. Ayurveda is the branch of medicine based on the everyday use of nutraceutical and functional foods in the Indian medicine system. Likewise, it is traditional medicine in China. Both the system includes medicinal herbs, spices, and tree species as preventive and protective medications. A nutraceutical provides physiological benefit as well as protection against lifestyle disease.[27] In recent years, metabolic syndrome is critical due to the socioeconomic burden caused by disease complications.[26] It is a network of metabolic disparities that increases the risks of various other co-morbidities.[67] Elevated plasma glucose and blood pressure, insulin resistance, and dyslipidemia are the standard features.[26]

Obesity, where the excess fat gets accumulated in various parts such as the abdomen, deteriorates the health condition.[18] This kind of metabolic disturbance also promotes cardiovascular risk, including acute coronary syndrome and acute myocardial infarction.[21] Cancer occurs through impaired cellular energy metabolism and extracts energy from oxidative phosphorylation or substrate-level phosphorylation and hence is also considered a metabolic disease.[53] Spices, though used widely as flavoring and coloring agents, are also recognized to possess medicinal properties. There has been much research on the chemistry and pharmacology of their bioactive compounds and investigations against various metabolic and lifestyle diseases. Their effects have potential therapeutic applications against disease and disorders with beneficial physiological effects.[58]

This chapter explores the potentiality of selected spices such as turmeric, ginger, fenugreek, coriander, and their intervention against lifestyle diseases such as diabetes, obesity, cancer, cardiovascular diseases, and safety.

## 12.2 SELECTED SPICES AND THEIR BIOACTIVE COMPOUNDS

Spices can be any plant part such as fruit, root, leaf, bark that are used as adjunct in food preparations. They host active principles that can modulate molecular markers that relate to their disease prevention or cure. The selected spices mentioned in this article are turmeric, ginger, fenugreek, and coriander (Figure 12.1).

**FIGURE 12.1** Selected spices: turmeric rhizomes (A), ginger rhizomes (B), fenugreek seeds (C), and coriander seeds (D).

Apart from their popularity in various cuisines, food and food formulations, spices like turmeric and fenugreek are widely used in cosmetology.[1] The most beautiful part is that these spices and their bioactive compounds can mitigate many diseases, disorders, irregular, and improper conditions. The significant properties that promote disease mitigation are antioxidant, antimicrobial, antithrombotic, antilithogenic, antimutagenic, and anti-inflammatory properties immuno-modulatory, would healing, digestive stimulative, and RBC protective effects.[58]

Any spice's total extract contains various active compounds that provide a synergistic effect in acting against any disorder. The details of the selected spice and their bioactive compounds are listed in Table 12.1.

**TABLE 12.1** Details of Spices and Their Bioactive Compounds.

| Spice | Scientific name | Family | Plant part | Bioactive compounds |
|---|---|---|---|---|
| Coriander[55] | *Coriandrum sativum* | Apiaceae | Seed, leaves | Rutin, linalool, geranyl acetate and γ-terpinene |
| Fenugreek[17] | *Trigonella foenum-graecum* L., | Fabaceae | Seed | Diosgenin, 4-hydroxyisoleucine, trigonelline, galactomannan and polyphenols |
| Ginger[59] | *Zingiber officinale* | Zingiberaceae | Rhizome | Gingerols, shaogals, paradols, galanals A and B, and diterpene compounds |
| Turmeric[33] | *Curcuma longa* | Zingiberaceae | Rhizome | Curcumin curcuminoids, glycosides, terpenoids, and flavonoids |

## 12.3 SELECTED SPICES: LIFESTYLE DISEASES AND NUTRACEUTICAL ACTIVITIES

Currently, nutraceuticals have gained popularity due to their natural origin, easy availability and as they are generally considered safe. They either include powders, whole extract or the pure compound. Food-based nutraceuticals are mostly obtained from grains, pulses, spices, dairy, egg, fermented products, etc. Molecular mechanisms of these spices against lifestyle diseases are shown in Table 12.2.

## 12.4 DOCUMENTATION BASED ON RESEARCH STUDIES ON SPICES

### *12.4.1 TURMERIC*

Turmeric and its compounds, especially curcumin, are well known for their health beneficial effects. Curcumin and its isomers protect pancreatic

**TABLE 12.2** Mechanism of Action of Selected Spices and Molecular Pathways of Action to Combat Metabolic Diseases.

| Disease condition | Spices and their mechanisms of action | Bioactive compounds responsible for the activity |
|---|---|---|
| **Turmeric** | | |
| Diabetic[10,16,32,33] | ↑ Insulin production and β-cell function | Curcumin (diarylheptanoid) |
| | ↑ Antioxidant defense system | |
| | ↑ Homeostasis model assessment of β-cell function (HOMA-β) | |
| | ↓ C-peptide levels | |
| | ↓ Lipid peroxidation | Curcuminoids |
| | ↓ Human Pancreatic Amylase (HPA) enzyme | Isopropanol and acetone extract of *Curcuma longa* |
| Hypolipidemic[23,70] | ↓ Oxidative stress | Curcumin |
| | ↓ Hepatic enzymes 3-hydroxy-3-methylglutaryl-CoA (HMG-CoA) reductase and Acyl-CoA cholesterin acyltransferase (ACAT) | |
| | ↓ Hepatic cholesterol, total cholesterol (TC), non-high-density lipoprotein cholesterol (HDL)-C levels | |
| | ↓ Hepatic fatty acid synthase (FAS) activity | |
| | ↑ β-oxidation of fatty acids | |
| Cancer[14,33,66] | ↓ Interleukins- 6 (IL-6), IL-8, Tumor necrosis factor (TNF)-α and inhibitory kappa β (Iκβ) *kinase* β (IKKβ kinase) | Curcumin |
| | ↓ Nuclear factor kappa-light-chain-enhancer of activated B cells (NF-κB), cyclooxygenase (COX)-2 and Signal transducer and activator of transcription 3 (pSTAT3) | |
| CVD[69] | ↑ HDL-C | Curcuminoids |
| | ↓ Low-density lipoprotein Cholesterol (LDL-C) | |
| | ↓ Histone acetyltransferase p300 (p300-HAT) | |

**TABLE 12.2** *(Continued)*

| Disease condition | Spices and their mechanisms of action | Bioactive compounds responsible for the activity |
|---|---|---|
| **Ginger** | | |
| Diabetic[3,36,37,49,51] | ↑ Antioxidant enzyme capacity | 1–2% ginger powder |
| | ↓ Lipid peroxidation | Ginger extract |
| | ↑ Protects the pancreatic β-cells | |
| | ↑ Insulin synthesis | |
| | ↑ Insulin sensitivity | |
| | ↑ Glucose uptake by peripheral tissues | |
| | ↓ Oxidative stress | |
| | ↓ Advanced glycation end (AGE) products | Ethyl acetate extract of ginger; 05–3% ginger |
| | ↑ Glucose transporter (GLUT) 4 receptors expression | Gingerols (phenol); Ethyl acetate extract of ginger |
| Hypolipidemic[3,20,38] | ↓ Serum total cholesterol | Ginger oleoresin |
| | ↓ LDL-C | Zingiber cfficinale |
| | ↓ Triacylglycerides (TG) | |
| | ↑ TC/HDL-C ratio | |
| | ↓ HMG-CoA reductase | |
| | ↑ LDL-receptors | |
| | ↑ Peroxisome proliferator-activated receptor (PPAR)-γ signaling pathway in adipocytes | Ginger extract; 6-gingerol; 6-shagaol (phenol) |
| | ↑ Fat catabolism | |
| | ↑ PPAR-δ pathway | |

**TABLE 12.2** *(Continued)*

| Disease condition | Spices and their mechanisms of action | Bioactive compounds responsible for the activity |
|---|---|---|
| | ↓ Reduced sterol regulatory element-binding transcription factor 1 (SREBP-1), PPAR-γ and CCAAT enhancer-binding proteins Homologous protein (c/EBP)-α | |
| Cancer[8,54] | ↑ Reactive oxygen species (ROS) generation | Zerumbone (sesqui-terpenoid and cyclic ketone), ginger extract |
| | ↑ Antioxidant *N*-acetyl cysteine attenuated cell death | |
| | ↓ Tumor growth | |
| | ↑ Caspases 8, 9, 3, and 7 and cleavage of poly ADP-ribose polymerase (PARP) | |
| | ↓ Extracellular-signal-regulated kinase (ERK1/2)/c-Jun N-terminal kinase (JNK)/(Activator protein 1) AP-1 pathway | |
| | ↓ NF-kB, TNF-α, COX-2, B-cell lymphoma 2 (Bcl-2), caspases | |
| | ↑ G2/M cell cycle arrest | |
| CVD[59] | ↑ Hypocholesterolemic property | Ginger extract |
| | ↓ Cardiac markers of infraction such as troponin | |
| | ↓ Enzyme activities of creatine kinase-MB isoenzyme, lactate dehydrogenase (LDH), alanine transaminase (ALT) | |
| | ↑ Antioxidant enzyme activity | |
| | ↑ Cell membrane integrity | |
| **Fenugreek** | | |
| Diabetic[17,31,39,61] | ↑ β-cell regeneration | Trigonelline (alkaloid) |
| | ↑ Blood lipid profile | 4-hydroxyisoleucine (non-protein aminoacid); Diosgenin (phytosteroid sapogenin), antihyperglycemic |

**TABLE 12.2** *(Continued)*

| Disease condition | Spices and their mechanisms of action | Bioactive compounds responsible for the activity |
|---|---|---|
| | | compound named GII purified from the water extract |
| | ↓ Hemoglobin A1c (HbA1c) | 30 g fenugreek seed powder |
| | ↑ Renal protection | 4-hydroxyisoleucine; Trigonelline |
| | ↑ Antioxidant activity | Diosgenin; Trigonelline |
| Hypolipidemic[39,40,61,63] | ↓ TG, LDL-C and TC | 25 g fenugreek seed powder; Diosgenin |
| | ↓ Fat accumulation | |
| | ↓ Transactivation of liver X receptor (LXR) | Diosgenin |
| | ↑ Glutathione peroxidase (GPx) | Fenugreek extract |
| | ↑ Catalase (CAT) in serum | |
| Cancer[28,35] | ↓ Proto-oncogene tyrosine-protein kinase Src (c-Src), JAK1 and JAK2 activation | Diosgenin |
| | ↓ STAT3-mediated gene expression | |
| | ↓ Cell proliferation | Fenugreek seed extract |
| | ↑ Apoptosis, cell cycle arrest at the G2/M phase | |
| | ↑ Caspase 3, 8, 9, p53, FAS, Fas-associated protein with death domain (FADD), Bcl-2-associated X protein (Bax) | |
| | ↓ Transforming growth factor beta 1 (TGF-β) induced phosphorylation of protein kinase B (Akt) | |
| CVD[41,44] | ↓ Thiobarbituric acid reactive substances (TBARS) levels | Fenugreek |
| | ↑ Superoxide dismutase (SOD), CAT, GPx, and reduced glutathione (GSH) activities | |
| | ↑ Serum levels of cardiac markers | Trigonelline |

**TABLE 12.2** *(Continued)*

| Disease condition | Spices and their mechanisms of action | Bioactive compounds responsible for the activity |
|---|---|---|
| **Coriander** | | |
| Diabetic[12,55] | ↑ Plasma insulin | 10% coriander seed powder |
| | ↓ Blood glucose and glycated hemoglobin | |
| | ↑ Free radical-scavenging activity | |
| | ↑ Liver glycogen content | |
| | ↑ Liver hexokinase | |
| | ↑ Phosphoglucomutase | |
| | ↑ Glycogen synthase | |
| | ↓ Glycogen phosphorylase | |
| | ↑ Glycolysis and glycogenesis | |
| | ↓ Glycogenolysis and gluconeogenesis | |
| | ↓ TNF-α, IL-6, TGF-β1 and NF-Kb | Linalool (two enantiomers of a naturally occurring terpene alcohol) |
| Hypolipidemic[9,24,25,47,57,60] | ↑ AMP-activated protein kinase (AMPK) | 70% methanolic extract of *C. sativum,* coriander seed oil |
| | ↑ Fatty acid oxidation | |
| | ↓ Fatty acid synthesis | |
| | ↓ Lowering of blood pressure and LDL | |
| | ↓ TL, TC, TG, and LDL-C | |
| | ↑ HDL-C | |
| | ↓ HMG-CoAR | Linalool |
| | ↑ Enzyme ubiquitin-dependent proteolysis | |

**TABLE 12.2** *(Continued)*

| Disease condition | Spices and their mechanisms of action | Bioactive compounds responsible for the activity |
|---|---|---|
| | ↓ SREBP-2 | |
| | ↑ Insig-1 expression | |
| | ↑ PPAR-α | |
| Cancer[11,15,22,62] | ↑ SOD activity | Coriander root extract |
| | ↑ Toxic $H_2O_2$ that kills cancer cells | |
| | ↑ Cell cycle arrest at G2/M phase | Ethyl acetate extract of coriander roots |
| | ↓ SOD, CAT GSH, and ↑ ROS | |
| | ↑ Sub-G1cells | |
| | ↑ Liver function | Ethanolic extract of coriander leaves |
| | ↑ Bilirubin (total and conjugated) | |
| | ↑ Total protein, globulin and albumin | |
| | ↑ Caspase-8, -9 and -3 | |
| | ↑ Alkaline phosphatase (ALP), ALT, Aspartate transaminase (AST) | |
| | ↑ Nrf-2 and p21 | Linalool |
| CVD[55] | ↑ Hypolipidemic potential | Coriander seed extract |
| | ↑ Antioxidant property | |

β-cell, improves the homeostasis of β-cell and deteriorate C-peptide levels.[10] They promote strong antioxidant activity and diminish lipid peroxidation through SOD, CAT, and GLP level.[16] Turmeric extract inhibits HPA enzyme, which lowers blood glucose levels by reducing starch hydrolysis.[32] Curcumin reduces collagenase-2 and matrix metalloproteinase (MMP)-13/collagenase-3 in diabetes, through which wound healing is progressed.[71] Curcumin intake elevates HDL-C and demotes LDL-C.[70] It reduces oxidative stress in hyperlipidemic patients and presents immunomodulatory effect through modulation of IL-1β, IL-4, and vascular endothelial growth factor (VEGF).[18,50] HMG-CoA reductase and acyl CoA cholesteryl acyltransferase (ACAT) is downregulated by curcumin to lower liver cholesterol, TC, and cholesterol other than HDL-C.[23]

Curcumin enhances fatty acid β-oxidation and suppresses fatty acid synthesis by inhibiting FAS enzyme.[7] It also activates AMPK and phosphorylate acetyl CoA carboxylase (ACC).[23] Apo-A-1, HDL, and paraoxonase (PON) increase in the plasma due to curcumin intake.[4] Curcumin has anticancerous properties, proving it to be antiproliferative in different cancer types through NFκB inhibition and downregulation of different interleukins such as IL-6, IL-8, inflammatory markers such as COX-2, TNF-α, cyclins, nitric oxide synthase (NOS), c-myc, Bcl-2, IKKβ kinase, phosphorylated STAT3, and MMP-9. It also involves mutagenesis, angiogenesis, metastasis, cell cycle, apoptosis, and tumor growth.[14,66] Curcumin reduces myocardial infarction by lowering the C-reactive protein, malondialdehyde and N-terminal pro-B-type natriuretic peptide levels. Thus, its antioxidant property also protects against cardiac issues.[69]

### 12.4.2 GINGER

Ginger is a highly pungent and potential spice. Gingerols and shaogals are the important molecules present in ginger. Since the molecules are pancreatic β-cell-protective, they reduce blood glucose levels due to their modulatory effect on insulin synthesis and insulin release and are proved to be insulin sensitive, enhancing the glucose uptake. They also have the potential to inhibit enzymes involved in carbohydrate and lipid metabolism. They activate the antioxidant mechanism and downstream lipid peroxidation.[3,36,51] Diabetic complications have been alleviated through the delay of glycation and inhibition of AGE products due to the antioxidant property of ginger bioactive compounds.[49,51] Glucose transportation due to GLUT4 is considered a significant reason for the hypoglycemic effect of ginger.[37] With the intake of ginger, there was a decrease in TC, LDL-C, and TG, an increase in HDL, activation of LDL-receptors, and HMG-CoA reductase inhibition in diabetic rats.[3,20] Ginger suppresses both the cholesterol level and the accumulation of lipids due to the stimulation of pancreatic lipase. It also modulated PPAR-γ to exert beneficial effects against obesity.[38] Ginger-induced cell death in cancerous condition and enhanced ROS generation with an increase in *N*-acetyl cysteine, an antioxidant.[54] 6-Gingerol was shown to cause apoptotic cell death due to activation of caspases 8, 9, 3, and 7 and cleavage of PARP.[15] Inhibition of the ERK1/2/JNK/AP-1 pathway is seen in colon cancer. In gastrointestinal cancer, 6-shogaol also plays a chemopreventive role along with 6-gingerol due to modulation in NF-kB, TNF-α, COX-2, Bcl-2, caspases etc. 6-Shogaol, in particular, contributes to G2/M cell cycle arrest and cell death.[8] Cardiovascular problems are directly linked to excessive cholesterol; hence the hypocholesterolemic property of ginger alleviates cardiovascular issues too. Ginger has shown to downregulate related cardiac markers such as troponin, enzymes such as LDH, ALT, AST and creatine kinase-MB isoenzyme to prevent myocardial infarction. Ginger also improves the integrity of the cell membrane through its antioxidant activity.[59]

### 12.4.3 FENUGREEK

TC, LDL-C, TAG, phospholipids, FFAs, and HbA1c in diabetic and obese conditions were lowered by fenugreek seed extract.[39,40,61] It reduced insulin resistance and glucose resorption, enhanced insulin secretion and β-cell regeneration, promoted lipid profile, renal protection, and antioxidant activity.[31] Glucose homeostasis was balanced due to enzymes involved in pathways such as glycolysis, gluconeogenesis, and lipogenesis.[46] Sterol regulatory element-binding protein-1 (SREBP-1), PPAR-γ, and c/EBP-α expressions were reduced to obstruct fat accumulation along with the deterioration of mRNA and protein expression of SREBP-1, reducing TG, LDL-C, and cholesterol.[64] Adipocyte differentiation, lipid accumulation through the antiproliferative effect was also inhibited.[19] It shows hypolipidemic effect by increasing HMG-CoA reductase activity and further increasing the excretion of bile acids and neutral sterols.[48] Fenugreek extract showed cell cycle arrest at the G2/M phase.[52] Diosgenin inhibited LXR transactivation and suppressed c-Src, JAK1, and JAK2 activation through which the STAT3 signaling pathway is modulated to inhibit cell proliferation.[35,63] Caspase 3, 8, 9, p53, FAS, FADD, Bax, and Bak dependent were highly expressed to induce apoptosis.[28] There was an increase in enzymatic and nonenzymatic antioxidant activity and decreased thiobarbituric acid reactive substances (TBARS) level.[41] Cardiac marker LDH, glutamate pyruvate transaminase, and creatine kinase-MB levels in serum were increased, giving a positive effect, and the overexpression of stress protein was restored to regular showing cardioprotective effect.[44]

### 12.4.4 CORIANDER

A decrease in blood glucose and increase in plasma insulin has been reported due to the coriander's efficient antioxidant activity. It also enhances glycogen synthesis through an increase in glycogenolysis and gluconeogenesis.[55] Oxidative stress is avoided through diminishing inflammatory markers such as TNF-α, IL-6, TGF-β1, and NF-κB.[12] There are reports showing AMPK activation and fatty acid synthesis suppression.[60] An increase in PUFA and MUFA were seen with a decrease in SFA.[43] TL, TC, LDL-C decreased, and HDL-C increased.[24,47] HMG-CoAR and SREBP-2 inhibition is seen..[9,57] PPAR-γ activation has reduced TG.[25] Cell proliferation has been inhibited and has shown antitumor and immunomodulating effect. There is an increase in SOD that increases $H_2O_2$ that can kill cancer cells.[62] Liver function, ALP, ALT, AST, are restored, and protein, globulin, albumin, and bilirubin are increased.[15] There was activation of caspase-8, Nrf-2 and p21 and downregulation of caspase-9 and -3.[11,22]

## 12.5 SAFETY ASPECTS

Spices are mostly considered and regarded as safe. There are studies that report safety of spices even at higher intake of up to 100 times the normal.[58] Yet safe dose, tolerance, bioavailability, mutagenicity, allergy and toxicity should be evaluated before its usage in any medical applications. Overall, the spices are well tolerated and highly safe. Precautions are advised in terms of toxicity, allergy, and pregnancy. Safety details of turmeric, ginger, fenugreek, and coriander are shown in Table 12.3.

## 12.6 SUMMARY

With the increasing trend of a sedentary lifestyle, lack of nutritious food and adequate exercise, metabolic diseases tend to increase at a higher pace.

**TABLE 12.3** Safety Aspects of Turmeric, Ginger, Fenugreek and Coriander.

| Parameter | Turmeric[2] | Ginger[5,42,65] | Fenugreek[6,13,56] | Coriander[34] |
|---|---|---|---|---|
| Allergy | No | Dust from ginger may produce IgE-mediated allergy | Chickpea allergy and its possible cross-reactivity is seen Patients with bronchospasm are allergic to curry powder with fenugreek seeds | Itching/stinging of lips and mouth |
| Maximum tolerated dose | 500–12,000 mg of curcumin | 1000 mg/kg body weight | MTD (9.77 g/kg), $LD_{50}$ dose of 2000 mg/kg | AOT ($LD_{50}$) in rats was reported as 4.13 g/kg up to 1000 g/kg/day; $LC_{50}$ value of 23 μg/mL |
| Mutagenicity | No | Ginger extract, gingerol, shogaol were mutagenic. Zingerone was nonmutagenic | Non mutagenic | Coriander drop, ethanolic extract, linalool showed non-mutagenicity; 8 mg/mL aqueous extract showed mutagenicity in two strains |
| Pregnancy | Safe | Ginger tea-early embryo loss increases along with growth in surviving fetuses | Not recommended due to its potential uterine stimulating properties | No developmental toxicity in rats at maternal doses of up to 1000 mg/kg/day |
| Safe Dose | 6 g/day in human | 250 mg and 1 g rhizome powder for four times daily | 2 g/kg in mice and 5 g/kg in rats | 160 mg/kg was NOEL for male rats |
| Tolerance | Well tolerated | Considered safe | Absolutely safe and well tolerated | Recognized safe (FDA and FEMA) safe as food and food ingredient |
| Toxicity | Poly(N-isopropyl acrylamide) and dipeptide nanoparticle are safe | No maternal and developmental toxicity in pregnant rats | Not seen | Acute oral toxicity and acute dermal toxicity due to coriander oil |
| Other effects | – | Consumption of 6 g and above causes heartburn and gastric irritation | No change in feed efficiency ratio, severe wheezing, transient diarrhea, flatulence and dizziness | – |

Nutraceutical should be incorporated into our daily food routine to develop a preventive mechanism for any illness or disorder. Although most spices are considered a part of food, they are used for medicinal benefits for ages. Experimental evidence regarding nutraceutical activity of selected spices such as turmeric, ginger, fenugreek, and coriander in combating lifestyle diseases such as diabetes, obesity, cancer, and cardiovascular diseases are available. Having huge scope and benefit to mankind, numerous studies have been undertaken, and the mechanisms of actions of different spices and their bioactive compounds were established. This chapter pools up the available reports and their safety aspects. In general, the selective spices chosen are considered to be safe. Overall, spice extracts and their bioactive compounds are highly potential and efficient in curbing any metabolic disorders.

## ACKNOWLEDGMENTS

Authors acknowledge: Director of CSIR-CFTRI at Mysore, India, for his encouragement & guidance; Uma Maheshwari is thankful to DBT, New Delhi, for the Junior Research Fellowship.

## KEYWORDS

- **coriander**
- **fenugreek**
- **ginger**
- **lifestyle diseases**
- **safety**
- **spices**
- **turmeric**

## REFERENCES

1. Aburjai, T.; Natsheh, F. M. Plants Used in Cosmetics. *Phytother. Res.* **2003**, *17* (9), 987–1000.
2. Ahmad, R. S.; Hussain, M. B.; Sultan, M. T.; Arshad, M. S.; Waheed, M.; Shariati, M. A.; Plygun, S.; Hashempur, M. H. Biochemistry, Safety, Pharmacological Activities, and

Clinical Applications of Turmeric: A Mechanistic Review. *Evid.-Based Complement. Altern. Med.* **2020**, *2020* (article ID 7656919), 1–14.

3. Akash, M. S. H.; Rehman, K.; Tariq, M.; Chen, S. *Zingiber officinale* And Type 2 Diabetes Mellitus: Evidence from Experimental Studies. *Crit. Rev. Eukaryotic Gene Expression* **2015**, *25* (2), 91–112.
4. Alappat, L.; Awad, A. B. Curcumin and Obesity: Evidence and Mechanisms. *Nutr. Rev.* **2010**, *68* (12), 729–738.
5. Ali, B. H.; Blunden, G.; Tanira, M. O.; Nemmar, A. Some Phytochemical, Pharmacological and Toxicological Properties of Ginger (*Zingiber officinale* Roscoe): A Review of Recent Research. *Food Chem. Toxicol.* **2008**, *46* (2), 409–420.
6. Al-Yahya, A. A. Reproductive, Cytological and Biochemical Toxicity of Fenugreek in Male Swiss Albino Mice. *Afr. J. Pharm. Pharmacol.* **2013**, *7* (29), 2072–2080.
7. Asai, A.; Miyazawa T. Dietary Curcuminoids Prevent High Fat Diet Induced Lipid Accumulation in Rat Liver and Epididymal Adipose Tissue. *J. Nutr.* **2001**, *131*, 2932–2935.
8. Badr, O. M.; Sakr, S. A.; Abd-Eltawab, H. M. Ameliorative Effect of Ginger Extract Against Pathological Alterations Induced in Mice Bearing Solid Tumors. *J. Biosci. Appl. Res.* **2016**, *2*, 185–196.
9. Cho, S. Y.; Jun, H. J.; Lee, J. H.; Jia, Y.; Kim, K. H.; Lee, S. J. Linalool Reduces The Expression of 3-Hydroxy-3-Methylglutaryl CoA Reductase via Sterol Regulatory Element Binding Protein-2-And Ubiquitin-Dependent Mechanisms. *FEBS Lett.* **2011**, *585* (20), 3289–3296.
10. Chuengsamarn, S.; Rattanamongkolgul, S.; Luechapudiporn, R.; Phisalaphong, C.; Jirawatnotai, S. Curcumin Extract for Prevention of Type 2 Diabetes. *Diab. Care* **2012**, *35* (11), 2121–2127.
11. Day, T. W.; Huang, S.; Safa, A. R. c-FLIP Knockdown Induces Ligand-Independent DR5-, FADD-, Caspase-8-, and Caspase-9-Dependent Apoptosis in Breast Cancer Cells. *Biochem. Pharmacol.* **2008**, *76* (12), 1694–1704.
12. Deepa, B.; Anuradha, C. V. Linalool a Plant Derived Monoterpene Alcohol, Rescues Kidney from Diabetes-Induced Nephropathic Changes via Blood Glucose Reduction. *Diabetologia Croatica* **2011**, *40* (4), 121–137.
13. Deshpande, P.; Mohan, V; Thakurdesai, P. Preclinical Safety Assessment of Glycosides Based Standardized Fenugreek Seeds Extract: Acute, Sub-chronic Toxicity and Mutagenicity Studies. *J. Appl. Pharma. Sci.* **2016**, *6* (9), 179–188.
14. Dhillon, N.; Aggarwal, B. B.; Newman, R. A.; Wolff, R. A.; Kunnumakkara, A. B.; Abbruzzese, J. L. Phase-II Trial of Curcumin in Patients with Advanced Pancreatic Cancer. *Clin. Cancer Res.* **2008**, *14*, 4491–4499.
15. Fahmy, H. A.; Shreif, N. H.; Gharib, O. A. The Protective Effect of *Coriandium Sativum* Extract on Hepato-Renal Toxicity Induced in Irradiated Rats. *Eur. J. Med. Plants* **2014**, *4* (2), 196–205.
16. Faizal, I. P; Suresh, S.; Satheesh Kumar, R.; Augusti, K. T. A Study on The Hypoglycemic and Hypolipidemic Effects of An Ayurvedic Drug Rajanya-malakadi in Diabetic Patients. *Ind. J. Clin. Biochem.* **2009**, *24*, 82–87.
17. Fuller, S.; Jacqueline, M. S. Diosgenin, 4-Hydroxyisoleucine, and Fiber from Fenugreek: Mechanisms of Actions and Potential Effects on Metabolic Syndrome. *Adv. Nutr.* **2015**, *6* (2), 189–197.
18. Ganjali, S.; Sahebkar, A.; Mahdipour, E.; Jamialahmadi, K.; Torabi, S.; Akhlaghi, S.; Ferns, G.; Parizadeh, S. M. R.; Ghayour-Mobarhan, M. Investigation of the Effects of

Curcumin on Serum Cytokines in Obese Individuals: A Randomized Controlled Trial. *Sci. World J.* **2014**, *2014* (ID: 898361), 1–6.
19. Ghorbani, A.; Hadjzadeh, M. A.; Rajaei, Z.; Zendehbad, S. B. Effects of Fenugreek Seeds on Adipogenesis and Lipolysis in Normal and Diabetic Rats. *Pak. J. Biol. Sci.* **2014**, *17* (4), 523–528.
20. Giri, J.; Sakthidevi, T. K.; Meerarani, S. Effect of Ginger on Serum Cholesterol Levels. *Ind. J. Nutr. Dietetics* **1984**, *21* (12), 433–436.
21. Golden, S. H.; Maguire, A.; Ding, J.; Crouse, J. R.; Cauley, J. A.; Zacur, H.; Szklo, M. Endogenous Postmenopausal Hormones and Carotid Atherosclerosis: A Case Study of The Atherosclerosis Risk in Communities Cohort. *Am. J. Epidemiol.* **2002**, *155* (5), 437–445.
22. Jana, S.; Patra, K.; Sarkar, S.; Jana, J.; Mukherjee, G.; Bhattacharjee, S.; Mandal, D. P. Antitumorigenic Potential of Linalool is Accompanied by Modulation of Oxidative Stress: An *In Vivo* Study in Sarcoma-180 Solid Tumor Model. *Nutr. Cancer* **2014**, *66* (5), 835–848.
23. Jang, E. M.; Choi, M. S.; Jung, U. J.; Kim, M. J.; Kim, H. J.; Jeon, S. M.; Shin, S. K.; Seong, C. N.; Lee, M. K. Beneficial Effects of Curcumin on Hyperlipidemia and Insulin Resistance In High-Fat-Fed Hamsters. *Metabolism* **2008,** *57* (11), 1576–1583.
24. Joshi, S. C.; Sharma, N.; Sharma, P. Antioxidant and Lipid Lowering Effects of *Coriandrum sativum* in Cholesterol Fed Rabbits. *Int. J. Pharm. Pharma. Sci.* **2012**, *4* (3), 231–234.
25. Jun, H. J.; Lee, J. H.; Kim, J.; Jia, Y.; Kim, K. H.; Hwang, K. Y.; Yun, E. J.; Do, K. R.; Lee, S. J. Linalool is a PPAR-A Ligand That Reduces Plasma TG Levels and Rewires The Hepatic Transcriptome and Plasma Metabolome. *J. Lipid Res.* **2014**, *55* (6), 1098–1110.
26. Kaur, J. Comprehensive Review on Metabolic Syndrome. *Cardiol. Res. Pract.* **2014**, *2014*, 23–37.
27. Keservani, R. K.; Kesharwani, R. K.; Vyas, N.; Jain, S.; Raghuvanshi, R.; Sharma, A. K. Nutraceutical and Functional Food as Future Food: A Review. *Der Pharmacia Lettre* (The Pharmacy Letters), **2010**, *2* (1), 106–116.
28. Khoja, K. K.; Shaf, G.; Tarique, N. H.; Naveed, A. S.; Abdrohman, S. A.; Abdullah, H. A.; Ali, A. Fenugreek, A Naturally Occurring Edible Spice, Kills MCF-7 Human Breast Cancer Cells Via an Apoptotic Pathway. *Asian Pac. J. Cancer Prev.* **2011**, *12* (12), 3299–3304.
29. Kim, S. G.; Veena, M. S.; Basak, S. K.; Han, E.; Tajima, T. Curcumin Treatment Suppresses IKK-Beta Kinase Activity of Salivary Cells of Patients with Head and Neck Cancer: A Pilot Study. *Clin. Cancer Res.* **2011**, *17*, 5953–5961.
30. Kim, J.; Ha, H. L.; Moon, H. B.; Lee, Y. W.; Cho, C. K.; Yoo, H. S.; Yu, D. Y. Chemopreventive Effect of *Curcuma Longa* Linn on Liver Pathology in Hbx Transgenic Mice. *Integr. Cancer Ther.* **2011**, *10*, 168–177.
31. Koupý, D.; Kotolova, H.; Rudá, K. Effectiveness of Phytotherapy in Supportive Treatment of Type 2 *Diabetes Mellitus*, II: Fenugreek. *Ceska a. Slovenska Farmacie* **2015**, *64* (3), 67–71.
32. Krup, V.; Prakash, L. H.; Harini, A. Pharmacological Activities of Turmeric (*Curcuma Longa* Linn): A Review. *J. Homeopathy Ayurvedic Med.* **2013**, *2* (133), 2167–1206.
33. Kunnumakkara, A. B.; Bordoloi, D.; Padmavathi, G.; Monisha, J.; Roy, N. K.; Prasad, S.; Aggarwal, B. B. Curcumin, the Golden Nutraceutical: Multitargeting for Multiple Chronic Diseases. *Br. J. Pharmacol.* **2017**, *174* (11), 1325–1348.
34. Laribi, B.; Kouki, K.; M' Hamdi, M.; Bettaieb, T. Coriander (*Coriandrum sativum* L.) and its Bioactive Constituents. *Fitoterapia* **2015**, *103*, 9–26.

35. Li, F.; Fernandez, P. P.; Rajendran, P.; Hui, K. M.; Sethi, G. Diosgenin, A Steroidal Saponin, Inhibits STAT3 Signaling Pathway Leading to Suppression of Proliferation and Chemo-sensitization of Human Hepatocellular Carcinoma Cells. *Cancer Lett.* **2010**, *292* (2), 197–207.
36. Li, Y.; Tran, V. H.; Duke, C. C.; Roufogalis, B. D. Preventive and Protective Properties of *Zingiber officinale* (Ginger) in *Diabetes Mellitus*, Diabetic Complications, and Associated Lipid and Other Metabolic Disorders: A Brief Review. *Evid-Based Complement. Altern. Med.* **2012**, *2012* (ID: 516870), 1–10.
37. Li, Y.; Tran, V. H.; Duke, C. C.; Roufogalis, B. D. Gingerols of *Zingiber Officinale* Enhance Glucose Uptake by Increasing Cell Surface GLUT4 in Cultured L6 Myotubes. *Planta Medica*, **2012**, *78* (14), 1549–1555.
38. Misawa, K.; Hashizume, K.; Yamamoto, M.; Minegishi, Y.; Hase, T.; Shimotoyodome, A. Ginger Extract Prevents High-Fat Diet-Induced Obesity in Mice Via Activation of the Peroxisome Proliferator-Activated Receptor Δ Pathway. *J. Nutr. Biochem.* **2015**, *26* (10), 1058–1067.
39. Moorthy, R.; Prabhu, K. M.; Murthy, P. S. Anti-Hyperglycemic Compound (GII) From Fenugreek Seeds, its Purification and Effect in *Diabetes Mellitus. Ind. J. Exp. Biol.* **2010**, *48* (11), 1111–1118.
40. Moosa, Abu, S. M.; Mamun, U. R.; Asadi, A. Z. S.; Nazma, A. M.; Mojib, U.; Ferdaus, A. Hypolipidemic Effects of Fenugreek Seed Powder. *Bangladesh J. Pharmacol.* **2006**, *1* (2), 64–67.
41. Murugesan, M.; Ramalingam, R.; Vaiyapuri, M. Cardioprotective Effect of Fenugreek on Isoproterenol-Induced Myocardial Infarction in Rats. *Ind. J. Pharmacol.* **2011**, *43* (5), 516.
42. Nagabhushan, M.; Amonkar, A. J.; Bhide, S. V. Mutagenicity of Gingerol and Shogaol and Antimutagenicity of Zingerone in Salmonella/Microsome Assay. *Cancer Lett.* **1987**, *36* (2), 221–233.
43. Nyakudya, T.; Siyanda, M.; Nompilo, M.; Kennedy, E. Dietary Supplementation with Coriander Seed: Effect on Growth Performance, Circulating Metabolic Substrates, and Lipid Profile of The Liver and Visceral Adipose Tissue in Healthy Female Rats. *Int. J. Agric. Biol.* **2014**, *16*, 1–9.
44. Panda, S.; Sagarika, B.; Anand, K. Trigonelline Isolated from Fenugreek Seed Protects Against Isoproterenol-Induced Myocardial Injury Through Down-Regulation of Hsp27 and αβ-crystallin. *Nutrition* **2013**, *29* (11), 1395–1403.
45. Prasad, S.; Tyagi, A. K. Ginger and its constituents: role in prevention and treatment of gastrointestinal cancer. *Gastroenterol. Res. Pract.* **2015**, *2015* (ID: 142979), 1–11.
46. Raju, J.; Gupta, D.; Rao, A. R.; Yadava, P. K.; Baquer, N. Z. *Trigonella foenum graecum* (fenugreek) Seed Powder Improves Glucose Homeostasis in Alloxan Diabetic Rat Tissues by Reversing the Altered Glycolytic, Gluconeogenic and Lipogenic Enzymes. *Mol. Cell. Biochem.* **2001**, *224* (1–2), 45–51.
47. Ramadan, M. F.; Amer, M. M. A.; Awad, A. E. S. Coriander (*Coriandrum sativum* L.) Seed Oil Improves Plasma Lipid Profile in Rats Fed A Diet Containing Cholesterol. *European Food Res. Technol.* **2008**, *227* (4), 1173–1182.
48. Ramulu, P.; Nappan, V. G.; Paruchuri, U. Hypolipidemic Effect of Soluble Dietary Fiber (Galactomannan) Isolated from Fenugreek Seeds in WNIN (GR-Ob) Obese Rats. *J. Med. Plants Res.* **2011**, *5* (19), 4804–4813.

49. Rani, M. P.; Krishna, M. S.; Padmakumari, K. P.; Raghu, K. G.; Sundaresan, A. *Zingiber officinale* Extract Exhibits Antidiabetic Potential Via Modulating Glucose Uptake, Protein Glycation and Inhibiting Adipocyte Differentiation: An In Vitro Study. *J. Sci. Food Agric.* **2012**, *92* (9), 1948–1955.
50. Sahebkar, A.; Mohammadi, A.; Atabati, A.; Rahiman, S.; Tavallaie, S.; Iranshahi M. Curcuminoids Modulate Pro-Oxidant-Antioxidant Balance but Not the Immune Response to Heat Shock Protein 27 and Oxidized LDL in Obese Individuals. *Phytother. Res.* **2013**, *27*, 1883–1888.
51. Saraswat, M.; Suryanarayana, P.; Reddy, P. Y.; Patil, M. A.; Balakrishna, N.; Reddy, G. B. Anti-glycating Potential of *Zingiber Officinalis* and Delay of Diabetic Cataract in Rats. *Mol. Vis.* **2010**, *16*, 1525–1537.
52. Sebastian, K. S.; Raghava, V. T. Differential Effects of Soybean and Fenugreek Extracts on the Growth of MCF-7 Cells. *Chemico—Biological Interactions,* **2007**, *170* (2), 135–143.
53. Seyfried, T. N.; Shelton, L. M. Cancer as a Metabolic Disease. *Nutr. Metabol.* **2010**, *7* (7), 1–22.
54. Shamoto, T.; Matsuo, Y.; Shibata, T.; Tsuboi, K.; Nagasaki, T.; Takahashi, H.; Funahashi, H.; Okada, Y.; Takeyama, H. Zerum Bone Inhibits Angiogenesis by Blocking NF-Kb Activity in Pancreatic Cancer. *Pancreas* **2014**, *43* (3), 396–404.
55. Singletary, K. Coriander: Overview of Potential Health Benefits. *Nutr. Today* **2016**, *51* (3), 151–161.
56. Smith, M. Therapeutic Applications of Fenugreek. *Altern. Med. Rev.* **2003**, *8* (1), 20–27.
57. Song, B. L.; De Bose-Boyd, R. A. Insig-Dependent Ubiquitination and Degradation Of 3-hydroxy-3-methylglutaryl Coenzyme A Reductase Stimulated by δ-and γ-tocotrienols. *Journal of Biological Chemistry*, **2006**, *281* (35), 25054–25061.
58. Srinivasan, K. Role of Spices Beyond Food Flavoring: Nutraceuticals with Multiple Health Effects. *Food Rev. Int.* **2005**, *21* (2), 167–188.
59. Srinivasan, K. Ginger (*Zingiber officinale*): A Spice with Multiple Health Beneficial Potentials. *Pharma Nutr.* **2017**, *5* (1), 18–28.
60. Suchankova, G. M.; Tekle, A. K.; Saha, N. B.; Ruderman, S. D.; Clarke; Gettys, T. W. Dietary Polyunsaturated Fatty Acids Enhance Hepatic AMP-Activated Protein Kinase Activity in Rats. *Biochem. Biophys. Res. Commun.* **2005**, *326*, 851–858
61. Suchitra, M. R.; Parthasarathy, S. Effect of Administration of Fenugreek Seeds on Hba1c Levels in Uncontrolled Diabetes Mellitus–A Randomized Controlled Trial. *Int. J. Pharm. Tech Res.* **2015**, *8* (2), 180–182.
62. Tang, E. L.; Rajarajeswaran, J.; Fung, S. Y.; Kanthimathi, M. Antioxidant Activity of Coriandrum Sativum and Protection Against DNA Damage and Cancer Cell Migration. *BMC Complement. Altern. Med.* **2013**, *13*, 347–360.
63. Uemura, T.; Tsuyoshi, G.; Min-Sook, K.; Noriko, M.; Shizuka, H.; Joo-Yong, L.; Yuki, N. Diosgenin, the Main Aglycon of Fenugreek, Inhibits LXRα Activity in HepG2 Cells and Decreases Plasma and Hepatic Triglycerides in Obese Diabetic Mice–3. *J. Nutr.* **2010**, *141* (1), 17–23.
64. Vijayakumar, M. V.; Vimal, P.; Gyan, C. M.; Manoj, K. B. Hypolipidemic Effect of Fenugreek Seeds is Mediated Through Inhibition of Fat Accumulation and Up Regulation of LDL Receptor. *Obesity* **2010**, *18* (4), 667–674.
65. White, B. Ginger: An Overview. *American Family Physician*, **2007**, *75* (11), 1689–1691.

66. Wilken, R. Curcumin: A Review of Anti-Cancer Properties and Therapeutic Activity in Head and Neck Squamous Cell Carcinoma. *Mol. Cancer* **2011**, *10* (12), 1–19.
67. Wilson, P. W.; D' Agostino, R. B.; Parise, H.; Sullivan, L.; Meigs, J. B. Metabolic Syndrome as a Precursor of Cardiovascular Disease and Type 2 Diabetes Mellitus. *Circulation* **2005**, *112* (20), 3066–3072.
68. Wongcharoen, W.; Phrommintikul, A. The Protective Role of Curcumin in Cardiovascular Diseases. *Int. J. Cardiol.* **2009**, *133* (2), 145–151.
69. Wongcharoen, W.; Jai-Aue, S.; Phrommintikul, A.; Nawarawong, W.; Woragidpoonpol, S.; Tepsuwan, T. Effects of Curcuminoids on Frequency of Acute Myocardial Infarction After Coronary Artery Bypass Grafting. *Am. J. Cardiol.* **2012**, *110*, 40- 44.
70. Yang, Y. S.; Su, Y. F.; Yang, H. W.; Lee, Y. H.; Chou, J. I.; Ueng, K. C. Lipid-Lowering Effects of Curcumin in Patients with Metabolic Syndrome: A Randomized, Double-Blind, Placebo-Controlled Trial. *Phytother. Res.* **2014**, *28*, 1770–1777.
71. Zhang, Y.; Yang, C.; Wang, W.; Liu, J.; Liu, Q.; Huang, F.; Chu, L.; Gao, H.; Li, C.; Kong, D.; Liu, Q. Co-Delivery of Doxorubicin and Curcumin by pH-Sensitive Prodrug Nanoparticle for Combination Therapy of Cancer. *Sci. Rep.* **2016**, *6* (1), 1–12.

## CHAPTER 13

# HEALTH BENEFITS OF OREGANO EXTRACT

ANUPAM JAISWAL, MEGHA VERMA, STUTI PANDEY, SHREYA CHITNAVIS, REKHA PATHAK, KAMAL SHAH, DURGESH NANDINI CHAUHAN, and NAGENDRA SINGH CHAUHAN

### ABSTRACT

The parts of oregano (*Origanum vulgare*: Spanish thyme or wild marjoram) are widely used for medicinal purpose and as food ingredients. Oregano consists of thymol, carvacrol, *p*-cymene, and *γ*-terpinene. Oregano is used as folk medicine for upset stomach, cold, cardiovascular disorder, nervous disorder, general health issues; and it has anti-inflammatory, antioxidant, antifungal, antiviral, and antiasthmatic activities, etc. Oregano extract and dry leaves are used in Italian and Mexican dishes to increase flavor/taste of vegetables, sauce, chili baked foods, and pasta.

### 13.1 INTRODUCTION

Medicinal plants are thought to be a rich source of components for drug discovery and synthesis. It is also suggested for its medicinal properties as a natural source of novel antibacterial, antioxidant, and cytotoxic agents.[51] Oregano (*Origanum vulgare*) is a Mediterranean herb that has been found to have a variety of bioactive qualities, including antioxidant, antimicrobial, anti-inflammatory, and analgesic effects. Oregano's antioxidant and antimicrobial qualities make it an excellent choice for use as a natural food preservative. The

Plant-Based Bioactive Compounds and Food Ingredients: Encapsulation, Functional, and Safety Aspects.
Junaid Ahmad Malik, Megh R. Goyal, Preeti Birwal, & Ritesh B. Watharkar (Eds.)

oregano extract is a potential natural food preservative. Additionally, oregano is also being studied for its beneficial effects on gastrointestinal health, as well as it is implied to be a good preservative as an extra benefit.[67] There are several species of oregano that may also contribute in culinary art apart from its medicinal uses.

*O. vulgare* belongs to the family Lamiaceae (Labiatae) also known as Spanish thyme and wild marjoram. Its light green leaves are used as a culinary flavoring, either "dry" or fresh (Figures 13.1 and 13.2) and it is used to induce sensory responses like "pungent," "pleasantly bitter," "herbaceous," and "aromatic."[56] Approximately, 60 plant species and 17 genera are commonly labeled as "oregano" because they have a similar flavor and color.[21] The phytochemicals present in oregano have been linked to and attributed to the herb's health benefits (Figure 13.3). Phytochemicals extracted from oregano are a diverse group of molecules produced during secondary metabolism, the majority of which do not appear to play a role in key metabolic functions.[45] Consumers are becoming increasingly concerned about the use of chemical additives in foods.[46]

**FIGURE 13.1** Oregano plant with leaves.

**FIGURE 13.2** Dry leaves of oregano.

Oregano is mostly known for giving Mediterranean cuisine an "Italian flavor," but it can also be used to enhance the flavor of baked products, vegetables, beans, fish, pizza, spaghetti sauce, and chili. It has been used in aromatic teas and is a common ingredient in Mexican recipes.[56]

Oregano has been used for medicinal purposes since the ancient Greek and Roman empires, when the leaves were used to cure skin ulcers, soothe aching muscles and as an antiseptic. It is used as traditional medicine for the treatment of asthma, cramping, diarrhea, and indigestion.[55] Various extracts of herb are available in market which have important component used for the study of comparison.[11]

Carvacrol and thymol are two important components of oregano contributing to warmth sensation in the mouth by acting on temperature-sensitive structures known as ion channels.[43,69] The herb consists 1–2.5% of essential oil, tannins, phenolic acid, flavonoids, triterpene acids,[54,66] and polysaccharides. The herb extracts have positive outcomes in gastric conditions[4] and exhibit antimicrobial[31,35] and antioxidant properties,[26,44] they may also have beneficial effects in cardiovascular diseases.[70]

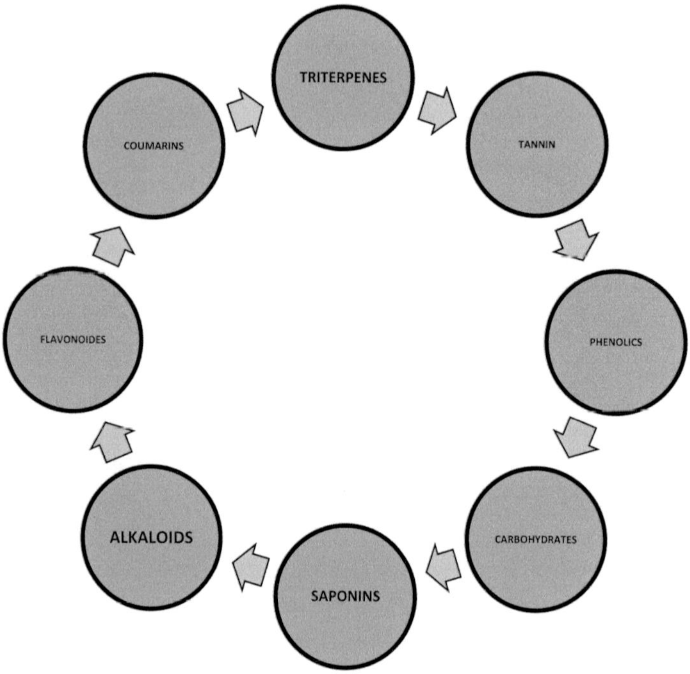


**FIGURE 13.3** Phytochemicals in parts of oregano plant.

This chapter explores potential adaptability and benefits of *O. vulgare* in food preservation.

## 13.2 NATURAL DISTRIBUTION AND MORPHOLOGY

*Origanum vulgare*, a perennial herbaceous plant belonging to the Lamiaceae family, its taxonomy is complicated[13,16] thrives in Mediterranean countries and is also found in Asia, the Arabian Peninsula, Africa, America, and Europe. It is a plant with soft gray-green leaves, a dense, shallow root system, and a bushy form. It has tiny fruits, light brown nuts with a distinctive aromatic odor.[15] *O. vulgare* is a perennial plant that grows to a height of 20–80 cm and has a woody base and herbaceous stalks. The leaves of this plant are opposite and egg shaped that are 1–4 cm long and 0.5–2.5 cm broad, with a smooth leaf edge and a rounded to pointy tip. They have tiny flowers and arranged in paniculate inflorescences on the terminal and lateral sides. The corolla is made up of five joined petals that are 0.4–0.8 cm long and range in color from white to purple. The sepals are also joined;

the flower's inner half consists of four stamens, and two fused carpels are made the pistil.[57]

The plants consist of dense spikes, and tubular five-toothed calyces, never becoming turbinate in fruit (Figure13.1). *O. vulgare* is divided into six subspecies based on changes in indumentums, the number of sessile glands on the leaves, bracts, and calyces, and the size and color of the bracts and flowers. The three subspecies of *O. vulgare* that are "rich" in essential oils are found in the southernmost range, while those that are "deficient" in essential oils are located in the northernmost range.[29] Six subspecies are:

1. *O. vulgare* L. subsp. *vulgare*
2. *O. vulgare* L. subsp. *glandulosum* (Desfontaines) Ietswaart
3. *O. vulgare* L. subsp. *gracile* (Koch) Ietswaart
4. O. vulgare L. subsp. *hirtum* (Link) Ietswaart
5. *O. vulgare* L. subsp. *viridulum* (MartrinDonos)Nyman
6. *O. vulgare* L. subsp. *virens* (Hoffmannsegg & Link) Ietswaart

It grows in a variety of habitats and ecologies.[58] *O. vulgare* grows best in a climatic life zone of 5–28°C, with annual precipitation ranging from 0.4 to 2.7 m and soil pH ranging from 4.5 to 8.7.[41] The cultivation and distribution of oregano found all over Europe, Central, and West Asia up to Taiwan.[25] This demonstrates the crop's monetary value in various communities and groupings across several countries. The average performance throughout time of leaf length was 2.21 cm, leaf width 1.18 cm (Figure13.1), plant height 37.18 cm, leaf-to-stem ratio 2.68, leaf yield/plant 159.27 g and leaf yield/ha 8.91 t. The maximum plant height value was found at Wondo Genet, while the lowest was found at Holleta.[28]

## 13.3 CHEMICAL COMPOSITION

The aerial portions of the plant are rich in essential oil with varying composition, in addition to a wide range of flavonoids, tannins, phenolic glycosides, and terpenoids.[49] The primary flavonoid derivatives discovered in the hydroalcoholic extracts, decoctions, and infusions of *O. vulgare* are luteolin-*O*-glucuronide and luteolin-7-*O*-glucoside, whereas rosmarinic acid is the important phenolic acid.[40] It is reported that caffeic acid, protocatechuic acid, vanillic acid, and *o*-coumaric acid are also available in plant.[30] There are variations in the chemical composition in different species of *O. vulgare* available.[38] Monoterpene and sesquiterpene hydrocarbons, as well as phenolic compounds are found in the volatile oil.[17] The primary constituents are terpenes like thymol, carvacrol, *p*-cymene, $\gamma$-terpinene, and linalool. Various chemotypes have been identified based on the primary identification.[37] Based on the concentrations of acyclic linalool/linalyl acetate, cymyl, and sabinyl compounds in 502 plants from 17 European countries, three primary *O. vulgare* chemotypes were identified.[38]

**TABLE 13.1** Solubility of Oregano Extract in Different Solvents.

| Phyto-constituents | Petroleum ether extract | Diethyl ether extract | Ethyl acetate extract |
|---|---|---|---|
| Alkaloids | – | – | – |
| Carbohydrates | – | – | + |
| Flavanoids | – | + | + |
| Glycosides | – | + | + |
| Proteins | – | – | + |
| Terpenes | – | + | + |

## 13.4 POTENTIAL USES OF OREGANO

### *13.4.1 FOOD PRESERVATION*

*Cypriot pastrami* (also known as samarella) is a sort of sun-dried meat product that uses oregano as a preservative.[65] The use of oregano essential oils in additive-free, dried, and cured meats as a food preservative. Essential oils are reported as safe food additives at concentrations less than 2 mg/kg body weight/day this approach has been recognized in the European Union.[1] Oregano extracts and oregano essential oils were used in several researches to determine the efficacy of food preservation. Aqueous extract of oregano was reported to considerably lower the microbial growth TBARS readings in raw beef samples compared to controls over a period of 15 days.[32] Oregano extract oil made by an O/W nanoemulsion prevented the growth of *L. monocytogenes*, *S. typhimurium*, and *E. coli* on commercial iceberg and romaine lettuce leaves at 0.05% and 0.1% concentrations with excellent efficacy was observed in romaine lettuce leaves.[7]

Incorporating a combination of extract to improve preservation quality is another approach of food preservation. Oregano's preservation properties have been observed to improve when combined with other natural extracts. In pork meat, chitosan (1%) and oregano extract oil (2–4%) combination was found to be beneficial in suppressing bacterial growth, protecting from lipid oxidation and increasing shelf life with low sensory impact.[48] In minced beef, oregano extract oil was found to have a stronger antimicrobial effect when combined with caprylic acid at 0.2% and 0.5%, respectively, compared to extract alone.[23] In a research work, in cooked chicken, oregano extract oil in combination with sage essential oil and honey helps in decreasing lipid oxidation as evaluated by TBARS after 48 and 96 h.[52] The researchers suggest that oregano extract and essential oil can be used alone or in combination with other herbs might improve the food preservation property.[67]

### 13.4.2 ANTIMICROBIAL ACTIVITY

The essential oil of oregano, as well as other extracts can inhibit the growth of Gram-positive and Gram-negative bacteria, yeast and some fungi. The range of effective concentration of oregano as antimicrobial agent was found to be 0.4 to 4000 μg/ml. Oregano oil was found to be beneficial in reducing microbial development in beef and sausage products at concentrations ranging from 0.1 to 2.0% w/w.[9,27] Oregano, on the other hand, did not inhibit the growth of *Yersinia enterocolitica* or *Listeria monocytogenes* in barbeque chicken in another investigation.[18] Oregano has been shown to suppress *Helicobacter pylori*, a bacteria linked to gastritis in humans, to kill human intestinal parasites in vitro (at a concentration of 175 μg/ml) and to effectively treat colibacillosis in newborn calves at a dose of 10 mg oregano oil per kg body weight.[6,39,47,53] In a short-term trial of persons infected with enteric parasites, daily use of 600 mg of emulsified oregano oil resulted in significant reductions in the prevalence of several parasites.[19]

### 13.4.3 INFLAMMATION AND IMMUNOLOGICAL PROPERTIES

Oregano extracts or ingredients have been shown to reduce inflammation in vitro and in vivo by a number of researchers. Thymol (at dosages ranging from 2.5 to 20 g/mL) reduced the release of elastase, an inflammatory disease marker, from human neutrophils elicited by a chemotactic peptide.[8] Similarly, rosmarinic acid lowered thrombin activity and prevented neutrophil elastase release.[43] In a chorioallantois membrane experiment, neither an essential oil of oregano nor the phytochemicals thymol, carvacrol, *p*-cymene, or γ-terpinene had anti-inflammatory effects whereas, thymol (10 μg) was proven to be an irritant.[12] Rosmarinic acid promoted lymphocyte apoptosis and was able to destroy T and natural killer cells both of which are ways to reduce detrimental immune responses.[3,24]

### 13.4.4 ANTIOXIDANT ACTIVITY

Oregano extracts have been shown to exhibit radical scavenging properties, as well as the ability to decrease lipid peroxidation, reduce nitric oxide activity and protect DNA from $H_2O_2$-induced oxidative damage.[2,14,64,72] Carvacrol, γ-terpinene, rosmarinic acid, ursolic acid, protocatechuate-glycosides, and thymol have all been suggested as components responsible for antioxidant activity of oregano.[56] Oregano or its components (42.5 mg/kg body weight) improved overall antioxidant status and aging brain antioxidant activity in rats.[71] Methanolic oregano extract had an antioxidant capacity of $13.59 \pm 0.87$ nmol equivalent of vitamin E per g of fat at a dosage of 5 g extract, which was more than twice the control group in a research of antioxidant herbs added to animal fat.[68]

### 13.4.5 BENEFITS FOR NERVOUS SYSTEM

In preliminary research, oregano and several components (such as carvacrol, thymol, and γ-terpinene) were shown to alter nervous system chemistry and functions, including olfactory stimulation responses.[20,22,50,73] Animal studies show that rosmarinic acid (2 mg/kg body weight i.p.) has antidepressant properties and may help to prevent an emotional aberration caused by stress.[59–61]

### 13.4.6 REGULATION OF BLOOD GLUCOSE AND LIPIDS

The evidence of benefit of oregano extract in the management of diabetes and cardiovascular disease are limited. Water extracts of oregano and rosmarinic acid showed significant in-vitro inhibiting action of glucosidase.[33] The activity of porcine pancreatic amylase was likewise inhibited by rosmarinic acid containing oregano extracts.[42] An aqueous extract of *O. vulgare* given orally at a dose of 20 mg/kg body weight showed antihyperglycemic activity in two animal studies utilizing diabetic rats.[36,63] Origanum species extract and its individual constituents like γ-terpinene and carvacrol reported to inhibit aggregation, adhesion of blood platelet, reduce biosynthesis of cholesterol and decrease total serum cholesterol and triglyceride levels, as observed in various studies.[5,10,33,59,70] Administration of carvacrol (100 μg/kg ip) lowered systolic and diastolic blood pressures in normotensive rats.[5]

## 13.5 SUMMARY

Phytochemical screening of *O. vulgare* revealed the presence of many types of biologically active naturally occurring chemicals. The main chemical constituents found in *O. vulgare* include: essential oils, glycosides, flavonoids, carbohydrates, terpenes, and proteins. Phytochemical screening revealed that a majority of components are found in ethyl acetate extract. The pharmacognostical study is one major criterion for the identification of plant. The present study will provide information about the benefits of *O. vulgare* in food preservation and health-promoting properties. The demand for natural alternatives has been increased on public interest for clean labeling of food products and alternatives to synthetic food preservatives. The presence of phytochemical content is essential in understanding the advantages of different forms tested for antioxidant and antimicrobial properties. A combination of various extracts or essential oils, in addition to employing a single herb like oregano, may create an even greater effect on food preservation.

## KEYWORDS

- **baked goods**
- **Carvacrol**
- ***Origanum vulgare***
- ***p*-cymene**
- **Spanish thyme**
- **thymol**
- **wild marjoram**
- **γ-terpinene**

## REFERENCES

1. Aguilar, F.; Autrup, H.; Barlow, S.; Castle, L.; Crebelli, R.; Dekrant, W. Use of Rosemary Extracts as a Food Additive—Scientific Opinion of the Panel on Food Additives, Flavorings, Processing Aids and Materials in Contact with Food. *Eur. Food Saf. Authority J.* **2008**, *721*, 1–29.
2. Aherne, S.; Kerry, J; O'Brien, N. Effects of Plant Extracts on Antioxidant Status and Oxidant-Induced Stress in Caco-2 Cells. *Br. J. Nutr.* **2007**, *97*, 321–328.
3. Ahn, S.; Oh, W.; Kim, B.; Kang, D.; Kim, M.; Heo, G.; Ahn, J. Inhibitory Effects of Rosmarinic Acid on Lck SH2 Domain Binding to Synthetic Phosphopeptide. *PlantaMedica: Journal of Medicinal Plant and Natural Product Research,* **2003**, *69* (7), 642–646.
4. Alhowiriny, T.; Alsheikh, A.; Alqasoumi, S.; Alyahya, M.; Eltahir, K.; and Rafatullah, S. Protective Effect of *Origanum majorana* on Various Models of Gastric Mucosal Injury in Rats. *Am. J. Chinese Med.* **2009**, *37* (3), 531–545.
5. Aydin, Y.; Kutlay, O.; Ari, S.; Duman, S.; Uzuner, K.; Aydin, S. Hypotensive Effects of Carvacrol on the Blood Pressure of Normotensive Rats. *PlantaMedica: J. Med. Plant Nat. Product Res.* **2007**, *73* (13), 1365–1371.
6. Bampidis, V.; Christodoulou, V.; Paneri, F. P.; Christaki, E. Effect of Dried Oregano Leaves Versus Neomycin in Treating Newborn Calves with Colibacillosis. *J. Vet. Med A* **2006**, *53* (3), 154–156.
7. Bhargava, K.; Conti, D. S.; Rocha, S. R.; Zhang, Y. Application of Oregano Oil Nanoemulsion to the Control of Foodborne Bacteria on Fresh Lettuce. *Food Microbiol.* **2015**, *47*, 69–73.
8. Braga, P.; DalSasso, M.; Culici, M.; Bianchi, T.; Bordoni, L.; Marabini, L. Anti-Inflammatory Activity of Thymol: Inhibitory Effect on the Release of Human Neutrophil Elastase. *Pharmacol. J.* **2006**, *77* (3), 130–136.
9. Busatta, C.; Vidal, R.; Popiolski, A.; Mossi, A. J.;Dariva, C.; Rodrigues, M. R. A,; Corazza, F. C. Corazza, L.; Oliveria, J. V.; Cansian, R. L. Application of Origanum majorana Essential Oil as an Antimicrobial Agent in Sausage. *Food Microbiol. J.* **2008**, *25*, 207–211.
10. Case, G.; He, H.; Elson, C. Induction of Geranyl Pyrophosphate Pyrophosphatase Activity by Cholesterol-Suppressive Isoprenoids. *J. Lipid Res.* **1995**, *30* (4), 357–359.
11. Clark, S. J; Dodds, C. J.; Handerson, I. F.; Martin, A. P. A Bioassay for Screening Materials Influencing Feeding in the Field Slug *Deroceraus Reticulam* (Muller) (Mollucsca Pulmonate). *Ann. Appl. Biol.* **1997,** *130* (2), 379–385.
12. Demrci, F.; Paper, D.; Franz, G.; Baser, K. Investigation of the *Origanum onites* L. Essential Oil Using the Chorio allantoic Membrane (CAM) Assay. *J. Agric. Food Chem.* **2004**, *52* (2), 251–254.
13. Derwich, E.; Benziane, Z.; Manar, A.; Boukir, A.; and Taouli, R. Phytochemical Analysis of and In Vitro Antibacterial Activity of the Essential Oil of *Origanum vulgare* from Morocco. *American Eurassian J. Sci. Res.* **2010**, *5* (2), 120–129.
14. Dragland, S.; Senoo, H.; Wake, K.; Holte, K.; Blomhoff, R. Several Culinary and Medicinal Herbs are Important Sources of Dietary Antioxidants. *J. Nutr.*, **2003**, *133* (5), 1286–1290.
15. El-Ashmawy, I. M.; Amal, S.; and Salama, O. M. Acute and Long Term Safety Evaluation of *Origanum majorana* Essential Oil. *Alexandria J. Pharma. Sci.* **2007**, *21* (1), 29–35.

16. Fatma, A. T.; Suer, Y.; Bayram, E.; Refika, R. A. G.; Sonmeze, C.; Isa, T.; Mehmet, A. F. Chemical and Genetic Variability of Selected Turkish Oregano. *Plant System. Evol.* **2010**, *288*, 157–165.
17. Fikry, S.; Khalil, N.; Salama, O. Chemical Profiling, Biostatic and Biocidal Dynamics of *Origanum vulgare* L. Essential Oil. *AMB Express* **2019**, *9*, 41.
18. Firouzi, R.; Shekarforoush, S.; Nazer, A.; Borumand, Z.; Jooyandeh, A. Effects of Essential Oils of Oregano and Nutmeg on Growth and Survival of *Yersinia enterocolitica* and *Listeria monocytogenes* in Barbequed Chicken. *J. Food Protect.* **2007**, *70* (11), 2626–2630.
19. Force, M.; Sparks, W.; Ronzio, R. Inhibition of Enteric Parasites by Emulsified Oil of Oregano In Vivo. *Phytother. Res.* **2000**, *14* (3), 213–214.
20. Fukumoto, S.; Sawasaki, E.; Okuyama, S.; Miyaki, Y.; Yokogoshi, H. Flavor Components of Monoterpenes in Citrus Essential Oils Enhance the Release of Monoamines from Rat Brain Slices. *Nutr. Neurosci.* **2006**, *9* (1–2), 73–80.
21. Grijalva, G. E. P.; Salas, P. M. A.; Lopez, L. N.; Mendoza, C. M. S.; Olivo, V. G.; Heredia, J. B. Flavonoids and Phenolic Acids from Oregano: Occurrence, Biological Activity and Health Benefits. *Plants (Basel)*, **2018**, *7* (1), 2–10.
22. Huang, M. H.; Wu, S. N.; Shen, A. Y. Stimulatory Actions of Thymol, a Natural Product, on Ca(2+)-Activated K(+) Current in Pituitary GH(3) cells. *Planta Medica* **2005**, *71* (12), 1093–1098.
23. Hulankova, R.; Borilova, G.; Steinhauserova, I. Combined Antimicrobial Effect of Oregano Essential Oil and Caprylic Acid in Minced Beef. *Meat Sci.* **2013**, *95* (2), 190–194.
24. Hur, Y. G.; Yun, Y.; Won, J. Rosmarinic acid Induces p56Lck-Dependent Apoptosis in Jurkat and Peripheral T Cells via Mitochondrial Pathway Independent from Fas/Fas Ligand Interaction. *J. Immunol.* **2004**, *172* (1), 79–87.
25. Ietswaart, J. H. A. (Ed.) (1980). *Taxonomic Revision of the Genus Origanum (Labiatae)*; Leiden University Press: Leiden, 1980; pp 3–5.
26. Jin Jun, W.; Kyung Han, B.; Won Yu, K.; Sung Kim, M.; Seop Chang, I.; Yun Kim, H.; Yon Cho, H. Antioxidant Effects of *Origanum majorana* L. on Superoxide Anion Radicals. *Food Chem.* **2001**, *75* (4), 439–444.
27. Juneja, V.; Thippareddi, H.; Friedman, M. Control of Clostridium perfringens in Cooked Ground Beef by Carvacrol, Cinnamaldehyde, Thymol or Oregano Oil During Chilling. *J. Food Protect.* **2006**, *69* (7), 1546–1551.
28. Kassahun, B. M.; Tilahun, S.; Zigene1, Z. D.; Teferi1, Z.; Mekonnen, M.; Melka, B. Morpho—agronomic Characteristics, Essential Oil Content and Essential Oil Yield of Oregano (*Origanum vulgare* L.) in Ethiopia. *Scholarly J. Agric. Sci.* **2014**, *4* (12), 565–571.
29. Kokkini, S. Taxonomy, Diversity and Distribution of *Origanum* Species. *Int. Plant Genetic Resour. Inst.* **1997**, 8–12.
30. Kosakowska, O.; Czupa, W. Morphological and Chemical Variability of Common Oregano (*Origanum vulgare* L. Subsp. Vulgare) Occurring in Eastern Poland. *Herba Polonica* **2018**, *64* (1), 11–21.
31. Kozlowska, M.; Laudy, A. E.; Starosciak, B. J.; Napiorkowski, A.; Chomicz, L.; and Kazimierczuk, Z. Antimicrobial and Antiprotozoal Effect of Sweet Marjoram (*Origanum majorana L.*). *Acta Scientarium Polonorum Hortorum Cultus* **2010**, *9* (4), 133–141.
32. Krishnan, K. R.; Babuskin, S.; Babu, P. A. S.; Fayidh, M. A.; Sabina, K.; Archana, G.; Sivaranjan, M.; Sukumar, M. Bio-Protection and Preservation of Raw Beef Meat Using Pungent Aromatic Plant Substances. *J. Sci Food Agric.* **2014**, *94* (12), 2456–2463.

33. Kwon, Y. I.; Vattem, D. A.; Shetty, K. Evaluation of Clonal Herbs of Lamiaceae Species for Management of Diabetes and Hypertension. *Asia Pac. J. Clin. Nutr.* **2006**, *15* (1), 107–118.
34. Lee, S. P.; Buber, M. T.; Yang, Q.; Cerne, R.; Cortes, R. Y.; Sprous, D. G.; Bryant, R. W. Thymol and Related Alkyl Phenols Activate the TRPA-1 Channel. *Br. J. Pharmacol.* **2008**, *153* (8), 1739–1749.
35. Leeja, L; Thoppil, J. E. Antimicrobial Activity of Methanol Extract of Origanum majorana L. (Sweet marjoram). *J. Environ. Biol.* **2007**, *28* (1), 145–146.
36. Lemhadri, A.; Zeggwagh, N. A.; Maghram, M.; Jouad, H.; Eddouks, M. Anti Hyperglycemic Activity of the Aqueous Extracts of *Origanum vulgare* Growing Wild in Tafilet Region. *J. Ethnopharmacol.* **2004**, *92* (2–3), 251–256.
37. Lopez, N. L; Grijalva, E. P. G.; Olivo, G. V.; Heredia, J. B. Essential Oils of Oregano: Biological Activity beyond Their Antimicrobial Properties. *Molecules (Basel, Switzerland)*, **2017**, *22* (6), 989.
38. Lukas, B.; Schmiderer, C.; Novak, J. Essential Oil Diversity of European *Origanum vulgare* L. *Phytochemistry* **2015**, *119*, 32–40.
39. Macotela, P. M.; Gonzalez, R. Y.; Maciel, G. A.; Robies, R. R.; Gordillo, M. M. Oregano Kills *Giardia Intestinalis* Trophozoites In Vitro: Anti-giardiasic Activity and Ultrastructural Damage. *Parasitol. Res.* **2006**, *98* (6), 557–560.
40. Martins, N.; Barros, L.; Buelga, S. C.; Henriques, M.; Silva, S.; Ferreira, I. C. F. R. Decoction, Infusion and Hydroalcoholic Extract of *Origanum vulgare* L.: Different Performances Regarding Bioactivity and Phenolic Compounds. *Food Chem.* **2014**, *158*, 73–80.
41. Marzi, V.; Fortunato, I. M.; Circella, G.; Picci, V.; Melegari, M. Oregano Results Obtained in the Context of Cultivation improvement of Medicinal Plants. *Agric. Res.* **1992**, *132*, 71–89.
42. Mccue, P, P.; Shetty, K. Inhibitory Effects of Rosmarinic Acid on Porcine Pancreatic Amylase In Vitro. *Asia Pac. J. Clin. Nutr.* **2004**, *13* (1), 101–106.
43. Melzig, M. F.; Henke, K. Inhibition of Thrombin Activity by Selected Natural Products in Comparison to Neutrophil Elastase. *PlantaMedica* **2005**, *71* (8), 787–789.
44. Mossa, A. T. H.; and Nawwar, G. A. M. Free Radical Scavenging and Anti-acetylcholinesterase Activities of *Origanum majorana* Essential Oil. *Human Exp. Toxicol.* **2011**, *30* (10), 1501–1513.
45. Oniga, I.; Puscas, C.; Dumitrescu, S. R.; Olah, N. K.; Sevastre, B.; Marica, R. Marcus, I.; Berghian, A. C. S.; Benedec, D.; Pop, E. C.; Hanjanu, D. *Origanum vulgare*: Chemical Composition and Biological Studies. *Molecules (Basel, Switzerland)*, **2018**, *23* (8), 2077–2082.
46. Oswell, N. J.; Thippareddi, H.; Pegg, R. B. Practical Use of Natural Antioxidants in Meat Products in the U.S.: A Review. *Meat Sci.* **2018**, *145*, 469–79.
47. O'Mahony, R.; AlKhtheeri, H.; Weerasekera, D.; Fernando, N.; Varira, D.; Honton, J.; Basset, C. Bactericidal and Anti-Adhesive Properties of Culinary and Medicinal Plants Against *Helicobacter Pylori*. *World J. Gastroenterol.* **2005**, *11* (47), 7499–7507.
48. Paparella, A.; Mazzarrino, G.; Lopez, C. C.; Rossi, C.; Sacchetti, G.; Guerrieri, O.; Serio, A. Chitosan Boosts the Antimicrobial Activity of *Origanum vulgare* Essential Oil in Modified Atmosphere Packaged Pork. *Food Microbiol.* **2016**, *59*, 23–31.
49. Pezzani, R.; Vitalini, S.; Iriti, M. Bioactivities of *Origanum vulgare*: An Update. *Phytochem. Rev.* **2017**, *16*, 1253–1268.

50. Priestley, C. M.; Williamson, E. M.; Wafford, K. A.; Sattelle, D. B. Thymol: A Constituent of Thyme Essential Oil is a Positive Allosteric Modulator of Human GABA(A) Receptors and a Homo-Oligomeric GABA receptor from *Drosophila Melanogaster*. *Br. J. Pharmacol.* **2003**, *140* (8), 1363–1372.
51. Rasool, H. B. A. Medicinal Plants (Importance and Uses). *Pharmaceut Analytica Acta* **2012**, *3* (10), 139–145.
52. Sampaio, G. R.; Saldanha, T.; Soares, R. A. M.; Torres, E. A. F. S. Effect of Natural Antioxidant Combinations on Lipid Oxidation in Cooked Chicken Meat During Refrigerated Storage. *Food Chem.* **2012**, *135* (3), 1383–1390.
53. Santoro, G. F.; Cardoso, M. D. G.; Guimaraes, L. G. L.; Salgado, A. P. S. P.; Barreto, R. F. S. M.; Soares, M. J. Effect of Oregano and Thyme Essential Oils on *Trypanosomacruzi* (Protozoa: Kinetoplastida) Growth and Ultrastructure. *Parasitol. Res.* **2007**, *100* (4), 783–790.
54. Sellami, I. H.; Maamouri, E.; Chahed, T.; Wannes, W. A.; Kchouk, M. E.; Marzouk, B._Effect of Growth Stage on the Content and Composition of the Essential Oil and Phenolic Fraction of Sweet Marjoram. *Ind. Crops Product.* **2009**, *30*, 395–402.
55. Shan, B.; Cai, B.; Sun, M.; Corke, H. Antioxidant Capacity of 26 Spice Extracts and Characterization of Their Phenolic Constituents. *J. Agric. Food Chem.* **2005**, *53* (20), 7749–7759.
56. Singletary, K. W. Oregano Overview of the Literature on Health Benefits. *Nutr. Today* **2010**, *45* (3), 129–138.
57. Skoufogianni, E.; Solomou, A. D.; Danalatos, N. G. Ecology, Cultivation and Utilization of the Aromatic Greek Oregano: A Review. *Notulae Botanicae Horti Agrobotanici Cluj—Napoca* **2019**, *47* (3), 545–552.
58. Snogerup, S. Evolutionary and Plant Geographical Aspects of Chasm phytic Communities. *Plant Life South West Asia* **1971**, 157–170.
59. Takahashi, Y.; Inaba, N.; Kuwahara, S.; Kuki, W. Effects of γ-terpinene on Lipid Concentrations in Serum Using Triton WR1339—Treated Rats. *Biosci. Biotechnol. Biochem.* **2003**, *67* (11), 2448–2450.
60. Takeda, H.; Tsuji, M.; Matsumiya, T.; Kubo, M. Identification of Rosmarinic Acid as a Novel Anti-depressive Substance in the Leaves of *Perilla frutescens*. *Nihon Shin Seish Yakurig Zasshi* (*Japanese Journal of Psychopharmacology*), **2002**, *22* (1), 15–22.
61. Takeda, H.; Tsuji, M.; Miyamoto, J.; Matsumiya, T. Rosmarinic Acid and Caffeic Acid Reduce the Defensive Freezing Behavior of Mice Exposed to Conditioned Fear Stress. *Psycho Pharmacol.* **2002**, *164* (2), 233–235.
62. Takeda, H.; Tsuji, M.; Inazu, M.; Egashira, T.; Matsumiya, T. Rosmarinic Acid and Caffeic Acid Produce Anti-Depressive Like Effect in the Forced Swimming Test in Mice. *Eur. J. Pharmacol.* **2002**, *449* (3), 261–267.
63. Talpur, N.; Echard, B.; Ingram, C.; Bagchi, D.; Preuss, H. Effects of a Novel Formulation of Essential Oils on Glucose-Insulin Metabolism in Diabetic and Hypertensive Rats: A Pilot Study. *Diab. Obes. Metabol.* **2005**, *7* (2), 193–199.
64. Tsai, P.; Tsai, T.; Yu, C.; Ho, S. Evaluation of NO—Suppressing Activity of Several Mediterranean Culinary Spices. *Food Chem. Toxicology*, **2007**, *45* (3), 440–447.
65. Ulusoy, B.; Hecer, C.; Kaynarca, D.; Berkan, S. Effect of Oregano Essential Oil and Aqueous Oregano Infusion Application on Microbiological Properties of *Samarella Tsamarella* (a Traditional Meat Product of Cyprus). *Foods (Basel, Switzerland)*, **2018**, *7* (4), 43–50.

66. Vagi, E.; Rapavi, E.; Hadolin, M.; Vasarhelyine P. K.; Balazs, A.; Blazovics, A; and Simandi, B. Phenolic and Triterpenoid Antioxidants from *Origanum majorana* and Extracts Obtained with Different Solvents. *J. Agric. Food Chem.* **2005**, *53* (1), 17–21.
67. Veenstra, J. P.; Johnson, J. J. Oregano Extract for Food Preservation and Improvement in Gastrointestinal Health. *Int. J. Nut.* **2019**, *3* (4), 43–52.
68. Vichi, S.; Eglseer, Z. K.; Jugl, M.; Franz, C. Determination of the Presence of Antioxidants Deriving from Sage and Oregano Extracts Added to Animal Fat by Means of Assessment of the Radical Scavenging Capacity by Photo Chemiluminescence Analysis. *Die Nahrung* (The Food), **2001**, *45* (2), 101–104.
69. Xu, H.; Delling M.; Jun, J. C.; Clapham, D. E. Oregano, Thyme and Clove-Derived Flavors and Skin Sensitizers Activate Specific TRP channels. *Nat. Neurosci.*, **2006**, *9* (5), 628–635.
70. Yazdanparast, R.; Shahriyary, L. Comparative Effects of *Artemisia dracunculus*, *Satureja hortensis* and *Origanum majoranaon* Inhibition of Blood Platelet Adhesion, Aggregation and Secretion. *Vasc. Pharmacol.* **2008**, *48* (1), 32–37.
71. Youdim, K. A.; Deans, S. G. Effect of Thyme Oil and Thymol Dietary Supplementation on the Antioxidant Status and Fatty Acid Composition of the Aging Rat Brain. *Br. J. Nutr.* **2000**, *83* (1), 87–93.
72. Zheng, W.; Wang, S. Y. Antioxidant Activity and Phenolic Compounds in Selected Herbs. *J. Agric. Food Chem.* **2001**, *49* (11), 5165–5170.
73. Zibrowski, E. M.; Hoh, T. E.; Vanderwolf, C. H. Fast Wave Activity in the Rat Rhinencephalon: Elicitation by the Odors of Phytochemicals, Organic Solvents and a Rodent Predator. *Brain Res.* **1998**, *800* (2), 207–215.

# PART IV

# SAFETY ASPECTS OF FUNCTIONAL AND NATURAL FOODS

# CHAPTER 14

# SAFETY ASPECTS OF FUNCTIONAL FOODS IN THE INDUSTRY: AN OVERVIEW

RUBEKA IDRISHI and SAMEER AHMAD

## ABSTRACT

Food Safety and Quality Management Systems (FSQMS) form an integral part of the food industry, without which one cannot fully understand the concept of having safe and nutritious food. In a global context, components of WTO (World Trade Organization) in the form of agreements for TBT (technical barriers to trade) along with various SPS (sanitary and phyto sanitary measures) globally govern all the operational matters of food safety and quality management of food industry. In contrast, the main governing body in India is the centrally managed FSSAI (Food Safety and Standards Authority of India). In this chapter, various aspects of food safety and quality management in food industries have been discussed with relation to the functioning of the same, which are cohesive in nature and can be better understood by clarifying the concept of quality, food safety standards, and regulation and quality management systems. The reader will be able to understand few basic and practical approaches in the form of principles of food safety management systems, risk analysis, safe food concept, and sanitary and hygiene practices in the food industry. Tools for managing food safety such as TQM (Total Quality Management), HACCP (Hazard Analysis Critical Control Points), Accreditation and Auditing and various organizations like ISO (International Organization for Standardization), their series, various standards benchmarked by GFSI (Global Food Safety

Plant-Based Bioactive Compounds and Food Ingredients: Encapsulation, Functional, and Safety Aspects.
Junaid Ahmad Malik, Megh R. Goyal, Preeti Birwal, & Ritesh B. Watharkar (Eds.)

Initiative), and a brief understanding for other supportive management systems for ensuring overall safety and quality in food industry will also be discussed.

In order to maintain the safety and quality in food industries, a cohesive approach of all the tools discussed in the above paragraph is needed. Thus, food personnel must undergo specific program/curriculum/training to familiarize themselves with essential tools and techniques which are needed in order to develop specific skills to cater specific food industry services.

## 14.1 INTRODUCTION

As the world is adapting new ways to combat the rising issues in safe food handling practices, the management pedagogies in all the spheres of food world are also reinventing themselves by continuous improvements in the existing system. With the humongous expansion of food industry whether it is manufacturing, retail, distribution or hospitality sector, the importance of maintaining good quality and safety standards as per the globally acceptable laws is now given ~~is~~ priority by all the food handlers. In India, food safety regulations are continuously reaching to global standards with the introduction of 'FSSA 2006' in the year 2011. With this, the requirement of adequately trained professionals in food safety and quality management for food industries is very important in order to meet consumer expectations, increasing awareness, and safety and quality assurance from farm to fork level. To maintain a stable standard and management there are different methodologies, techniques, curriculums, tools, and management basics which should be understood well by a person who in anyway is handling or dealing with foods at industrial level.

In view of the same, this chapter will make the readers familiar with all the basics of quality and safety management needed to ensure safe and well managed practices in a food industry. This chapter will provide an outline of state-of-the-art theoretical information and practical approaches which are directly and indirectly related to some of the common scenarios which happens in a food industry and is thus framed for transmission of both the knowledge and techniques which are applied in food industry.

As the chapter will proceed, the reader will get familiarized with the importance of food safety and management basics, physical, chemical, and microbiological safety of foods, analytical quality assurance in food laboratories, food standards and quality control, and various safety management tools used in food industries for food standards and quality control such as HACCP, ISO series, TQM, OHSAS along with supportive management systems like GMP, GHP, GAP, etc.

## 14.2 PRINCIPLES OF SAFETY AND QUALITY MANAGEMENT IN FOOD INDUSTRIES

### *14.2.1 FSQMS*

To introduce FSQMS to any food handler it is very important to understand food safety, various food safety systems, and Total Quality Management (TQM).

#### *14.2.1.1 FOOD SAFETY*

According to FAO "*Food safety is all about handling, preparation and storing of food in order to prevent infection and diseases and to ensure that our food keeps sufficient nutrients for us to have a balanced and healthy diet because food and water which is unsafe means that it was exposed to any kind of external matter such as dirt and germs, or it may even be rotten, which can in turn cause various infections or diseases such as diarrhea, salmonellosis, meningitis etc.*" It further states that the ramifications of unsafe food however go a long way and that too beyond human suffering. The contaminated food affects socioeconomic development, healthcare system overloading and retards economic growth and trade. The threats of food safety cause an enormous pressure on economies from disturbances or restrictions in the trade facilitation whether global or local, food and associated income loss to wastage of natural resources.[8]

#### *14.2.1.2 FOOD SAFETY SYSTEMS*

The complete food safety system is governed by three sectors which are consumers who drive the need of food safety, the food industries, and the public health agencies which ensures that food processors have an obligation of assuring that they produce the safest food possible in all the sectors[10] (Figure 14.1).

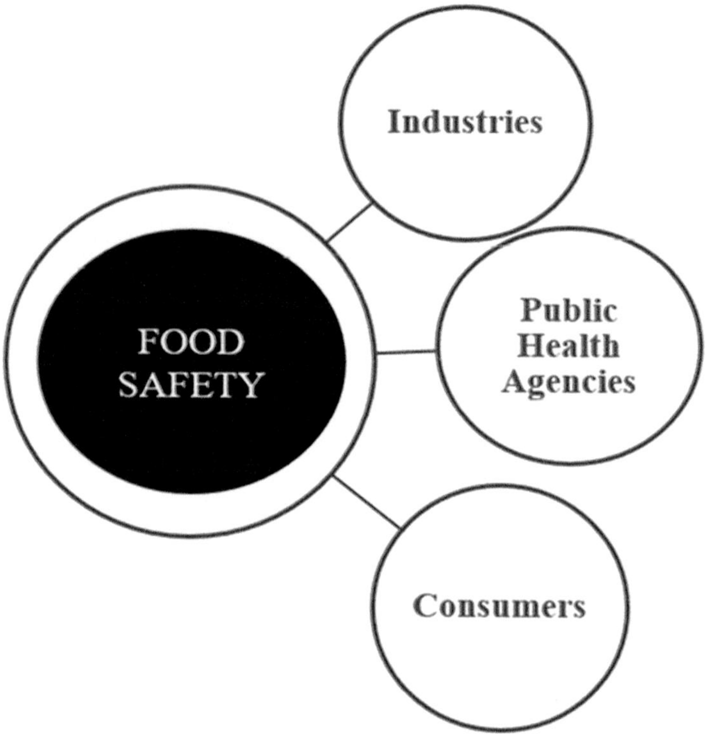


**FIGURE 14.1** Three sectors of a food safety system.

#### *14.2.1.3 TQM (TOTAL QUALITY MANAGEMENT)*

TQM is a holistic term used in maintaining quality in an organizational setup. The term "Quality" here refers to the ultimate customer satisfaction and target-based quality which result in the final product. TQM is a perfect depiction of life, attitude, and organization of a company which work for quality service to customers that fulfill their desired requirements. It is a collective approach of management strategies and organizational effort. It is primarily centered on quality of the management based on the effective and collective participation of all its members which together see a long term success with parameters decided for ultimate satisfaction of its customers as well as adding some value to the society.[19] TQM applies to any kind of industry and organization; and in food industry it is mostly centered towards maintaining the food safety and quality through KAIZEN approach which is a part of TQM only and focuses on making "small improvements" anywhere in the food industry. TQM clubs all its three words in execution as shown in Figure 14.2:

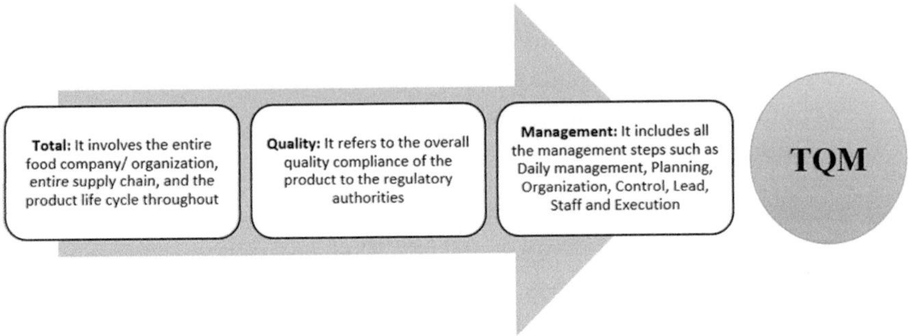


**FIGURE 14.2** Total quality management (TQM).

*Principles of TQM*

- Management commitment: In this step, PDCA cycle comes into play which is 1. Plan, 2. Do, 3. Check, and 4. Act
  **Plan:** This step involves lying down the objectives and processes to deliver a particular food product and service with desired results in accordance with the customer satisfaction.
  **Do:** Implementing the planned process
  **Check:** Continuous monitoring and measurement of the processes (e.g. HACCP) and products in compliance to the regulatory authorities, policies, objectives, requirements, and detailed documentation.
  **Act:** Taking various actions for continuous improvement of process performance.
- Employee empowerment: This step is followed in an organization to manage practice of sharing information, rewards, and power with all the employees in order to encourage them for better decision-making and problem-solving skills. It works on four key factors viz., Empower, Enhance, Enable, and Engage.
- Fact-based decision making: This step of TQM requires continuous collection and analysis of data for improvement of decision-making process with a higher degree of precision, prediction of challenges, and analysis of past mistakes.
- Continuous improvement: This step takes into account all the benchmarks of excellent practices and instilling a sense of employee ownership. It works on four key factors which are: Key matrices and

reporting, Analysis and understanding, Modeling and planning, and Decisive action.

- Customer focus: Lastly but most importantly, an organization aims to target the customer expectations by defining all its processes and functions with the point of view of customers. It includes maintaining a good relationship with all its partners such as the supplier, service providers, and customers and prioritizing their needs without compromising in the quality.

There are many factors which affects the TQM in any organization, such as top management commitment is needed to ensure smooth application of TQM in organization; similarly, other qualities like focus, relationship and rapport building with clients/customer, leadership and emotional intelligence are also needed for the same.[19] Manufacturing sectors are highly impacted by ownership and employee engagement.[1] Morath and Doluschitz[22] did a research project to analyze that to what extent the food companies in the food industries in Germany are applying different activities of the TQM and found that this approach is very useful for improving their business performance but its application is limited to some extent in food industries as compared to others; hence it is not able to fetch its full potential in all the industries. Heerden and Jooste[13] proposed various guidelines for the integration of TQM and physical asset management techniques in South African food and engineering institutes and their results identified that the system integration in TQM acts as a central phenomenon for improving the food safety and quality.

## 14.3 QUALITY AND SAFETY MANAGEMENT IN THE FOOD INDUSTRY

After understanding the operating principles of FSQMS, it is insured in a food industry that a food handler knows all the basic food safety management systems which governs the production of safe food and ensures that it travels safe from farm to fork. It includes various management approaches like Risk Analysis, HACCP, Auditing, Standard and Accreditation, ISO series application, Laboratory Quality Management Systems, and Retailer Standards.

### *14.3.1 RISK ANALYSIS*

CAC (Codex Alimentarius Commission) is a joint entity formed by WHO (World Health Organization) and FAO (Food and Agriculture

Organization) advisory committees of all the member states and subject committees in 1963. Risk analysis was introduced by CAC only, it is defined as a continuous process consisting of three components, such as: (1) to assess risk, (2) to manage risk, and (3) to communicate risk.[16] These three components are further broken down to their respective key factors as shown in Figure 14.3.

On the basis of this framework (Figure 14.3), food industries dealing with different commodities and products use these factors to design and assess their risk-based tree in order to minimize the losses and to avoid hazards. It is a framework which works on science-based standards requiring knowledge and understanding of modern food safety and quality practices along with public health institutions and infrastructure, and provides a cohesive environment which supports and values the risk analysis (RA) process. RA is integral to all food systems. It includes legislations, policies, analysis of food lab and premises, timely inspection, monitoring contamination by surveillance and communication, information and education of outbreaks and food-borne diseases.[7]

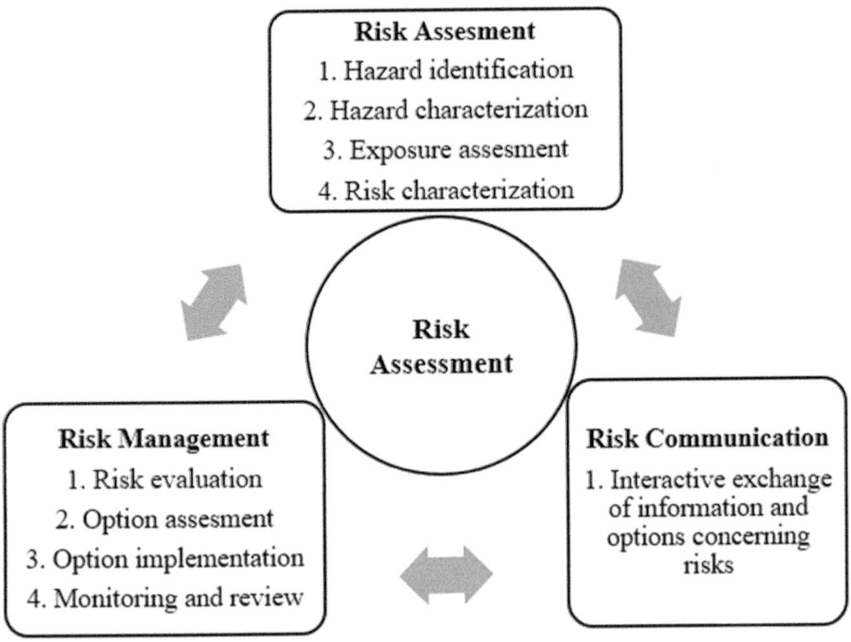


**FIGURE 14.3** Risk analysis and its components.

*Source:* Adapted from FAO/WHO (1997).

To manage safety risk in food industries various precautionary and uncertainty analyses are routinely done which often include industry-specific steps for precautionary measures. These measures should be specific to each hazard and should be well reflected by the various risk management options on scientific basis.[2]

#### 14.3.1.1 APPLICATIONS OF RISK ANALYSIS

- Formation of new standards, e. g., processing standards for seafood, eggs, poultry, etc.
- Incorporating the changes in already existing standards such as approval or deletion of food additives, adding novel foods, and removing genetically modified foods.
- Evaluation of existing standards.
- To evaluate current food-based approach in terms of technologies and its impact.
- To address all queries about safety and quality which can be a threat and arises from risks in imported and domestic foods. It helps in identifying and considering emerging health risks associated with food and in managing the health response to all domestic and international food incidents.
- Evaluation of existing and already proposed standards of food where various risks related to health and safety have changed over a period of time.[5]

### 14.3.2 HACCP—HAZARD ANALYSIS CRITICAL CONTROL POINT

It is one of the advanced FSMS. HACCP is defined as "a collaborative system which undergoes identification, evaluation, and control of hazards which are very significant for the food safety and quality of a process" (definition by CODEX).

More broadly HACCP is a system that gives full confidence to the management that food safety is being managed effectively because it is laid on scientific and rational basis of food safety, keeping all the potential hazards and critical control points (CCPs) in food processing and development of continuous monitoring procedures in real time which determines if the hazard identified can be effectively controlled or not. HACCP is a preventive, systematic, and proactive approach of food safety which is based upon seven principles developed to control the physical, chemical, and biological hazards right from farm to fork.[25]

#### *14.3.2.1 BASICS FOR APPLYING HACCP IN FOOD INDUSTRY*

There are various prerequisites to be understood and acted upon before establishing a complete HACCP team, therefore these should be understood as per the protocol (discussed below under blocks 1 & 2) and then a HACCP plan should be laid out provided that the plan is specific for each and every food industry; for example, Shi[26] proposed a management system for analyzing the hazards and critical control points which helped to tackle the issues in the FSMS of hotels where it played a significant role in reducing the microbial load during processing which was different for different services in the hotel. Hence, HACCP proved to be a very effective management system on the hotel's FSQM.

#### ***Block 1: Good Manufacturing Practices (GMP) and HACCP prerequisites (GMP discussed in Section 14.4—with other supportive FSQMS)***

Though GMP is a supportive FSQMS, it is usually coupled with HACCP to strengthen its applicability and to reduce the occurrence of hazards. Following are the steps included in HACCP in the first block:

#### ***Preparation and Planning***

1. Understanding and awareness of the concept of HACCP
2. Formation and training of team
3. Auditing of the available support systems and gap analysis
4. Planning of HACCP project

   HACCP plan will include the following:

   a) Product description
   b) HACCP team list
   c) HACCP control charts
   d) Hazard analysis charts
   e) Process flow charts

***Block 2: Application of seven principles of HACCP***

1. Analyze hazards
2. Acknowledge the CCPs (critical control points)
3. Establish range/limits for CCPs
4. Set up a monitor control of CCPs
5. Set up the corrective action procedures
6. Verification for checklisting that applied HACCP is working properly or not
7. Set up the record keeping and document procedures
    a) The record keeping includes the following:
    b) Monitoring records
    c) CCP record reviews
    d) Audit reports
    e) Training reports

### *14.3.2.2 BENEFITS OF HACCP IMPLEMENTATION IN FOOD INDUSTRIES*

| Consumers | Industries | Regulatory authority/Government |
|---|---|---|
| 1. Decreases the risk of food-borne illnesses<br>2. Strengthen food supply chain | 1. Reduces business risk<br>2. Cost cutting in losses which incurs due to spoils, wastes, and recalls | 1. Public health improvement, hence reduction in public health investment<br>2. Facilitates international trade |
| 3. Increases awareness and quality of life | 3. More exposure to market and value chain | 3. Strengthen the confidence to do bigger trades |

### *14.3.2.3 FACTORS AFFECTING HACCP IMPLEMENTATION*

Apart from the benefits of implementation of HACCP in the food industry there are many factors which make its implementation difficult in many scenarios. The risks and difficulties in HACCP implementation in Russian enterprises of food industries were researched and it was found that there were two kinds of risks: external and internal. External ones being more significant and permanent in nature as compared to internal such as harmony imbalance with regulatory framework in terms of HACCP requirements, complex presentation of information, complex use of terminologies, poor understanding of HACCP, not enough information materials for training, and application of HACCP in an organization; whereas the internal ones quoted were low level of awareness and education, mismanagement in personnel of catering enterprise, lack of appreciation and rewards, and non-compliance in record keeping and verification.[4]

Apart from these internal and external factors there are few others quoted by Milios et al.[21] who addressed different factors and technical barriers which affects HACCP implementation in food industry. It was found that lack of awareness, lack of training, owners not perceiving HACCP a benefit approach, management regressions, lack of specifications in product and production lines, and cost of implementation and maintenance were negatively impacting the performance and implementation of HACCP system.

### *14.3.3 AUDITING AND ACCREDITATION*

**Auditing** in food safety means a functionally independent and systematic series of examination of the existing food safety and quality measures taken up by production units to check whether they match the standards and claims which has been made by the company being audited.[9]

**Accreditation** is getting registered and certified by direct accreditation or by the respective authority designated by bodies like FDA which permits it as a third party which meets the desired requirement of manufacturing a certain product or to carry out a specific task. The process of accreditation involves a group of people forming a panel called **accreditation body** which possess the authority to perform accreditation of third parties. These third parties are called **certification bodies.**[28] This regulatory framework works in a specific manner as followed by FDA (Figure 14.4).

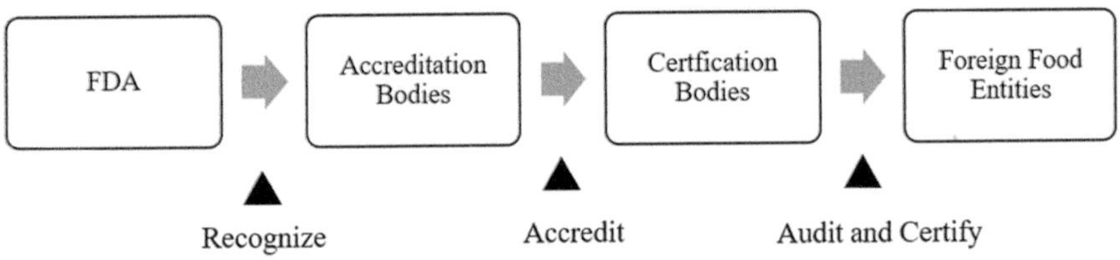


**FIGURE 14.4** The FDAs third party certification program working framework.[20]

There are various points, which are considered by a food industry while going for a certification by this method. Mostly the only points of contact for the food businesses are the scheme owner and the certification body. The points which are considered are[11]:

- Selecting the right scheme
- Selection of a certification body
- Applying for certification
- Scheduling
- Certification audit

Various audit tools are used by the auditing bodies so as to maintain the standards of certification throughout all the sectors of food industry. Considering certification and compliance with the same, it helps the industries and small businesses to repute their brands, maintain good customer relations and in keeping themselves safe from legal matters.[12]. Examples of tools used are matrices, Kaizen tools, process charts, and diagrams. They are now been amalgamated with upcoming software technology such as advanced food safety software which can customize the way a standard is given, customer or regulator requirement etc.

### *14.3.4 INTERNATIONAL ORGANIZATION FOR STANDARDIZATION (ISO) SERIES*

ISO is an autonomous, non-profit NGO with networks globally for which every country has a representative. They make international standards abbreviating them as ISO series. ISO have its coordination point as Central Secretariat in Geneva, Switzerland. Their financial developments are met by selling these practical tools and standards which is common between the interested parties. They have laid out many standards for foods to address the challenges in safe and sustainable ways of handling foods by mentoring and best practices for producing food products in hygienic environment, analysis, promotion of FSQMS through food and allied industries for the help of these internationally accepted standards[18] ISO have different series for food which are **ISO 9001:2015, ISO 22000:2018, ISO/TS 22004:2014, ISO/TS 22003:2013, ISO 22005:2007, ISO17021-10:2018, and ISO 15189:2012.**[27]

#### *14.3.4.1 ISO 9001:2015*

This standard provides the requirements for the QMS which are generic and is open to many organizations irrespective of its genre, size, production, services, and customers to which it caters. It demonstrates ability of an organization that it can consistently provide products and services exactly what it promises to deliver. It also aspires to increase the customer satisfaction through its improvement, implementation, and the compliance with the confirmatory standards to the intended user, the regulatory body, and statutory requirement.

#### *14.3.4.2 ISO 22000:2018*

FSMS–It gives standards and prerequisites for various organizations which are operating in food chain and its requirements for a FSMS to ensure quality and safety throughout the food chain, till it reaches to end consumer. It has been revised from the series 22000:2005 with lateral corrigendum number 1:2006.

#### *14.3.4.3 ISO/TS 22004:2014*

FSMS–It provides amended regulations for the erstwhile series ISO 22000:2005 which includes guidance to compliance with ISO 22000, and FSMS like HACCP and is designed in such a way that it is applicable all together with all the relevant standards published by ISO.

#### *14.3.4.4 ISO 22005:2007*

It provides the regulations and requirements for traceability (includes feed chain also) with guidelines for designing and applicability in the feed and food system.

#### *14.3.4.5 ISO/TS 22003:2013*

It provides guidelines for certification, auditing, and accreditation of different FSMS. It was last reviewed and confirmed in 2016 from its earlier version ISO/TS 22003/2007 and is under review to be replaced by ISO/CD 22003-1.

#### *14.3.4.6 ISO 17021-10:2018*

It provides standards for conformity assessment in which it provides standards and requirements for different management systems of the auditing and certification bodies, in which the part 10 contains competency guidelines for certification and auditing of facilities getting audited for OHSAS (Occupational Health and Safety Management System).

#### *14.3.4.7 ISO 15189:2012*

Laboratory Quality Management Systems–It introduces standards for lab quality, competence, processing lines, equipment and facilities, personnel hygiene, assessments: audit, accreditation and quality assessment, food personnel requirement, services and catering, record keeping, organization, and management. It has been reviewed from ISO 15189:2007 to ISO 15189:2012 and is under review to be replace by ISO/CD 15189.[29]

There are subsections of these series which covers food products, FSMS, microbial protocols, marine and fisheries, and carbohydrates and starches along with their byproducts and essential oils and these subsections further breakdown into vertical and horizontal standards of their major categories as followed[18]:

- ISO/TC 34 for Food products, ISO/TC 34/SC 5- Milk and milk products, ISO 20633 for Adult nutrition and infant food products, and ISO/TC 34/SC 4- Cereals and pulses.
- ISO 22000 for FSMS- Provides various conditions for any food organization to own a subdivision and ISO/TC 34/SC 17 provides various management tools and techniques for safe food.
- ISO/TC 34/SC 9 provides microbial protocols which have subdivisions, ISO 16140 provides requirements and microbial protocols for food and it's value chain and its validation methods, ISO 6579-1- provides various horizontal standards and methods to detect, enumerate, and serotype *salmonella*, ISO 10272-1 provides microbial protocol for animal feeds (horizontal method includes detection of campylobacter spp.), ISO 11290-1- provides horizontal standards for enumeration and detection of *L. monocytogenes*, ISO 15216-1- provides virus determination protocols, ISO 16654- provides method of detection for *E.coli* O157, ISO 19020- provides horizontal method for detecting immuno-enzymatic reactions of enterotoxins released by *Staphylococcus* spp., ISO 22964- provides microbiology of the food chain- horizontal method for the detection of *Cronobacter* spp., ISO 12875- provides traceability of finished products- specification for captured finfish and its value chain, and ISO 12878- provides sustainability and environmental aspects of fisheries and marine life.

**ISO**, apart from focusing purely on food safety looks at business processes and structures as well. At the same time, it is independent in nature, hence companies can decide for themselves whether they want to excel and embrace or not, unlike HACCP which is mandatory for food industries in order to assure food safety and quality to end customers. Franzosi and Estorilio[24] surveyed and analyzed various companies in food industry for its maturity in product development process and their obscurity to the status of their accreditations and certifications. The study highlighted that amongst all the companies surveyed, there was certain level of maturity in management, planning, and execution of FSMS which was required to get the ISO certification because the companies already possessing the ISO certification were more stable in market. The maturity level meant a proper management, an organization which is compatible with the products and the service requirements which are fulfilled by them is managed sufficiently and the processes involved are safe, well-planned, and operated in a controlled manner.

### 14.3.5 RETAILER STANDARDS

- British Retail Consortium Global Standard- BRCGS (Erstwhile BRC) Food and IOP
- International Food Standard (IFS)
- Safe Quality Food Standard (SQF: 1000 SQF: 2000)

The GFSI (Global Food Safety Initiative) recognizes and benchmarks the most preferred standards in retail, upon which the customers can rely and they are most looked upon when designing a Total Quality Management System (TQMS) and considered while considering standard identification. The GFSI have given approval to five standards namely: the British Retail Consortium Global Standard (BRCGS), the International Food Standard (IFS), the Safe Quality Food Standard (SQF: 1000 SQF: 2000), the Food Safety System Certification (FSSC) 22000; and the Dutch HACCP-based food safety system.[13]

#### *14.3.5.1 BRCGS (ERSTWHILE BRC) FOOD AND BRCGS INSTITUTE OF PACKAGING (IOP)*

BRCGS (which was earlier known as BRC) is a global standard which has specified requirements for retailers, manufacturing, and food service organization. It was established in 1992, as a trade facilitating authority for guidelines and legislation in UK which now represents the retail trading in

145 countries.[3] It has developed and released various standards for third party BRCGS certification in following sectors: Packaging of food and packaging material, food safety, distribution and storage facilities and methods, retail and wholesale trade, consumer products and services, ethical trading, broker and agents, appropriate sourcing of raw materials and goods, certification programs such as gluten free, toxic substances awareness, plant and plant product certifications. Both of these standards (BRCGS Food and BRCGS-IOP) had laid the standards and elements to be fulfilled as per BRC/IOP for food packaging and package production industries.[17]

BRC has six retailer standards, which are implied in wholesale and retail customer complaints and issues; it includes all the e-commerce retail and wholesale guidelines for food safety and security, legislations against frauds and crime. BRC— (Food standards) It also provides a globally accepted standard common for certifying the food industries trading into supply chain, provides assistance in fulfilling the legal demands of manufacturers, protects consumer rights and its standards, covers topics like HACCP, QMS, process and product control and factory environment standards. BRC (packaging standards) provides the criterion regarding the best practices in packaging and package manufacturing industries of feed, food, and other edible and non-edible products. An industry compliancing with the same gets the benefits of staying in good books of raw material suppliers, manufacturers and, the retailers. These guidelines under this sector covers technical and risk management standards, industry planning standards, controlling contamination and personnel hygiene and safety standards.[6]

#### 14.3.5.2 INTERNATIONAL FOOD STANDARD (IFS)

Under GFSI only, the members of German retail federation with its French and Italian counterparts have formed IFS which draws standards for safe and quality food for retail and wholesale branded food products. It applies to all the post-agricultural steps of food processing and the retailers of all over the world and is now accepted by many countries.[14] The purpose of IFS standard is very well emphasized in its objectives which are as followed[23]:

- Ensuring transparency and traceability throughout the supply chain.
- Improvement in confidence and self-protection of the consumers.
- Improving retail services.
- Novel products standardization and evaluation as self-form brands.
- Reduction of cost and time saving.

- Uniformity in evaluation system with a common standard.
- Unique formulation and implementation of the auditing in the setup.
- Working with accreditation organizations and personnel's skilled as evaluators.

#### *14.3.5.3 STANDARD FOR SAFE QUALITY FOOD (SQF: 1000 AND SQF: 2000)*

SQF standards were developed by AGWEST Trade and Development (the state of Western Australia) and SQF 1000 and SQF 2000 are the two different quality codes as per their application and utilization. On one hand, where SQF 1000 code has been designed primarily for food manufacturers and growers; SQF 2000 has a wider utilization approach reaching up to food manufacturing, distribution, and retail sectors. These standards are regularly updated by the SQF Institute Technical Committee. SQF emphasizes on the following key points[15]:

- Aims to minimize the frequency of auditing which are erratic and repeatable in nature.
- Enables manufacturers and producers to showcase "Due Diligence" along with compliance standards for traceability and regulatory requirements.
- It encourages cost cutting through skipping the spoilage, malfunctioning, and waste production part and enabling "one system, one audit."
- It provides a pathway for value chain facilitating personnel and retailers to get a lead over the non-authorized ones to encourage the gain in terms of profit as well as trade relations.
- This standard is globally accepted because of its suitability to all horizontal and vertical standards and for all kind of food producers and suppliers.

## 14.4 OTHER SUPPORTIVE FOOD SAFETY AND QUALITY MANAGEMENT SYSTEMS

Apart from basic FSQMS there are various management systems which are supportive FSMS such as Good Agricultural Practices (GAP), Good Hygiene Practices (GHP), Good Manufacturing Practices (GMP), Good Retail Practices (GRP), Good Transport Practices (GTP), Good Distribution

Practices (GDP), and Nutritional Labelling and Traceability. These management systems preach quality and safe handling practices throughout the food chain, thus ensures that the food reaches its intended consumer in the safest form possible with quality at par level needed.

These standards are extremely important and includes participants, who are responsible for food safety and quality from farm to consumer table including all the processes like agricultural production where Global GAP and country's own GAP standards are applied, transportation of goods in a supply chain requiring a good understanding of GTP, whereas GHP and GMP ensures quality and safety in the production, packaging, manufacturing, and inventory management system. Retail and wholesale of goods (unprocessed or processed) implies GRP and GDP at all the steps of sale to ensure safety and quality to the final consumer and to the supplier as well.

To support these management systems there is a role of nutritional labelling which speaks to the consumer about its type and nature on and off shelves without involving someone, to ensure that the customer picks up a right product for themselves. Traceability plays a huge role in the complete food chain so as to ensure a correct backup for every mishappening, to trace back any issues in the food chain or just for conducting a self-audit and inspection, if required.

## 14.5 SUMMARY

Now-a-days having a safe and quality food is a matter of great concern and with increasing global exchange of dietary diversity the problems like food fraud, food recalls, food-borne diseases and various other risks which are associated with food such as allergies, hazards in industries are on an escalating trajectory as well. To combat all these issues food industries worldwide look for a comprehensive solution which can cater all these issues way back before the food reaches the final fork. Standards and regulations are to be understood as different food safety along with all quality management systems which compliances with the respective regulatory framework mandatory for ensuring safety and quality to the commodities being traded and handled in the food industries; and therefore in order to implement the same each and every person in the food industry and in the supportive framework like farm, supply chain, wholesale, and retail needs to acknowledge the basic mechanism and application of various FSQMS of food industry. The concept of FSQMS starts with basic principles of food safety and quality management, understanding its systems which are: risk analysis, HACCP, auditing, and accreditation, ISO, retailer standards (BRCGS, IFS and SQF), TQM and few supportive systems such as GAP, GHP, GMP, GRP, GTP, and GDP. These supportive systems help in reducing the chances of failure of core systems of FSQM and hence, applying them should be prioritized before applying the core systems. These standards do require timely amendments because of the changing food safety laws which are often influenced by changing consumer behavior, dietary pattern, and globalization. Because all the systems of FSQMS are interconnected with their allied sectors such as nutritional labelling and traceability, an understanding of other sectors should also be taken into account before the application of any FSQMS.

## KEYWORDS

- **food handler**
- **food quality**
- **food safety**
- **FSSAI**
- **HACCP**
- **ISO series**
- **quality management**
- **TQM**

## REFERENCES

1. Arora, N.; Gupta, S. Total Quality Management for Employee Engagement: A Study. *Test Eng. Manage.* **2020**, *82* (1), 12769–12786.
2. Bruno, A. Risk Analysis and Science in Codex by FAO and WHO. *FAO/IAEA Int. Symp. Food Saf. Qual.* **2014**, *1*, 10–12.
3. CERMET working paper. *Conformity Assessment Service: "BRC—Global Standard Food" Product Certification—Basic Requirements*, **2007**, *1*, 1–7.
4. Chernova, E.; Bazhenova, I.; Tsupranova, L.; Bychenkova, V.; Bazhenova, T. Identification and Analysis of Difficulties in the Implementation of HACCP in Russian Food Industry Enterprises. *E3S Web Conf.* **2020**, *161*, 1–8. https://doi.org/10.1051/e3sconf/202016101076.
5. Commonwealth Department of Health and Aging National Food Incident Response Protocol. Report on Addressing Food-Related Health Risks. *Food Standards Australia New Zealand Risk Analy. Food Regul. Rep. 24*, **2012,** *24*, 1–7.

6. Egyan Kosh. *Manual: BRC Food and BRC/IOP Standards—An Overview* **2005**, *16*, 5–18.
7. FAO/WHO. *Food Safety Risk Analysis Part I: An Overview and Framework Manual Provisional Edition*; FAO: Rome, June 2005; pp 1–76.
8. Food and Agriculture Organization of the United Nations (FAO). *Report—The Future of Food Content*, 2019; pp 1–28.
9. FSSAI (Food Safety and Standards Authority of India). *Report—Guidelines for Accreditation of FSMS Assessment Agencies*, 2012; pp 1–37.
10. Gardner, S. Consumers and Food Safety: A Food Industry Perpective. *Food and Agriculture Organization of the United Nations—Rome*; 1993; *Report 1*; pp 1–5.
11. Global Food Safety Initiative. *Report—Certification and Accreditation Framework*, 2011; pp 1–8.
12. Global Food Safety Initiative. Food & Beverage Industry Standards. *Get a Quote*; 2020; pp 1–10.
13. Heerden, van M. A.; Jooste, J. L. Guide for Integrating Total Quality Management and Physical Asset Management in the Food Industry. *SA J. Ind. Eng.* **2018**, *29* (4), 155–170.
14. IFS Audit Portal. *Working Paper–IFS—Certification*; 2010; pp 1–4.
15. IGNOU. E—Gyankosh SQF 1000 and SQF 2000. *Food Saf. Manage. Syst.* **2000**, *18*, 37–51.
16. International Programme on Chemical Safety (IPCS). Risk Assesment and Its Role in Risk Analysis. *Principles Methods Risk Assess. Chem. Food* **2009**, *1* (2), 1–16.
17. International Trade Centre (ITC). *Manual–BRC Global Standards Food* **2013**, *1*, 4–6.
18. ISO. ISO and Food. *ISO-Standards*, **2017**, *1*, 1–16.
19. Kumar, V.; Singh, J.; Kumar, D.; Antil, M. Total Quality Management. *Natl. J. Adv. Res.* **2016**, *2* (3), 5–8.
20. Mermeistein, N. H. Auditing Food Safety. *Food Technol. Magaz. Article*, **2018**, *1*, 1–11.
21. Milios, K.; Drosinos, E. H.; Zoiopoulos, P. E. Factors Influencing HACCP Implementation in the Food Industry. *J. Hellenic Vet. Med. Soc.* **2012**, *63* (4), 283–290.
22. Morath, C.; Doluschitz, R. Total Quality Management in the Food Industry: Current Situation and Potential in Germany. *Appl. Stud. Agribusi. Comm.* **2009**, *3* (3–4), 83–87.
23. Nikolic, A.; Mujcinovic, A.; Memic, A., Eds. IFS (International Food Standards). *24th International Scientific—Expert Conference of Agriculture and Food Industry—Sarajevo 2013 Professional* **2013,** *1*, 301.
24. Oliveira-Franzosi, D. L.; Estorilio, C. C. A. Relationship between PDP Maturity and ISO Certification in the Food Industry. *Gestao e Producao* **2019**, *26* (4), 1–19.
25. Rotaru, G.; Sava, N.; Borda, D.; Stanciu, S. Food Quality and Safety Management Systems: Brief Analysis of the Individual and Integrated Approaches. *Agroalimentary Processes Technol.* **2005**, *XI* (1), 229–236.
26. Shi, Z. Study on Food Quality and Safety Management Based on Hotel. *De Gruyter* **2017**, *XXI* (2), 91–96.
27. Tranchard, S. All ISO Standards for Food Safety Management Systems. *Int. Org. Standard. (ISO)* **2007**, *1*, 1–4.
28. US-FDA. Report by Third Party Certification Body Accreditation for Food Safety Audits: Model Accreditation Standards. *Guidance for Industry and FDA Staff* 2016, 240–402.
29. World Health Organization (WHO). *Quality Management System Handbook* 2011.

## CHAPTER 15

# SAFETY ASPECTS AND ROLE OF FUNCTIONAL AND NUTRACEUTICAL FOODS

AMRITA POONIA

## ABSTRACT

Our health depends on the consumption of healthy, nutritious, functional, and nutraceutical foods that are helpful in combating many chronic diseases (such as: cancer, obesity, diabetes mellitus, disorders of gastrointestinal tract, etc.). Consumption of bioactive compounds in plant-based foods attracts attention of the consumers. Enough literature is not available on (1) the inclusion of these bioactive compounds derived from fruits and vegetables is helpful to fight diseases; and (2) to depict the relationship between health and diet; and to validate these sufficiently.

## 15.1 INTRODUCTION

According to Chumphukam et al.,[2] "*functional food is defined as the food which is consumed in form of normal diet and shows physiological benefits for reducing the risk of incessant disease apart from basic nutrition.*" A nutraceutical can be defined as a diet supplement that provides a concentrated form of bioactive compound of food in a non-food matrix to enhance health and prevent disease. Kailasapathy[6] defined "*nutraceutical as natural functional or bioactive compounds obtained from plants and used in the form of pills, capsules, powders & liquids which provides physiological benefits from incessant diseases.*" According to recent literature, about 82%

Plant-Based Bioactive Compounds and Food Ingredients: Encapsulation, Functional, and Safety Aspects.
Junaid Ahmad Malik, Megh R. Goyal, Preeti Birwal, & Ritesh B. Watharkar (Eds.)

of cardiovascular disease patients preferred nutraceuticals in form of health-enhancing compounds. Functional and nutraceutical foods play a critical role in reducing the burden on health system by their preventive mechanism of action. Knowledge and health trends are main causes of growth of functional food market.

## 15.2 BENEFITS OF FUNCTIONAL AND NUTRACEUTICAL FOODS

- Functional food is designed to minimize an existing health risk problem like high cholesterol and blood pressure.
- Latest food products enriched with health attributes have gained popularity among the masses as they are believed to offer consumers an enhanced ability to mitigate the risk of certain diseases.
- Ma-kinen-Aakula[7] reported that lactose-free and gluten-free functional foods are also beneficial to the consumers.
- To improve the regular stomach and colon functions by consuming prebiotics and probiotics. It also helps children to improve their learning capability and behavior.

## 15.3 SOURCES OF FUNCTIONAL AND NUTRACEUTICAL FOODS

### *15.3.1 PLANT-BASED FUNCTIONAL FOODS*

- **Primary metabolites:** Plant proteins, omega-3 fatty acids, and β-glucans are present in this category.
- **Secondary metabolites:** This category includes phytoestrogens, vitamins, antioxidants, steroids, gamma-linolenic acid (GLA), and tocopherols. Phase II enzyme inducers (found in Brassica vegetables) act as functional foods by glycosylating insoluble toxins to produce soluble compounds that are excreted, are also included in this category. Various sources of plant-based functional foods are rice bran, oats, flaxseed, soy, garlic, and wine (grape), tea (green tea and black tea).

### *15.3.2 ANIMAL SOURCE BASED FUNCTIONAL FOODS*

- **Zoochemicals:** Animal-derived functional foods include omega-3 and omega-6 fatty acids, conjugated linolenic acid (CLA), small peptides, whey and casein, and glucosamine. Omega-3 fatty acids include alpha-linolenic, DHA, and EPA fatty acids. Various sources of functional foods from animal sources are dairy products (probiotics) and fish.

### *15.3.3 MICROBIAL SOURCE BASED FUNCTIONAL FOODS*

- **Probiotics** are living microorganisms driven in adequate number to endure in the intestinal ecosystem such as *L. casei* or *Bifidobacter* species, which bolster health.
- **Prebiotics** are those dietary components which bolster growth of the probiotic bacteria.
- **Synbiotics** is a heterogenous mixture of probiotics and prebiotics taken in specific quantities which is beneficial to one another.

### *15.3.4 MISCELLANEOUS FUNCTIONAL FOODS*

- **Algae:** Algae works by providing omega-3 fatty acids, which boost immunity, regulate inflammation, and counter to neurodegenerative diseases.
- **Mushrooms:** Functional foods obtained from mushrooms consists of antiviral, sterile, and anti-inflammatory properties.

## 15.4 SIGNIFICANCE OF FUNCTIONAL FOOD AND NUTRACEUTICALS

- **Vast and growing market:** There is huge and expanding market of functional food and nutraceuticals which is eyed by the various pharmaceutical companies that are now trying to foray into the functional food and nutraceuticals business.
- **Consumer's faith:** The producers and consumers of functional foods and nutraceuticals are aware about the advantages of these products and have faith too. They are also aware that these products are non-toxic and harmless as compared to the conventional pharmaceuticals.
- **High health care costs:** The cost of the other medicines and pharmaceuticals is comparatively more and also has some side effects, novel regulations, and scientific discoveries.

- **Lifestyle diseases:** The main reason of the lifestyle disease is the wrong dietary habits and sedentary work. This may lead to diabetes, heart problems, cancer, and high blood pressure. In literature, the role of functional and nutraceutical food is speculated.
- **Consumer Knowledge and Popularity:** Functional foods and nutraceuticals are very popular among consumers because of their safety and health benefits. There are about 5,000 nutraceutical products that are available in the market worldwide. Sutar et al.[9] identified factors affecting the popularity of functional and nutraceutical foods (Figure 15.1).

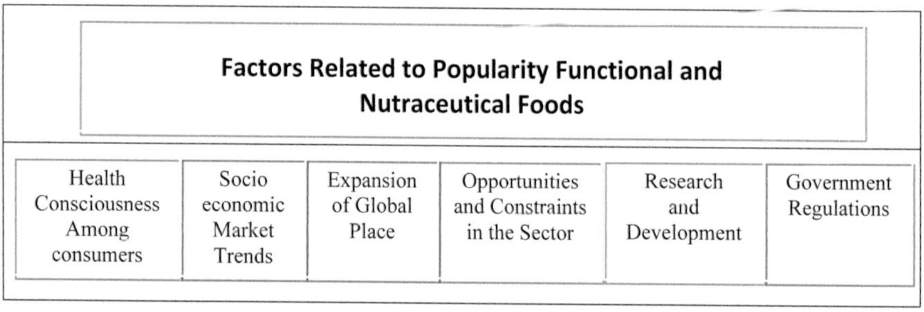


**FIGURE 15.1** Factors affecting the popularity of functional and nutraceutical foods.

- **Public acceptance and availability:** Functional and nutraceutical foods are well accepted by the consumers. Increased knowledge on nutritional aspects, medicinal plants, and biotechnology has brought a great change about food and health. Zhao[10] reported that nutrition therapy and phytochemicals are the new concepts in recent years.

## 15.5 NUTRITIONAL CONTENT CLAIM OF FUNCTIONAL AND NUTRACEUTICAL FOODS

Nutritional claim of functional and nutraceutical foods is given in Table 15.1. *Codex Alimentarius*—FAO[3] has defined two types of nutrition claims:

- **Nutrient content claim:** This claim represents the nutrient composition of a food. Nutrition claim characterizes any representation stating, suggesting or mentioning about particular nutritional properties of a food including but not only limited to the fat, energy value, protein content and carbohydrate, as well as the content of minerals and vitamins.

- **Nutrient comparative claim:** This parameter correlates the nutrient levels of two or more than two foods.
- **Health claims** represent, suggest, and signify that a link exists among a food category, a food or one of its constituents and health, that is,
- The impact of a particular nutrient to the growth and development of the organism function;
- The impact on psychological function, behavior;
- Weight control or weight, or reduce feelings of hunger or an increase satiety, reduce energy available. The impact on the health and development of children.
- Risk reduction of disease.

## 15.6 HEALTH ASPECTS OF FUNCTIONAL AND NUTRACEUTICAL FOODS

The health and wellness of human being depends upon the consumption of balanced and nutritious diet. In recent years, functional and nutraceutical foods have received much attention because of interests of scientific community, consumers, and food manufacturers. It has been revealed that the diet containing foods based on plants, that is, vegetables and fruits prevent and minimize the chances of chronic diseases like coronary heart disease, obesity, and diabetes mellitus and also promote human health.

Balsano and Alisi[1] reported that plant-based foods promote health due to presence of bioactive compounds. Inclusion of fiber in diet from fruits, vegetables as well as grains is firmly related to risk contraction of CVDs. Hence, there is need to change behavior and lifestyle which is the practical method for reduction of chronic diseases. There is lot of scientific literature which demonstrates the link between consumption of fruits and vegetable rich diets in reducing cancer, mainly epithelial cancers found in the respiratory and alimentary tracts.[4] Different bioactive compounds with their sources and health benefits are listed in Table 15.2.

## 15.7 PROBIOTICS

Probiotics are "*live microorganisms which provide health benefits on the host when administered in optimum amounts. Species of Lactobacillus and Bifidobacterium are commonly used as probiotics.*" The prerequisites of the strains used as probiotics should be nontoxic, nonpathogenic, immune to gastric acid, cling to gut epithelial tissue, resistant to bile acid, and immune to antibacterial mechanisms exploring the gut. Dairy products (namely: ice cream, cheese, yoghurt, and milk chocolates) are good matrices for probiotics because of the presence of whey proteins, thus increasing the viability of probiotics, high pH, fat, solids, buffering capacity, and low oxygen.

**TABLE 15.1** Claims for Nutritional Content of Functional and Nutraceutical Foods.

| Nutritional claim | Source |
|---|---|
| • Low in energy | • Reduced nutrient |
| • Very low sodium/salt | • High content of vitamins and/or mineral/s |
| • Sodium-free or salt-free | • Reduced nutrient |
| • Source of fiber | • Light/lite |
| • Low in fat | • Naturally high in fiber |
| • Fat-free | • Increased nutrient |
| • Low saturated fat | • High content of omega-3 fatty acids |
| • Saturated fat-free | • High content of unsaturated fat |
| • Sugars-free | • No added sodium or salt |
| • With no added sugars | • High content of monounsaturated fat |
| • High in fiber content | • Naturally linked to other nutrition claim |
| • Low in sugars | • High content of polyunsaturated fat |
| • Contains nutrients or other substances | |
| • Source of protein | |
| • Sodium-free salts | |
| • High protein | |

High pH (5.5–6.5) increases the survival rate of lactic cultures during storage; and higher total solids and lipids have protective effect to probiotics. Probiotics are very helpful in prevention of cancer, hepatic disease, antibiotic associated diarrhea, and heart disease. Role of probiotics in the prevention of cancer has been illustrated in Figure 15.2.

## 15.8 SAFETY ASPECTS OF FUNCTIONAL AND NUTRACEUTICAL FOODS

Safety issues of functional foods are becoming popular day by day throughout the world and the safety issues are also triggering brains of researchers as well

**TABLE 15.2** Bioactive Compounds of Functional and Nutraceutical Foods.

| Active compounds | Source | Health benefits |
|---|---|---|
| Alkaloids (Quinine, ergot alkaloids and coumarin, morphine) | Certain types of plants | Anti-obesity, antioxidant, antidepressant, and antidiabetic |
| Anthocyanidins | Fruits and flowers | Antioxidants and anti-mutagenic properties |
| Carotenoids (Leutin, lycopene, α- and β- carotene, zeaxanthin and fucoxanthin) | Fruits and vegetables | Antioxidants, protects against uterine, prostate, colorectal, lung, and digestive tract cancers, improvement of bone density |
| Fiber | A variety of fruits and vegetables | Protects against colorectal diseases |
| Glucosinolates and Saponins | Cruciferous | Activators of liver detoxification enzymes, inhibits the neoplastic effect of various carcinogens, and anti-inflammatory effects |
| Isothiocyanates Sulforaphane | Cruciferous vegetables | Anti-inflammatory |
| Limonoids | Citrus fruits | Inhibiting phase-I enzymes and inducing phase-II detoxification enzymes in liver, provide protection to lung tissue |
| Phenolic constituents (Flavones, isoflavones, flavonones, phenolic acids, anthocyanins, resveratrol, curcumin, flavonols, ellagic acid and ellagitannins) | Various plants and whole grains | Antioxidants, lowers the risk of CHD, Antidiabetic, immunomodulatory, anti-metastatic, anti-inflammatory, radioprotective and beneficial effects in brain functions, neuro- and geno-protective |
| Phytosterols | Various plants | Exhibit anti-inflammatory, antineoplastic, antipyretic and immune-modulating activities |
| Terpenes (Borneol, gossypol menthol and santonin) | | Anticarcinogenic, antihelminthic, prevention of gall stone formation, antimicrobial and antidiabetic |
| Tocotrienols and tocopherols | Grains | Suppressed the growth of diverse tumors cell lines via initiation of apoptosis |

as consumers. It is very important to determine the right doses of the bioactive compounds/phytochemicals to be incorporated into the food systems or to be administered as nutraceuticals. Functional and nutraceutical foods are very important in modern nutrition for promoting health and reducing the danger of chronic diseases. Therefore, it is of utmost importance that these foods must fulfill the standards and safety aspects.

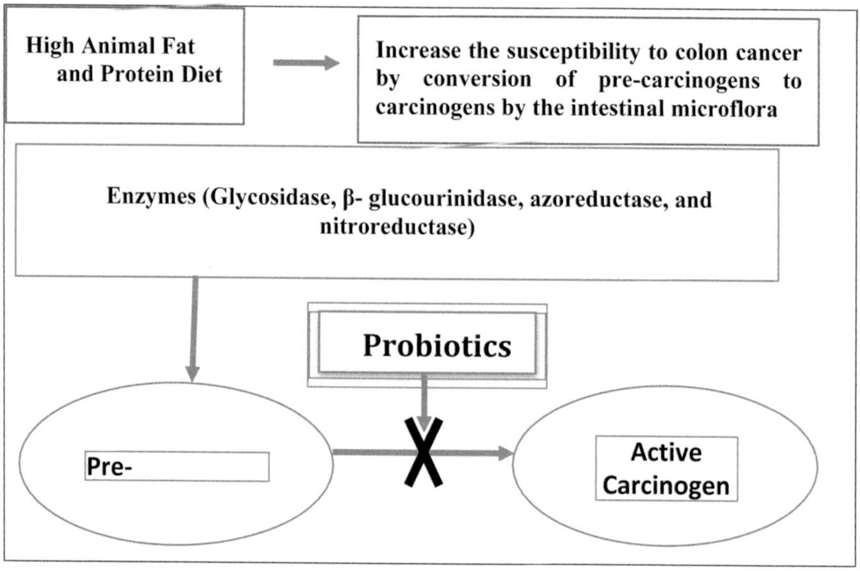


**FIGURE 15.2** Role of probiotics in cancer prevention.

Sufficient evidence is available in the literature on the prospects of consumption of functional foods for long time along with its attributed health benefits. The interactions may occur among the various components of functional foods; and biological processes have also been characterized. Some safety aspects of great concern are adverse metabolic effects, possibility of post manufacturing contamination, enhanced stimulation of immunological systems, and possible gene transfer. It is mandatory for the manufacturer of these foods to show clinical documentation for the lack of these negative effects.

In case of probiotics, the live microbes in use must not significantly endanger antibiotic resistance, nontoxic, and hemolytic activity.[10] According to FAO/WHO,[3] the manufacturer of functional foods should organize placebo-based clinical trials and carry out evaluation of their results in four phases, such as: efficacy, effectiveness, safety, and surveillance. Table 15.3 presents "International regulations for functional and nutraceutical foods" by the selected countries.

The commendations regarding the safety of functional and nutraceutical foods [by General Accounting Office (GAO) of US Government, 2000] are listed below:

- **Develop, promulgate regulation and guidance for industry** on documentary evidence of safety of ingredients, labels and enhance system for recording and analyzing reports related to health problems and correlated to functional and nutraceutical foods.
- **Functional foods** are monitored in the United States by the FDA (Food and Drug Administration) under the authority of following two laws:
- **Federal Food, Drug and Cosmetics Act** (FDAC) of 1938 **by the US Government** provides the regulation for all foods and food additives.
- **Dietary Supplement Health and Education Act (DSHEA) 1994 the US Government** amended the FDAC (1938) to cover dietary supplements and ingredients of dietary supplements. New product/ingredients in the market must explain its safety to a "*reasonable expectation of no harm (emphasis added)*" as part of the latest dietary ingredient notification to the FDA [US-CFR 190.6, 2008].

### 15.8.1 FOOD SAFETY AND STANDARD ACT (2006) OF FSSAI

In chapter IV, article 22 of FSSAI (Government of India) addresses on the functional food, dietary supplements. This chapter also focuses on the need to regulate these products which comprise of organic foods, novel foods, irradiated food, GM ingredients of food, functional foods, foods required for special dietary uses, nutraceuticals, and health supplements. Articles 23 and 24 of this act include guidelines on the packaging and labeling of foods and restrictions about advertising of these foods [www.commonlii.org/in/legis/num_act/fsasa2006234]. According to this act, the functional and nutraceutical foods are defined by those foods which are:

- Specially prepared from natural plants in the powder form, extracts, and concentrates (single or in combination);
- Substances from animal origin;
- Vitamins, proteins, and minerals (not more than the RDA or enzymes);
- Dietary substances consumed by human beings to supplement the diet by augmenting the total dietary intake;

**TABLE 15.3** International Regulations of Functional and Nutraceutical Foods.

| Country | Regulations | Remarks |
|---|---|---|
| Canada | "Health Canada" regulates the health food industry; and the Canadian Food Inspection Agency enforces these regulations. | In Canada, most of the nutraceuticals fall under the "Natural Health Products Regulations" of the Food and Drugs Act of January, 2004. |
| China | Food Hygiene Law of Peoples Republic of China; the Administrative Licensing Law of People's Republic of China. | http://www.npc.gov.cn/zgrdw/englishnpc/Law/2011- 02/15/ content_1620635.htm, Effective from July, 2005 |
| Japan | Foods for Specified Health Use (FOSHU) | Approve the statements made on food labels concerning the physiological effects of the food on the human body. |
| New Zealand & Australia | Food Standards Australia and New Zealand (FSANZ) | The Code is enforced by state and territory departments, agencies, and local councils in Australia, the Ministry for Primary Industries in New Zealand and Australian Department of Agriculture and Water Resources for food imported into Australia. https://www.foodstandards.gov.au/Pages/default.aspx |
| USA | US Food and Drug Administration (FDA) has regulated the nutraceutical industry since 1947 | Health claim may be authorized for a food based on the extensive review of the scientific literature and testing. |

- Products that are labeled as **"food for special dietary uses"** (functional food or nutraceutical dietary supplements), which are not described for use as conventional food, and whereby such products may be formulated in the form of powders, granules, tablets, capsules, liquids, jelly, and other dosage forms but not parenterally and are meant for oral administration; and
- Such product does not compromise a drug as defined in clause (b) and Siddha, Ayurvedic, and Unani drugs as defined in clauses (a) and (h) of Section 3 of the Drugs and Cosmetics Act 1940 and rules made therein.

## 15.9 SUMMARY

- Consumers are health conscious and looking for functional foods to control their health and well-being;
- There are various sources of functional foods, such as: plant-based, animal sources based, and microbes-based sources;
- Diets containing plant derived foods, that is, vegetables and fruits prevent and reduce the chances of chronic diseases like coronary heart disease, obesity, and diabetes mellitus and also promote human health;
- Probiotics are "referred to as live microorganisms which provide health benefits on the host when administered in optimum amounts." Examples of probiotics species are *Lactobacillus* and *Bifidobacterium.*
- According to FAO/WHO, the manufacturer of functional foods (a) should carry out placebo-based clinical trials; and (b) analyze their results in four phases, such as: effectiveness, efficacy, safety, and surveillance.

## ACKNOWLEDGMENTS

The author acknowledges the Department of Food Science and Technology, Institute of Agricultural Sciences, Banaras Hindu University, Varanasi for the help and support in the work related to this review article. Sincere thanks are extended to Institutions of Eminence (IoE) scheme at Banaras Hindu University, for the financial support under Incentive to Senior Faculties under IoE.

## KEYWORDS


- **bioactive compounds**
- **functional foods**
- **health benefits**
- **heart diseases**
- **nutraceutical**
- **probiotics**
- **safety aspects**


## REFERENCES


1. Balsano, C.; Alisi, A. Antioxidant Effects of Natural Bioactive Compounds. *Curr. Pharm. Des.* **2009**, *15*, 3036–3073.
2. Chumphukam, O.; Chaiwangyen, W.; Sivamaruthi, B. S.; Chaiyasut, C.; Natakankitkul, S.; Kangwan, N.; Suttajit, M. The Innovation of Functional Foods in Asia: IFFA2018. *Asia Pac. J. Clin. Nutr.* **2019**, *28*, 419–426.
3. *Codex Alimentaris* (FAO). Guidelines for Use of Nutrition and Health Claims. *Nutr. Health Claims* (CAC/GL 23-1997) **1997,** 24.
4. Denny, A.; Buttriss, J. Plant Foods and Health: Focus on Plant Bioactives; British Nutrition Foundation Synthesis Report No. 4, 2005. www.eurofir.net/temp/PLANTspFOODS spANDspHEALTHspFOCUSspONspBIOACTI VEShs1hs.pdf (accessed on Aug 31, 2021).
5. Joint FAO/WHO. Working Group Report on Guidelines for the Evaluation of Probiotics in Food (London, ON, Canada); FAO: Rome; April 30 and May 1, 2002.
6. Kailasapathy, K. Encapsulation Technologies for Functional Foods and Nutraceutical Product Development. *CAB Rev. Persp. Agric. Vet. Sci. Nutr. Nat. Resour.* **2009**, *4* (33), 1–19.
7. Makinen-Aakula M. Trends in Functional Foods Dairy Market. In *Proceedings of the Third Functional Food-net Meeting*, Liverpool, 2006; pp 18–19. http://www.functional-foodnet.eu/images/site/assets (accessed on Aug 31, 2021).
8. Nowicka, G.; Naruszewicz, M. Assessing Health Claims for Functional Foods. In *Functional Foods, Cardiovascular Disease and Diabetes*; Arnoldi, A., Ed.; Woodhouse Publishing: London, 2004; pp 7–10.
9. Sutar, N.; Sutar P.P.; Debabandya, M. New Horizons in Functional Food Sector: An Indian Perspective. *J. Dairy. Foods Home Sci.* **2010**, *29* (3), 166–172.
10. Zhao, J. Nutraceuticals, Nutritional Therapy, Phytonutrients, and Phytotherapy for Improvement of Human Health: A Perspective on Plant Biotechnology Application. *Recent Patents Biotechnol.* **2007**, *1*, 75–97.

CHAPTER 16

# SAFETY ASPECTS OF NANOMATERIALS IN NATURAL FOODS

FARHANA MEHRAJ ALLAI, KHALID GUL, Z. R. AZAZ AHMAD AZAD, INSHA ZAHOOR, SADAF NAZIR, and ARSHIED MANZOOR

## ABSTRACT

Nanotechnology has numerous promising applications and is associated with food technology in several ways, like in the formulation of nanocapsules, nanofibers, nanocoatings, nanodispersions, nanoencapsulation, nanoemulsions, etc. Nanomaterials are being used due to their tunable physical, chemical, and mechanical characteristics. Novel nanoparticles have been formulated for improving the safety and quality of food. As the human population is growing at a fast pace, and the production of food is not just at par to feed this increasing population. Therefore, this mismatch has evolved the idea of using nanomaterials to preserve food for long and deliver it globally. Food-derived nanomaterials have also significantly contributed in the development of nutraceutical and functional food products. Thus, nanoparticles, as well as nanomaterials, have attracted several aspects and interests due to their unique characteristics. This chapter outlines the most recent trends in nanomaterials, their role at several levels of the food industry, classification, potential applications, and the use of nanomaterials in food safety and toxicity.

## 16.1 INTRODUCTION

The rapid and increasing growth of nanotechnology over the last decades has found its way across the globe into almost all the major industries.

Plant-Based Bioactive Compounds and Food Ingredients: Encapsulation, Functional, and Safety Aspects.
Junaid Ahmad Malik, Megh R. Goyal, Preeti Birwal, & Ritesh B. Watharkar (Eds.)

Among these major sectors, food is one of the most promising domains that has revolutionized both traditional and commercial food products by using the latest approaches like contact of nanoparticles with food (engineered nanoparticles), precision processing and handling, smart and active packaging, food additives, preservatives, encapsulation, formulation of novel products, nanosensors, food safety and security, and so on.[5,35,58]

Nanotechnology has been focusing on the latest and most novel research, innovations, and concepts to develop nanoparticles at higher rate. The importance of nanoparticles comes in mind when researchers found that the tiny size of these materials can affect the optical and physicochemical characteristics of a substance, such as 20 nm platinum (Pt), palladium (Pd), gold (Au), and silver (Ag) nanoparticles, which have yellowish gray color, dark black color, red wine color, and black color, respectively. Nanotechnology has significantly benefited various important branches of the food industry, including food productivity and quality, development of functional food, developing and designing of instruments to enhance food storage, quality, traceability, and shelf-life. In food processing, nanotechnology has helped in the enhancement of bioavailability and absorption of essential micro- and macronutrients and other supplements with the use of nanodelivery techniques.[61]

Nano-based techniques are an integral part of research in the food industry but are facing various challenges in terms of cost, safety, and toxicity. The application of these novel technologies has gained great attention of researchers working in the field of food processing, preservation, storage, and transportation in ensuring safety of the food.[68] Also, several of latest nano-based and innovative food packaging materials have been developed in association with nanosensors to evaluate their beneficial effects on both the environment and human health. Nanostructured materials are used for packaging of food, beverages, and health supplements, etc. Moreover, nanoscale ingredients also boost the color retention, aroma, flavor, bioavailability of nutrients, extension of shelf-life, inhibition of microbial spoilage, and enhances the durability of packaging materials.[21] Researchers found that consumers accept more of nanoparticles in food packaging rather than in the matrices of food.

Engineered nanomaterials offer several benefits, like better physicochemical characteristics of material, surface charge, agglomeration state, good preservative properties, enhanced absorption capability of active compounds, and increased flavor or color retention. As the size of particle is reduced, the surface area per unit mass enhances in an indirect manner, that is, inversely proportional. Surface area is increased by manipulating the porosity of nanoparticle. Large surface area can be achieved by formulating clusters or tiny particles to increase the surface to volume ratio of each component, thus enhances the rate of dissolution, surface reactivity but modifies the toxicity profile. The potential applications of nanotechnology have been listed in Table 16.1.

This chapter illustrates the details of different types of nanotechnological applications in food sector with a greater emphasis on different types of nanomaterials used and their formulations. In addition, safety-related aspects are also discussed.

## 16.2 PROSPECTIVE AND CURRENT APPLICATIONS OF NANOMATERIALS IN THE FOOD SECTOR

Among industrial applications, nanomaterials have gained more importance in the food sector. These applications are usually associated with the improved characteristics as bioavailability or release of active substances. Nanotechnology can be employed at several stages during the production of food, like processing, packaging and labeling, transportation, and quality checks during or postproduction. (Figure 16.1). Nanoparticles are used generally in the form of nanocapsules, nanoemulsions, and nanocomposites in food products.[51] Nanotechnology is a vast field and has several different classifications, each having various functions as shown in Table 16.2.

Nanoencapsulation is one of the most attractive methods, having antimicrobial property regarding the specific food product. It was found that nanoencapsulation increases the inhibition rate and improves the antimicrobial characteristics for longer periods, such as nisin-loaded carrageenan nanocapsules have improved antibacterial activity against Salmonella, P. aeruginosa etc. Furthermore, samples of tomato juice showed similar effects for 6-months period as compared to the components evaluated separately. In another study, supplementation of essential oils of nanometric encapsulated in fruit juices increased their antimicrobial effect, with minimum effect on the quality parameters of the finished products.[6] Chitosan combined with zinc oxide of nanoscale, resulting in enhancement of antimicrobial activity by using support of polyvinyl alcohol and then were electrospun to develop composite nanofibers.[65]

**TABLE 16.1** Potential Application of Nanotechnology in Agriculture and Food Industry.

| Agriculture | Food processing and engineering | Food packaging | Supplements |
|---|---|---|---|
| Pesticides and other agro-chemicals of nano-formulated grade are used to enhance the efficacy in comparison to other formulations developed by conventional methods | During processing and storage nano-based food additives have an application to improve the stability and bioavailability of nutrients present in the food, to increase the beneficial properties of the food products and to enhance the potency | Detection of chemicals or food-borne pathogens by attaching antibodies to fluorescent nanoparticles | Nanosize powders to improve nutrient absorption |
| For delivery of agrichemicals, pesticides, and fertilizers nanocapsules are used | Bioengineered nanocapsulated flavor enhancer | Nano form materials covered with thin coating of metals such as metal oxides or aluminum in order to act as a barrier against the gases and vapors | Cellulose nanocrystal composites as drug carriers |
| Nanocapsules filled with tebuconazole and carbendazim are used as fungicide | Nanofiltration in which nanopores can be used for the purification and softening of water | Micro and nano size beads of polylactic and polycaprolactone incorporated into phase change materials to show enhanced thermal buffer capability of the packaging material. | For better stability, better absorption or targeted delivery nanoencapsulation of nutraceuticals is done. |
| For monitoring crop growth and soil conditions, nanosensors are used | Nanomaterial additives have been used like in beverages, synthetic amorphous silica (SAS) is used as a clarifying agent. SAS, calcium silicate, and dicalcium phosphate is also used in powdered food items as anticaking and free flow agent | Nanocomposite of polymer nanoclay has beneficial effect in enhancing the thermal characteristics and tensile strength of packaging material | Delivery of nutrients to cells more efficiently without affecting taste or color of food is done by coiled nanoparticles known as Nanocochleates |
| Nanoparticles of inorganic metals and ashes of nanoscale have been developed for antifungal, antimicrobial, and insecticidal properties | Nanoparticles are used to bind and selectively remove the pathogens or any chemicals from food. | For plastic beer bottles, nanocrystals of silicate enriched with nanocomposite layers showed enhanced barrier characteristics | For better absorption, active molecules of vitamin sprays are dispersed into nanodroplets |

**TABLE 16.1** *(Continued)*

| Agriculture | Food processing and engineering | Food packaging | Supplements |
|---|---|---|---|
| Plant and animal pathogens are identified by nanosensors | Iron of nanostructured form can prove to be beneficial for fortification of iron in certain foods | Silicate nanoparticles are used for stronger, lighter, and more heat-resistant films | |
| Vaccines are delivered as nanocapsules | Smart delivery system of nano-based food additives for monitoring safety and quality of food | Modified permeation behavior of foils like silver nanoparticles | |
| During storage, pests are controlled by using NP loaded with garlic essential oil and coated with polyethylene glycol | Nanotechnology is also used in diary industry for separation of milk proteins, standardization of milk, and increasing the quality and safety of diary fluids | Nano-based food packaging material like biodegradable packaging, smart, active, intelligent packaging, and nanolaminates to protect food from microorganism and thus extends the shelf-life and quality of food. | |

**TABLE 16.2** Classification of Nanomaterials and the Functions They Perform in Food Industry.

| Type of nanomaterials | Functions | References |
|---|---|---|
| **Nanoparticles** | Biodegradable nanoparticles of polylactic acid are often used for drug delivery. | |
| | Nanoparticles are also used for encapsulation and delivery of micronutrients, such as zinc, vitamins, iron, and protein. | [50] |
| | For better efficiency, polylactic acid is used in combination with polyethylene glycol and functional ingredient of nanoscale may be encapsulated and can be released at a specific site. | |
| | Organic and inorganic nanomaterials, especially silver nanoparticles, are used mostly due to their antimicrobial property. | |
| | Gold nanoparticles are widely used as a detector or sensor. | [68] |
| | Nanoparticles of titanium dioxide are also recommended as food additive, disinfecting agent, and flavor enhancer. | |
| **Nanoemulsions** | Nanoemulsions have large surface area due to which it can interact with various biologically active components like lipase enzymes found in gastrointestinal tract that readily digests the tiny droplets of nanoemulsions. | [72] |
| | Nanoemulsions are used for the formulation of flavored oils, sweeteners, and are known to possess antimicrobial property that are very effective on gram positive bacteria and thus can be used as disinfectant in case of packaging material. | [66] |
| | Soybean and tributyl phosphate are used for the development of nanoemulsions. These nanoemulsions were then applied to reduce the growth of microorganisms and spoilage of food. | [56] |
| | It aids the uniform texture in ice-cream and is used to improve the consistency, texture, aroma, and flavor in food product. | [34] |
| **Nanolaminates** | Nanolaminates can be used in combination with edible films and coatings of various fruits and vegetables for packaging purpose. | [52] |

**TABLE 16.2** *(Continued)*

| Type of nanomaterials | Functions | References |
|---|---|---|
| | It also serves as carrier for various bioactive and functional components like antioxidants and antimicrobial agents. | [58] |
| | It improves the color, retains the flavor, and texture of food. | |
| **Nanoencapsulation** | During processing, packaging, storage, and transportation nanoencapsulation increase the bioavailability, optimal hydration, quality, and stability of the unstable food product. | [7] |
| | It releases the bioactive compounds to the specific site at desired rate and time. | |
| | Unwanted flavors in food are masked by encapsulation process. | [3, 40, 54] |

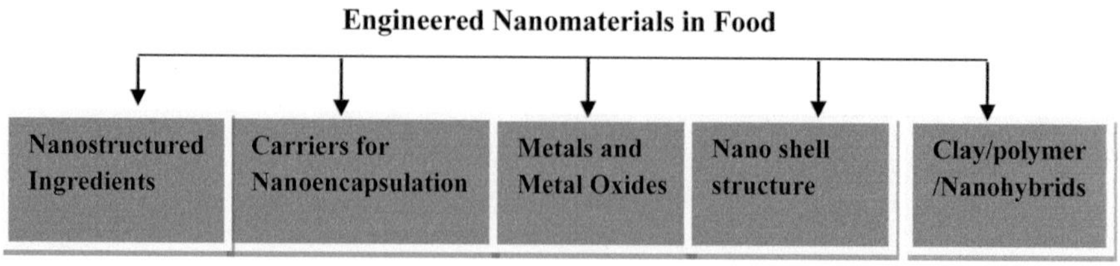


**FIGURE 16.1** Overview of nanomaterials in food.

## 16.3 ROLE OF NANOMATERIALS IN FOOD PROCESSING, PRESERVATION, AND PACKAGING

Nano-sized particles are naturally found in the food system, like fats and carbohydrates.[50] These ingredients can thus be used for the preparation of nanoemulsions, food grade polymers, and nanoencapsulates. Moreover, industrialists and scientists have already identified various applications of nanotechnology in almost each and every sector of the food industry, starting from the field practises in agriculture (such as vaccine delivery, identification of plant and animal pathogen, pesticides, and fertilizers) till the product reaches to the consumer. In between food processing, preservation and packaging are the most active areas of the food industry.[33]

In food processing, nanotechnology contributes for several functional ingredients including nanoencapsulation, food additives, nanoemulsions to develop aromas, flavor, consistency, and texture to enhance the bioavailability of ingredients having nutritional benefits and thus making food products more healthy, hygienic and nutritive for end user[16,20] or they may mask the undesirable flavors produced from the healthy food like food oils. Nanofiltration is another potential application to develop products which have the same characteristics as in their fresh produce with less processing like in juices, dairy industry for separation of lactic acid, lactose recovery, and whey protein concentrate.[55] Titanium dioxide nanoparticles are mostly used for sterilization of food product surfaces, drinking water, and as a whitening agent in candy and dairy products.[17] Additionally, nanomaterials have a thermal insulating characteristic known as nanosulate that reduces the loss of heat and saves energy costs.

Nanocomposites materials having specifically been used for packaging of food are LDPE-low density polyethylene-silver/LDPE, CuO/LDPE, Zinc oxide/LDPE, polylactic acid, isotactic polypropylene, and gelatin. Many researchers reported that polystyrene, chitosan, and polyvinyl chloride packages blended with zinc oxide and Cu nanoparticles inactivate pathogens. Other nanomaterials which are used during the processing, preservation, and packaging of food materials are reported in Table 16.3. The major relations between nanotechnology and food product industry are: (i) increasing shelf-life, (ii) extending food security, and (iii) improving delivery of nutrients and flavors. Hence, nanotechnology reformulated traditional food product packaging as intelligent and active packaging that includes the usage of different engineered nanoparticles such as metal oxides, nanosilver food containers, PET bottles containing nano-titanium nitride, and plastic beer bottles containing nano-clay, which provides protection to the product and brand presentation.[27]

The functional ingredients and encapsulation of nutraceuticals are important for the food preservation, as it is prime step in the food processing, storage, and passage through the GIT. Research has shown that nanoscale-edible coating is alternative method for extending the shelf-life of foods as well as eliminating or reducing the microbial spoilage. Coatings made from nano-silica, chitosan, gelation incorporating nanocrystals, lysozyme nano laminate, or alginate have been found to have significant effect on the quality of fresh produce and thus prolonged the storage life.[13]

## 16.4 RESEARCH ADVANCES FOR USE OF NANOMATERIAL IN FOODS

In the twenty-first century, the major challenges of growth of population, health of consumers, food supply, its safety and sustainability of bioresources, across the globe are rapidly increasing. Latest techniques and novel approaches in science and technology are very much important for identifying the solutions to the above challenges like use of different nanomaterials as discussed below. Nanotechnology is one of the latest emerging techniques for adding nutritive ingredients to the food chain. Nanotechnology regulations are themselves challenging as the word incorporates wide range of diverse technologies, materials, and products with a large spectrum of potential health benefits and risks.[47]

Klebsiella and Raoultella bacterial genera are mostly present in environment and foods, are associated with human health hazards. So, two types of strip-based techniques were formulated to detect bacteria. Lateral flow test strips labeled with palladium nanoparticles and linked with anti-klebsiella, an antibody that binds to specific bacteria. In another, antibodies were immobilized on nitrocellulose membrane, resulting in effective on-site monitoring of food, thereby increasing food safety.[63] Gold or graphene oxide nanoparticle composites together with an electrochemical immune sensing platform, are used for the detection of *E. coli* bacterial cell and Cronobacter Sakazakii bacterium that is poisonous to

**TABLE 16.3** Nanoparticles in Food Processing, Preservation, and Food Packaging.

| Domain | Nanomaterial | Function | References |
|---|---|---|---|
| Food processing | Nanoemulsions | It can be used in sauces, processed meat, puddings, and mayonnaise to improve texture. | [23] |
| | Nanoscale salt particles | Less quantity of salt gives same taste to tongue receptor without compromising the taste. | [23] |
| | Zinc oxide | UV sensing | [4] |
| | Titanium dioxide | Gives color to the food | [19] |
| | Nano-sized micelles | Used in milk protein casein to develop health promoting ingredient such as vitamin $D_2$. | [57] |
| | Nanoparticle-encapsulated lycopene | Used to enhance the bioavailability. Used in Jam, Jelly, paste sauce, tomato juice. | [12] |
| | Nanoparticles based on starch or protein | Delivery of bioactive substances to enhance efficacy. Encapsulating materials. | [23] |
| | Nanotea enriched with nanoselenium | Increases bioavailability and uptake of selenium. | [6] |
| | Silicon dioxide | Flavor enhancer Anticaking agent | [31] |
| | Nanometric delivery system of essential oils- mixture of terpenes and d-limonene | In fruit juices, it is used to improve the antimicrobial activity. | [24] |
| | Chitosan nanoparticle | Increased microbial activity in fish finger. | [1] |
| | Nanocapsulated oleoresin of paprika | Enhances sensory characteristics in chicken breast fillet. | [71] |
| | PLGA nanoparticles | Improves antimicrobial activity in meat products. | [53] |
| Food Preservation | Gold-based nanoparticle | In milk, B1 aflatoxin can be detected. | [45] |
| | Lipid-based nanoencapsulation | It is used widely to enhance the antioxidant activity by improving the solubility and bioavailability. | [48] |
| | Nano-silica with chitosan | During prolonged storage, it is used to preserve the fresh food. | [70] |
| | omega 3 nanoencapsulated | Used in bread making acrylamide (degradation byproduct) was decreased strongly. | [29] |

**TABLE 16.3** *(Continued)*

| Domain | Nanomaterial | Function | References |
|---|---|---|---|
| Food Packaging | Nanoparticles and nanocomposites of silver | Act as an antimicrobial Zeolites-silver containing material are used for the disinfection purpose. | [64] |
| | Active packaging coated with nano zinc oxide | Improves shelf-life of fresh cut Fuji apples. | [44] |
| | Nanoemulsions prepared from tributyl phosphate, nonionic, or soybean | Microbial growth is being checked and thus decreasing the spoilage of food. | [56] |
| | Zinc oxide nanoparticle | Used in juice like orange and liquid egg albumin | [26] |
| | | Significantly decreases the yeast and mold counts, Salmonella without affecting quality parameters. | |
| | Iron | Magnetic sensing when embedded in silicon dioxide. | [42] |
| | Silver oxide | Decreases growth of microorganisms in apples. | [73] |
| | Magnesium oxide in combination with polylactic acid polymer | It protects the material against the bacterial spoilage. | [62] |
| | Zerovalent iron particles | It has an application as oxygen scavengers. | [30] |
| | Polyethylene with nano powder like kaolin, silver, anatase, and rutile titanium dioxide | Used to preserve fruits like strawberries. | [69] |
| | Silicate nanoparticles | Limit flow of oxygen in packaging container and also check the moisture leakage so that product remains fresh for long time. | [38] |
| | Silver | Antimicrobial agent | [60] |
| | | When meat starts to spoil, the chemical reaction between the hydrogen sulfide and silver nano layer takes place and thus changes the clolor of the label. | |
| | Nanosensors | Smart sensing chips for safety analysis. | [23] |
| | Nanobarcodes | Authenticity of food, antimicrobial agent for packaging of food. | [23] |
| | Nickel oxide | Co-catalyst | [18] |

infants as it can be found in formula milk powder.[59] The synthesis of MgO nanoparticles was used for the elimination of heavy metal ions and bacterial disinfection from the aqueous solution.[15]

For the consumers, there has been a big issue regarding the food adulteration. As the population is growing at fast pace, the provision of standard and healthy food products is a main concern in today's world. Adulteration is wide ranging from addition of melamine and detergent to milk, adding beet invert sugar in fruit juices, mixing color or dye to spices, and adding duck or horse meat to beef. The scope of food adulteration is a source of huge income, easy way to gain profit. Unfortunately, thousands of consumers die each year because of illness caused due to the adulterated foods. Thus, nanomaterials or nanosensors are used to detect the unhealthy ingredients in food products that include graphene, metal oxide nanoparticles, and carbon nanotubes and its derivatives. An electrospun nanofiber plays an essential role in the design and development of biosensors to track adulteration, spoilage, and contamination in foods. These sensors are faster, more accurate, and cheaper to authenticate the safety and quality of food.[11]

Nanotechnology synergistically combined with other recent research areas to develop nanosensors for detection of metabolites, innovative processing techniques to produce nanoenabled products, encapsulation systems, to develop nanostructured foods, and intelligent delivery systems for nutraceuticals.[23] Bitterness in food is another challenging aspect as consumers demand for functional foods with decreased amounts of fat, sodium, and sugar, although the bitterness can be suppressed by these substances. In functional foods, astringent and bitter polyphenols exert health promoting benefits due to the presence of phytochemicals. During in mouth processing, bitter taste can be minimized by reducing contact between taste receptors and bitter molecules by forming nanostructure such as water insoluble gels, multiple emulsions O/W/O or W/O/W type to inhibit the diffusion and retard the release of bitter substances. Advanced food research is still under intense search to ensure development of novel products, food quality, and safety while satisfying demand of consumers for novel, hygienic, and healthy products.

## 16.5 SAFETY AND QUALITY OF FOOD NANOMATERIALS

Food safety is a global issue, due to the presence of various chemical additives that are being used in the production of food products. In different industrial sectors, the increasing demand of nanomaterials has raised several questions on the safety and quality concerns and their adverse effects on human health.

The food-borne pathogen-related diseases caused due to contaminants and additives that poses several challenges to the food processing industries. It is necessary to understand the action of nanoparticles and nanotoxicity (Figure 16.2). Research has been done to study the mechanism responsible for the detection of nanotoxicity and to overcome the limitations of nanoparticles.[41] The generation of reactive free radical oxygen scavengers and oxidative stress is one of the essential mechanisms of nanotoxicity. Nanoparticles, like Ag nanoparticles, are used to kill or inactivate the microbes in food chain without causing any deterioration to the food products.[32]

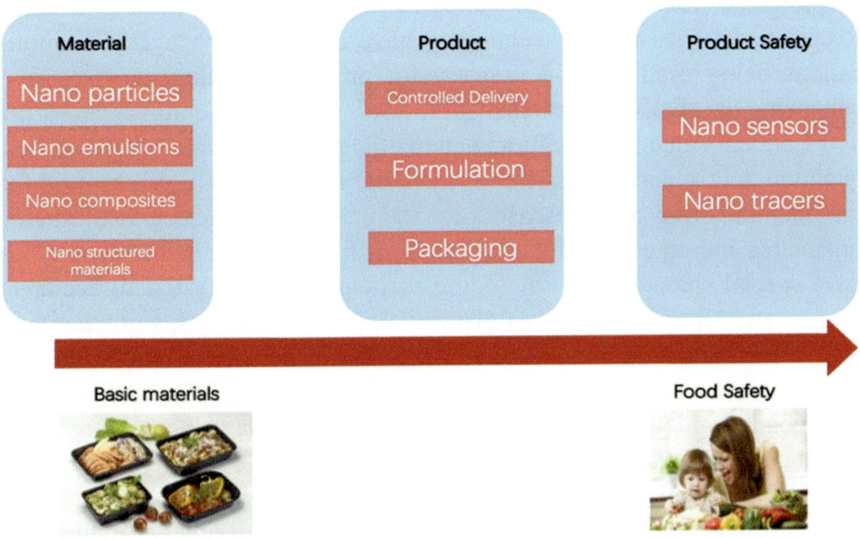


**FIGURE 16.2** Nano interventions in foods for food safety.

Nanomaterial-based assays and nanosensors have applications to diagnose the pathogens and contaminants present in the food that were caused due to the nanoparticles.[25] Advanced nanoparticles have helped to develop the biosensors for detection of food safety. Biosensors linked with detectors and biological components, possess several benefits for analysis and detection of various toxins and pathogens in food, as biosensors are highly specific and sensitive, resulting in the precise recognition such as fluorometric, electrochemical, optical, colorimetric, and bioluminescent.[43]

Electrochemistry is one of the efficient and rapid analytical techniques that is desirable for the food quality and safety assessment process over conventional methods. High-performance electrochemical sensing devices are used for the detection of some contaminants and additives, such as malachite green, caffeic acid, bisphenol, hydrazine, and nitrite. Also, the fabrication, design, and characterization of nanoparticles such as gold nanoparticles, metal, and bimetallic nanocomposites were also addressed by electrochemical sensors.[46] Another technique is surface-enhanced Raman scattering. With this technique, aggregated silver nanoparticles were used to analyze bacterial isolates.[37,39] A plasmonic biosensor-based device in association with self-assembled gold nanoparticles was developed to detect Salmonella typhimurium in pork meat samples.[49]

During processing, packaging, storing, and distribution, the food must be protected from any sort of biological, chemical, and physical hazards. Recent techniques in nanotechnology have drastically changed the applications being used in food industry for processing, safety and security, enhancing the shelf-life, decreasing the packaging waste, and increasing the nutraceutical value.[67] In addition, nanoparticles including carbon nanotubes, quantum dots, metal nanoparticles, and other nanomaterials, can be used for the manufacturing of nanosensors for the analysis and detection of microorganisms and other tests for application of food safety. Food nanoparticles have special characteristics including particle shape and size, surface area, charge, and topography, rate of dispersion, purity, solubility, and reactivity, etc.,[2,8–10,22,29] and have various advantages like nanomaterial in conjugation with biomolecules form the bases for the manufacturing of novel nano-based biorecognition devices,[14] microbial detection technique is incorporated with biofunctionalized nanomaterials that leads to real time, sensitive, and quick mode of detection of cells of microorganisms, etc.

Nanotechnology can enhance the functionality of food either by reducing or completing the removal of chemical toxicants or by increasing the nutritional supplements of nanoscale, for example, nutraceuticals such as lycopene, phytosterols, and β-carotenes can act as carriers to avoid accumulation of cholesterol in body. Green tea is another example that has a several health benefits as it contains nanoselenium, which has a significant role in the human body.[36] Nanomaterials have also been employed to control, protect, and manage food allergens such as $SiO_2$ nanoparticles. Toxins can also be produced by the release of heavy metals from nanomaterials. These metals can have adverse effect on health of humans that include Ag, ZnO, and CuO, enhances the intracellular reactive oxygen species level and finally causes damage of DNA.[14] It is expected in future that nanoscale particles/materials, nanoscale science, and nanotechnology, all together will contribute to the development of new techniques, tools, methods, and devices for identification, supervision, detection, tracking, and monitoring along the entire food chain.

## 16.6 SUMMARY

Food grade nanomaterials have promising applications in the food processing and preservation, including their use as food additives, protection of nutraceuticals, enhancement of the bioavailability, production of smart packaging, and development of biosensors for detecting microbes, pathogens, and toxins. Public awareness about the benefits, challenges, and opportunities that come in collaboration with novel nanotechnological techniques will bring better acceptance of the emerging technology. At the same time, the intervention of nanomaterials in the food industry has brought several changes in food processing, preservation, and packaging. These modifications and changes have considerably enhanced the shelf-life of foods with reduced the rate of food spoilage and deterioration. Knowing the properties of the nanoparticles in foods is an essential aspect for producers, as they significantly influence the food safety and quality. Despite of enhancing opportunities, the prime concern is about the health of a consumer who certainly consumes the food containing nanoparticles, which could gain entry into the human tissues, resulting in the accumulation of contaminants and toxic substances that produce unintentional effects to the health of the human body. The use of several nanoparticles having bioengineered characteristics might pose harmful effects on the human health and the environment. That is the reason why more and more scientific research has been carried out in order to formulate the new, latest, and novel techniques for detecting and controlling the negative aspects of nanomaterials.

## KEYWORDS

- **analysis**
- **delivery system**
- **food safety**
- **nanoparticles**
- **preservation**
- **packaging**
- **toxicity**

## REFERENCES


1. Abdou, E. S.; Osheba, A. S.; Sorour, M. A. Effect of Chitosan and Chitosan-Nanoparticles as Active Coating on Microbiological Characteristics of Fish Fingers. *Int. J. Appl. Sci. Technol.* **2012,** *2*, 158–169.
2. Akbari-Alavijeh, S; Shaddel, R.; Jafari, S. M. Nanostructures of Chitosan for Encapsulation of Food Ingredients. In *Biopolymer Nanostructures for Food Encapsulation Purposes*; Jafari. S. M., Ed., Vol. 28, Section 4; Academic Press Inc.: New York, 2019; pp 381–418.
3. Akhavan, S.; Assadpour, E., Katouzian, I., Jafari, S. M. Lipid Nano Scale Cargos for the Protection and Delivery of Food Bioactive Ingredients and Nutraceuticals. *Trends Food Sci. Technol.* **2018,** *74*, 132–146.
4. Allen, M. W.; Zemlyanov, D. Y.; Waterhouse, G. I. N.; Metson, J. B.; Veal, T. D.; Mcconville, C. F.; Durbin, S. M. Polarity Effects in the X-Ray Photoemission of ZnO and Other Wurtzite Semiconductors. *Appl. Phys. Lett.* **2011,** *98* (10), 1–3.
5. Alshammari, F. H.; Han, J. H.; Kim, B. C.; Rather, I. A. The Intertwine of Nanotechnology with the Food Industry. *Saudi J. Biol. Sci.* **2018,** *25*, 27–30.
6. Ami, P.; Falguni, P.; Nihir, S.; Chandraprakash, K. Application of Nanotechnology in the Food Industry: Present Status and Future Prospects. In *Handbook of Food Bioengineering*; Grumezescu, A. M., Maria, A., Eds., Vol. 12, Chapter 1; Elsevier Inc., 2018; pp 1–27.
7. Assadpour, E.; Jafari, S. M. Systematic Review on Nanoencapsulation of Food Bioactive Ingredients and Nutraceuticals by Various Nanocarriers. *Crit. Rev. Food Sci. Nutr.* **2018,** *59* (19), 3129–3151.
8. Assadpour, E.; Jafari, S. M. An Overview of Biopolymer Nanostructures for Encapsulation of Food Ingredients. In *Biopolymer Nanostructures for Food Encapsulation Purposes*; Jafari, S. M., Ed.; Academic Press Inc.: Boston, 2019; pp 1–35.
9. Assadpour, E.; Jafari, S. M. A Systematic Review on Nanoencapsulation of Food Bioactive Ingredients and Nutraceuticals by Various Nanocarriers. *Crit. Rev. Food Sci. Nutr.* **2019,** *59* (19), 3129–3151.
10. Assadpour, E.; Jafari, S. M. Nanoencapsulation: Techniques and Developments for Food Applications. In *Nanomaterials for Food Applications*; Lopez Rubio, A.; Fabra Rovira, M. J.; Martınez Sanz, M.; Gomez-Mascaraque, L. G., Eds., Chapter 3; Elsevier Inc., 2019; pp 35–61.
11. Assadpour, E.; Dima, C.; Jafari, S. M. Fundamentals of Food Nanotechnology. In *Handbook of Food Nanotechnology, Applications and Approaches*; Jafari, S. M., Eds., Chapter 1; Elsevier Inc.: Boston, 2020; pp 1–35.
12. Auweter, H.; Bohn, H.; Haberkorn, H.; Horn, D.; Luddecke, E.; Rauschenberger, V. Production of Carotenoid Preparations in the Form of Coldwater-Dispersible Powders, and the Use of the Novel Carotenoid Preparations. *US Patent* **1999,** *5*, 968, 251.
13. Bajpai, V. K., Kamle, M., Shukla, S., Mahato, D. K., Chandra, P., Hwang, S. K., Han, Y. K. Prospects of Using Nanotechnology for Food Preservation, Safety, and Security. *J. Food Drug Analys.* **2018,** *26* (4), 1201–1214.
14. Bata-Vidacs, I.; Adanyi, N.; Beczner, J.; Farkas, J.; Szekacs, A. Nanotechnology and Microbial Food Safety. In *Microbial Pathogens and Strategies for Combating Them: Science, Technology and Education*, Chapter 1; Formatex: Badajoz, Spain, 2013; pp. 155–159.

15. Cai, Y.; Li, C.; Wu, D.; Wang, W.; Tan, F.; Wang, X.; Wong, P. K.; Qiao, X. Highly Active MgO Nanoparticles for Simultaneous Bacterial Inactivation and Heavy Metal Removal from Aqueous Solution. *Chem. Eng. J.* **2017,** *312*, 158–66.
16. Chaudhry, Q.; Castle, L.; Watkins, R. Nanotechnologies in Food: What, Why and How? In *Nanotechnologies in Food*, Chapter 1; Royal Society of Chemistry: London, 2017; pp 1–19.
17. Chen, W. T.; Chan, A.; Jovic, V.; Sun-Waterhouse, D.; Murai, K.; Idriss, H.; Waterhouse, G. I. N. Effect of the $TiO_2$ Crystallite Size $TiO_2$ Polymorph and Test Conditions on the Photo-Oxidation Rate of Aqueous Methylene Blue. *Topics Catalys.* **2015,** *58* (2–3), 85–102.
18. Chen, W.-T.; Chan, A.; Sun-Waterhouse, D.; Moriga, T.; Idriss, H.; Waterhouse, G. I. N. $Ni/TiO_2$: A Promising Low Cost Photocatalytic System for Solar $H_2$ Production from Ethanol-Water Mixtures. *J. Catalys.* **2015,** *326*, 43–53.
19. Code of Federal Regulations (CFR). Electronic Code of Federal Regulations. Title 21: Food and Drugs. Part 73- Listing of Color Additives Exempt from Certification. The United States office of the Federal Register (OFR) and the United States; Government Publishing Office: Washington, DC, 2018.
20. Cushen, M.; Kerry, J.; Morris, M.; Cruz-Romero, M.; Cummins, E. Nanotechnologies in the Food Industry—Recent Developments, Risks and Regulation. *Trends Food Sci. Technol.* **2012,** *24* (1), 30–46.
21. Dasgupta, N.; Ranjan, S.; Mundekkad, D.; Ramalingam, C.; Shanker, R.; Kumar, A.; Nanotechnology in Agro-Food: From Field to Plate. *Food Res. Int.* **2015,** *69*, 381–400.
22. Dima, S.; Dima, C. Protection of Bioactive Compounds. In *Frontiers in Bioactive Compounds: Natural Sources, Physicochemical Characterization and Applications*; Apetrei, C., Ed., Vol. 1; Bentham eBooks, 2016; pp 255–301.
23. Dongxiao, S. W.; Geoffrey, I. N. W. Recent Advances in the Application of Nanomaterials and Nanotechnology in Food Research. In *Novel Approaches of Nanotechnology in Food Nanotechnology in the Agri-Food Industry*, Chapter 2, Vol. 1; Academic Press Inc.: Boston, 2016; pp 21–66.
24. Donsì, F.; Annunziata, M..; Sessa, M..; Ferrari, G. Nanoencapsulation of Essential Oils to Enhance their Antimicrobial Activity in Foods. *LWT—Food Sci. Technol.* **2011,** *44*, 1908–1914.
25. Duncan, T. V. Application of Nanotechnology in Food Packaging and Food Safety: Barrier Materials, Antimicrobials and Sensors. *J. Colloid Interface Sci.* **2011,** *363* (1), 1–24.
26. Emamifar, A.; Kadivar, M.; Shahedi, M.; Soleimanian-Zad, S. Effect of Nanocomposite Packaging Containing Ag and ZnO on Inactivation of *Lactobacillus Plantarum* in Orange Juice. *Food Control* **2011,** *22* (3–4), 408–413.
27. Enescu, D., Cerqueira, M. A., Fucinos, P., Pastrana, L. M. Recent Advances and Challenges on Applications of Nanotechnology in Food Packaging: A Literature Review. *Food Chem. Toxicol.* **2019,** *134*, 1–24.
28. European Commission (EC). Regulation (EC) No. 1333/2008 of the European Parliament and of the Council of 16 December **2008** on Food Additives. The European Parliament and The Council of The European Union, 2008.
29. Faridi, E. A.; Jafari, S. M..; Assadpour, E. Preparation of A Multiple Emulsion Based on Pectin-Whey Protein Complex for Encapsulation of Saffron Extract Nano Droplets. *Food Chem.* **2017,** *221*, 1962–1969.

30. Foltynowicz, Z.; Bardenshtein, A.; Seangerlaub, S.; Antvorskov, H.; Kozak, W. Nanoscale, Zerovalent Iron Particles for Application as Oxygen Scavenger in Food Packaging. *Food Pack. Shelf-Life* **2017,** *11*, 74–83.
31. Gokmen, V.; Mogol, B.; Lumaga, R..; Fogliano, V.; Kaplun, Z.; Shimoni, E. Development of Functional Bread Containing Nanoencapsulated Omega-3 Fatty Acids. *J. Food Eng.*, **2011,** *105*, 585–591.
32. Handford, C. E.; Dean, M.; Henchion, M.; Spence, M.; Elliot, C. T.; Campbell, K. Implications of Nanotechnology for the Agri-Food Industry: Opportunities, Benefits and Risks. *Trends Food Sci. Technol.* **2014,** *40* (2), 226–241.
33. Henchion, M.; McCarthy, M.; Dillon, E. J.; Greehy, G.; McCarthy, S. N. Big Issues for a Small Technology: Consumer Trade-Offs in Acceptance of Nanotechnology in Food. *Innov. Food Sci. Emerg. Technol.* **2019,** *58*, 1–9.
34. Hogan, S. A.; McNamee, B. F.; O'Riordan, E. D.; O'Sullivan, M. Microencapsulating Properties of Sodium Caseinate. *J. Agric. Food Chem.* **2001,** *49* (4), 1934–1938.
35. Huang, Q.; Yu, H.; Ru, Q. Bioavailability and Delivery of Nutraceuticals Using Nanotechnology. *J. Food Sci.* **2010,** *75* (1), R50–R57.
36. Inbaraj, B. S., & Chen, B. H. Nanomaterial-Based Sensors for Detection of Foodborne Bacterial Pathogens and Toxins as well as Pork Adulteration in Meat Products. *J. Food Drug Analys.* **2016,** *24* (1), 15–28.
37. Jarvis, R.M; Goodacre, R. Discrimination of Bacteria Using Surface Enhanced Raman Spectroscopy. *Analy. Chem.* **2004,** *76* (1), 40–47.
38. Jones, N.; Ray, B.; Ranjit, K. T.; Manna, A. C. Antibacterial Activity of ZnO Nanoparticle Suspensions on a Broad Spectrum of Microorganisms. *FEMS Microbiol. Lett.* **2008,** *279* (1), 71–76.
39. Kahraman, M.; Yazıcı, M. M.; Sahin, F.; Culha, M. Convective Assembly of Bacteria for Surface-Enhanced Raman Scattering. *Langmuir* **2008,** *24* (3), 894–901.
40. Katouzian, I.; Faridi Esfanjani, A.; Jafari, S. M.; Akhavan, S. Formulation and Application of a New Generation of Lipid Nano-Carriers for the Food Bioactive Ingredients. *Trends Food Sci. Technol.* **2017,** *68* (Supplement C), 14–25.
41. Klaine, S. J.; Alvarez, P. J. J.; Batley, G. E.; Fernandes, T. F.; Handy, R. D.; Lyon, D. Y.; Mahendra, S.; Mclaughlin, M. J.; Lead, J. R. Nanomaterials in the Environment: Behavior, Fate, Bioavailability, and Effects. *Environ. Toxicol. Chem.* **2008,** *27* (9), 1825–1851.
42. Leveneur, J., Waterhouse, G. I. N., Kennedy, J., Metson, J. B., Mitchell, D. R. G. Nucleation and Growth of Fe Nanoparticles in $SiO_2$: A TEM, XPS, and Fe L-edge XANES investigation. *J. Phys. Chem.* C, **2011,** *115* (43), 20978–20985.
43. Lv, M., Liu, Y., Geng, J., Kou, X., Xin, Z., & Yang, D. Engineering Nanomaterials-Based Biosensors for Food Safety Detection. *Biosens. Bioelectr.* **2018,** *106*, 122–128.
44. Li, M.; Zhu; Lin, D. Toxicity of ZnO Nanoparticles to *Escherichia Coli*: Mechanism and the Influence of Medium Components. *Environ. Sci. Technol.* **2011,** *45* (5), 1977–1983.
45. Mao, X.; Huang, J.; Fai Leung, M.; Du, Z.; Ma, L.; Huang, Z.; Gu, L. Novel Core-shell Nanoparticles and Their Application in High-Capacity Immobilization of Enzymes. *Appl. Biochem. Biotechnol.* **2006,** *135* (3), 229–239.
46. Manikandan, V. S., Adhikari, B., & Chen, A. Nanomaterial Based Electrochemical Sensors for the Safety and Quality Control of Food and Beverages. *Analyst* **2018,** *143* (19), 4537–4554.
47. Marchant, G. E.; Sylvester, D. J.; Abbott, K. W. What Does the History of Technology Regulation Teach Us About Nano Oversight? *J. Law Med. Ethics* **2009,** *37* (4), 724–731.

48. Mozafari, M. R.; Flanagan, J.; Matia-Merino, L.; Singh, H. Recent Trends in the Lipid-Based Nanoencapsulation of Antioxidants and Their Role in Food. *J. Sci. Food Agric.* **2006,** *86* (13), 2038–2045.
49. Oh, S. Y.; Heo, N. S.; Shukla, S.; Cho, H. Y.; Ezhil Vilian, A. T.; Kim, J.; Lee, S. Y.; Han, Y. K.; Yoo, S. M.; Huh, Y. S. Development of Gold Nanoparticle Aptamer-Based LSPR Sensing Chips for the Rapid Detection of *Salmonella Typhimurium* in Pork Meat. *Sci. Rep.* **2017,** *7* (1), 1–10.
50. Peters, R. J.; Bouwmeester, H.; Gottardo, S.; Amenta, V.; Arena, M.; Brandhoff, P. Nanomaterials for Products and Application in Agriculture, Feed and Food. *Trends Food Sci. Technol.* **2016,** *54*, 155–164.
51. Purohit, K.; Khitoliya, P.; Purohit, R. Recent Advances in Nanotechnology. *Int. J. Sci. Eng. Res.* **2012,** *3* (11), 1–11.
52. Ranjan, S.; Dasgupta, N.; Chakraborty, A. R.; Samuel, S. M.; Ramalingam, C.; Shankar, R, Kumar, A. Nanoscience and Nanotechnologies in Food Industries, Opportunities and Research Trends. *J. Nanopart. Res.* **2014,** *16* (6), 161–23.
53. Ravichandran, M.; Hettiarachchy, N. S.; Ganesh, V.; Ricke, S. C.; Surendra, S. Enhancement of Antimicrobial Activities of Naturally Occurring Phenolic Compounds by Nanoscale Delivery Against *Listeria monocytogenes*, *Escherichia coli* O157,H7 and *Salmonella typhimurium* in Broth and Chicken Meat System. *J. Food Saf.* **2011,** *31*, 462–471.
54. Rezaeia, A.; Fathib, M.; Jafari, S. M. Nanoencapsulation of Hydrophobic and Low-Soluble Food Bioactive Compounds within Different Nanocarriers. *Food Hydrocoll.* **2019,** *88*, 146–162.
55. Salehi, F. Current and Future Applications for Nanofiltration Technology in the Food Processing. *Food Bioprod. Process.* **2014,** *92* (2), 161–177.
56. Sanguansri, P.; Augustin, M. A. Nanoscale Materials Development—A Food Industry Perspective. *Trends Food Sci. Technol.* **2006,** *17* (10), 547–556.
57. Semo, E.; Kesselman, W.; Danino, D.; Livney, Y. D. Casein Micelle as a Natural Nano-Capsular Vehicle for Nutraceuticals. *Food Hydrocoll.* **2007,** *21* (6), 936–942.
58. Shukla, K. Nanotechnology and Emerging Trends in Dairy Foods, the Inside Story To Food Additives and Ingredients. *Int. J. Nanosci. Nanotechnol.* **2012,** *1* (1), 41–58.
59. Shukla, S.; Haldorai, Y.; Bajpai, V. K.; Rengaraj, A.; Hwang, S. K.; Song, X.; Kim, M.; Huh, Y. S.; Han, Y. K. Electrochemical Coupled Immuno-sensing Platform Based on Graphene Oxide/Gold Nanocomposite for Sensitive Detection of Cronobacter Sakazakii In Powdered Infant Formula. *Biosens. Bioelectron.* **2018,** *109*, 139–149.
60. Smolander, M.; Hurme, E.; Koivisto, M.; Kivinen, S. PCT International Patent Application WO 2004/102185 A1, 2004.
61. Sophiya, D. L.; Thangjam, R. Progress and Challenges of Nanotechnology in Food Engineering. In *Impact of Nanoscience in the Food Industry—Handbook of Food Bioengineering*, Chapter 4, Vol. 12; Elsevier Inc.: Boston, 2018; pp 87–112.
62. Swaroop, C.; Shukla, M. Nano-Magnesium Oxide Reinforced Polylactic Acid Biofilms for Food Packaging Applications. *Int. J. Biol. Macromol.*, **2018,** *113*, 729–736.
63. Tominaga, T. Rapid Detection of *Klebsiella Pneumoniae*, *Klebsiella Oxytoca*, *Raoultella Ornithinolytica* and Other Related Bacteria in Food by Lateral-Flow Test Strip Immunoassays. *J. Microbiol. Methods* **2018,** *147*, 43–49.
64. U.S. FDA. Environmental Decision Memo for Food Contact Notification No. 1569. Biologist, Regulatory Team 2, Division of Biotechnology and GRAS Notice Review (HFS-255), 2015.

65. Vikesland, P. J.; Wigginton, K. R. Nanomaterial Enabled Biosensors for Pathogen Monitoring—A Review. *Environ. Sci. Technol.* **2010,** *44* (10), 3656–3669.
66. Wang, Y.; Zhang, Q.; Zhang, C. L.; Li, P. Characterization and Cooperative Antimicrobial Properties of Chitosan/ Nano-ZnO Composite Nanofibrous Membranes. *Food Chem.* **2012,** *132* (1), 419–427.
67. Wesley, S. J.; Raja, P.; Sunder, A. A.; Tiroutchelvamae, D. Review on Nanotechnology Applications in Food Packaging and Safety. *Int. J. Eng. Res. Technol.* **2014,** *3*, 645–651.
68. Xiaojia, H.; Hua, D.; Huey-min, H. The Current Application of Nanotechnology in Food and Agriculture. *J. Food Drug Analy.* **2019,** *27*, 1–21.
69. Yang, F. M.; Li, H. M.; Li, F.; Xin, Z. H.; Zhao, L. Y.; Zheng, Y. H.; Hu, Q. H. Effect of Nano-Packaging on Preservation Quality of Fresh Strawberry (*Fragaria Ananassa* Duch.) During Storage at 4°C. *J. Food Sci.* **2010,** *75* (3), 236–240.
70. Yu, Y. W.; Zhang, S. Y.; Ren, Y. Z.; Li, H.; Zhang, X. N.; Di, J. H. Jujube Preservation Using Chitosan Film with Nano-Silicon Dioxide. *J. Food Eng.* **2012,** *113*, 408–414.
71. Yusop, S. M.; O'Sullivan, M. G.; Preub, M.; Weber, H.; Kerry, J. F.; Kerry, J. P. Assessment of Nanoparticle *Paprika Oleoresin* on Marinating Performance and Sensory Acceptance of Poultry Meat. *LWT—Food Sci. Technol.* **2012,** *46*, 349–355.
72. Zarif, L. Nanocochleate Cylinders for Oral and Parenteral Delivery of Drugs. *J. Liposome Res.* **2003,** *13* (1), 109–110.
73. Zhou, L.; Lv, S.; He, G.; He, Q.; Shi, BI. Effect of PE/AG2O Nanopackaging on the Quality of Apple Slices. *J. Food Qual.* **2011,** *34* (3), 171–176.

# CHAPTER 17

# SAFETY AND QUALITY ASPECTS OF FOODS: MONITORING METHODS

RAMANDEEP KAUR, MANINDER KAUR, and
RAMANDEEP SINGH SIDHU

## ABSTRACT

This chapter describes statistical methods and their applications for the evaluation of food quality and the nature of food. These methods comprise descriptive statistics, statistical process control, acceptance sampling, and other tools with tables and figures used as standards. Food quality is a characteristic that significantly affects product acceptability. Different statistical methods play important aids to detect experimental data, summaries, and concluding the optimal conditions. Different statistical designs for analysis and optimization have also been described. The fundamental requirements for using a particular statistical test and its significance and specificities have also been reviewed along with optimized or in-limits conditions.

## 17.1 INTRODUCTION

Food is the lifeline of human existence, and good quality food is required to meet the body requirements and promote and lead a healthy life. Good quality food can encourage proper development of the body and is the prime requirement of consumers. Therefore, the quality assessment is of great significance to assure food safety to the consumers. Various chemical and experimental biochemical methods are used both at the industrial and academic level for food safety assessment which generates a considerable

Plant-Based Bioactive Compounds and Food Ingredients: Encapsulation, Functional, and Safety Aspects.
Junaid Ahmad Malik, Megh R. Goyal, Preeti Birwal, & Ritesh B. Watharkar (Eds.)

amount of data. Different statistical techniques are used to analyze the food quality data to conclude the safety level of food. Before going into the further evaluation of data, it should be ensured that results of experimental data fall within the statistical limits. Also, the quality of data to be tested should be excellent and complete.

The application of statistical tools on food quality began in the late 1700s after introducing the concept of standardization by Eli Whitney. Basic statistical techniques as the size of the sample, sample space, size variance of sample, distribution, mean tests, hypothesis, and standard deviation, are required to assess food quality to develop safe food of good quality for consumers. The usages of statistical methods in food research are continuously growing with advancements in novel statistical techniques. Statistical analysis was introduced as a subject two decades ago for providing training to food technologists.[5] For statistical analysis, different methods have been explained in this chapter. Food quality includes physical, bio-chemical, rheological, and sensory characteristics assessed under descriptive analysis.

Further, for improving food quality during the process, statistical process control methods used in the food industry and consumer acceptance are analyzed with acceptance analysis methods. General statistical techniques and optimization of process parameters for better food quality are also explained in this chapter. The importance of application is highlighted with statistical information regarding the food technology section to certify consistent knowledge by persistent usage.[7]

This chapter focuses on the functioning of statistical quality control, understand the basic statistical process control methods, and find out the differences between attributes and variables.

## 17.2 FOOD QUALITY

Food may be defined as a nutritious substance originating from plant or animal sources to sustain life, which is considered one of the basic needs. It provides energy to the body as it has all essential nutrients such as proteins, carbohydrates, fats, vitamins, and minerals.[1] These are further classified according to the composition of food products. Quality refers to the characteristics of food that the consumers perceive. These may be categorized into physical, physico-chemical, biochemical, microbial, rheological, and sensory characteristics. Apart from this, it also includes the standard norms and quality standards that FSSAI formulates (Food Safety and Standards Authority of India), Prevention of Food Adulteration Act (1954), Grading and Marking of schemes like ISI mark and BIS (Bureau of Indian Standard) and AGMARK (Agricultural Produce Grading and Marketing Act), FSSAI Act 2006, ISO (International Organization for Standardization), GMP (Good Manufacturing Practice), GHP (Good Hygiene Practice), and HACCP food management systems.[8]

## 17.2.1 QUALITY PARAMETERS

### 17.2.1.1 PHYSICAL PARAMETERS

Physical parameters are the parameters that could be judged from the food product's appearance, color, and weight. These are size, shape, weight, volume, surface area, density, porosity, color, appearance, etc.

The **size of food products** varies according to the type, variety, and maturity of food, especially in raw fruits and vegetables. Determination of size for solid food products is done based on its spherical and non-spherical parameters. For granular food products, sieve analysis is done to separate the grains into various fractions depending upon the size of the granular material.

**Shape** is the factor in determining the grade characteristics of food products. Regular shaped food products fetch a higher price as compared to distorted shapes. The shape could be expressed as round, oblate, oblong, conic, ovate, elliptical, robbed, regular, or irregular.[13] The different methods to determine shapes and shapes are given in Tables 17.1 and 17.2, respectively.

**TABLE 17.1** Methods to Determine the Shape Characteristics of Food Products.

| Food product | Method |
|---|---|
| Granular or particle size food | Sieve analysis |
| Non—spherical object | Multiple lengths, i.e., length, breadth and thickness, which are mutually perpendicular to each other. |
| Spherical solid object | Diameter |

**Weight of food product** is another characteristic that can be determined physically with weighing balance or comparing with the object of a known weight.

**Volume** is usually used to determine the quality for non-porous food materials. Volume is of high importance in case of baked products and liquid packed food products. Generally, the platform scale method is used to determine the volume, which can be calculated by using the formula:

$$\text{Volume} = \frac{\text{Mass of displaced water(kg)}}{\text{density of water(kg / m3)}} \quad (17.1)$$

**TABLE 17.2** Description of the Shape of an Irregular Object.

| Shape | Description |
|---|---|
| Elliptical | Approaching ellipsoid round |
| Lopsided | Axis connecting stem and apex slanted |
| Oblate | Flattened at the stem end and apex |
| Oblong | Vertical diameter larger than the horizontal diameter |
| Oblong Conic | Tapered toward the apex |
| Ovate | Egg-shaped and broad at the stem end |
| Round Approaching spheroid | Approaching spheroid |
| Roundness | Roundness = $A_p/A_c$ |
| Sphericity | Sphericity = $D_i/D_c$ |

**Density** is represented by the symbol **(ρ),** which is expressed as mass per unit volume. Density is further of two types: particle or true density and bulk density. In processed food products, density determines the weight of the package and the amount of material filled in the package. In terms of granular material, it is useful to convert the weight of grains into volume.

Porosity can be expressed as the percentage of void space in the material, allowing the flow of air and gases through the mass product. Porosity is of high importance in determining the thermal and freezing parameters of food products.

**Color and appearance:** Color itself constitutes the appearance characteristics of food products. Color is determined in terms of characteristics of light, that is, intensity and wavelength. Also, depending upon the visual perception, it is classified as light, dark, or colorless. Color measurement is done either with a spectrophotometer or colorimeter to determine the values of hue and chroma.

#### *17.2.1.2 BIOCHEMICAL PARAMETERS*

Biochemical parameters include determining moisture content, pH, acidity, ascorbic acid content, and other bio-active components, determining the nutritional value of food products. Their nature and amount vary according to the variety and form of food product. However, it is usually observed that some amount of them is lost during pre-processing and processing operations.

**Moisture content:** The amount of water present in the food products may be free or inbound form. Free water is available for subsequent processing operations of freezing and drying. It also enhances various physical and biochemical reactions, ultimately resulting in microbial growth and spoilage of food products. It influences the storability of foods and reduction in water activity retards the growth of micro-organisms and slow down various enzyme-catalyzed reactions. Moisture content is usually determined by Hot Air Oven or Vacuum Oven Method. Also, moisture meters are available to measure the moisture content of grains.

**pH:** It represents the amount of acid or base present in a food. Lower the pH value, indicates acidic nature, while higher values correspond to basic or alkaline nature. Preservation quality of food is linked with pH of food. Very low or very high pH is effective in reducing microbial growth, especially yeasts and molds. pH can be determined by pH meter.

**Brix value** reflects the soluble solids concentration. It is measured by a refractometer and is equivalent to the percentage by weight of sugar. Brix is an essential parameter in controlling micro-organisms in food, for example, fruit juices, jams, and jellies. More the brix value, lower water content available for microbial growth as the sugars are bound to free water.

**Ascorbic acid** measures the vitamin C content of food products, which is a very effective antioxidant. To prevent spoilage and preservation of freshness of food products, a low amount of ascorbic acid value. Ascorbic acid is commonly used as a preservative, including bread, cured meats, jams, jellies, and other sauces. The titration method is employed for determining the ascorbic acid content of food. Some of the other bioactive compounds are listed in Table 17.3.

#### *17.2.1.3 RHEOLOGICAL PROPERTIES*

Designing flow processes, quality control, storage, processing stability measurements, predicting texture, and rheological properties play an essential role. Cereals, dairy products, meats, fruits, and vegetables are some of the food commodities for which rheological properties are tested. For example, viscosity, elastic modulus, and tensile strength are major parameters in determining the behavior of wheat flour dough. Similarly, compressibility, crispness, and breaking strength of baked products play a role in the acceptability by consumers.

**TABLE 17.3** Important Bioactive Compounds in Food Products.

| Parameter | Importance | Instruments and chemicals | Products |
|---|---|---|---|
| Anthocyanin | Offer anti-inflammatory, anti-viral, and anti-cancer benefits | Spectrophotometric analysis | Fruits and vegetables rich in red color |
| Anti-oxidants activity | Helps to maintain level of nutrients | Chemical & Spectrophotometric analysis | Fruits and vegetables |
| Lycopene | Provides health and lowers the risk of cancer | Spectrophotometric analysis | Red colored vegetables and fruits |
| Total carotenoids and β-carotene | Dietary source of vitamin A | Spectrophotometric analysis | Yellow color rich foods |

The hardness of food products, especially in the case of grains, is of high importance. Plasticity and mobility are applicable in the case of dough-like food materials and liquid products, respectively.[2] In meat products, tenderness is the most important factor affecting consumer evaluation of meat quality and acceptability. The various instruments to measure rheological properties are listed in Table 17.4.

**TABLE 17.4** Instruments to Determine the Rheological Parameters of a Food.

| Parameter | Food product | Instrument |
|---|---|---|
| Firmness | Cheese, milk, and curd | Torsiometer, Ball compressor, Hardness tester |
| Hardness | Fruits and vegetables | Penetrometer |
| Plasticity | Dough and dough like materials | Barbenderfarinograph, Mixograph |
| Tenderness | Meat | Warner Bratzler shear press, Grinder tenderometer |
| Texture | Fruits and vegetables | Compressimeter, Penetrometer, Shear testing device |

#### *17.2.1.4 SENSORY CHARACTERISTICS*

Sensory evaluation of food is a discipline that analyzes and measures human responses to food composition and its characteristics. For example, appearance, touch, taste, flavors, and odors characteristics are due to the food composition.

**Flavor** is the combined effect of taste and aroma. Taste is recognized by taste buds classified as bitter, salty, sour, savory, and sweet. Aroma is the odor of food.

**Odor** plays a paramount role in quality evaluation of food products as the aroma perception is the first perception made by the consumer than the taste. Different tests, rating tests, sensitivity tests, and descriptive tests are the tools to measure the sensory characteristics of food products.

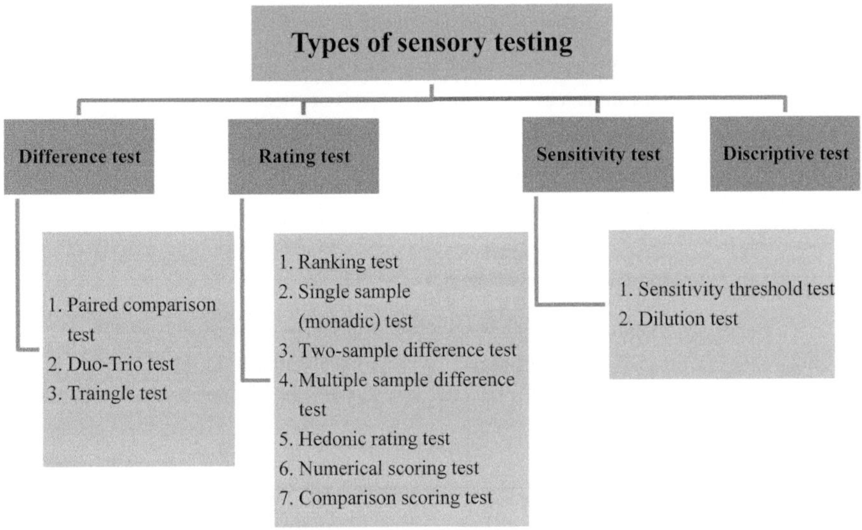


**FIGURE 17.1** Sensory testing methods.

## *17.2.2 METHODS TO MONITOR QUALITY*

The monitoring of quality refers to certify a specific quality level of a food item. However, this is done by following certain quality levels or other production methods, examinations, and undertaking processes or services. It also includes the assessment of various quality attributes which judge that the food should either be accepted or rejected. A sampling of food products is the primary operation performed for conducting various quality analysis tests. These tests may employ instrumentation techniques or chemical analysis, or novel food assessment technologies like biosensors, electric nose, and IOT-based technologies. According to the type of product and to the degree of accuracy required, the quality monitoring method could be applied. Basic techniques for quality monitoring involve specific equipment like some physical testing tests executed using modest instruments, for example, thermometers, refractometers, and spectrometers.[11] By contrast, chemical and microbiological tests are more complicated, requiring proper procedures to perform the test and experienced personals. Various quality assessment parameters for different food products are described in Table 17.5.

**TABLE 17.5** Quality Assessment Parameters for a Food.

| Type of product | Quality assessment parameters |
|---|---|
| Dairy products | Determination of shelf life of milk |
| | Determination of ripeness of cheese |
| | Distinguish between different types of cheese |
| Fruits and vegetables | Assessment of maturity and ripening index |
| | Assessment of damaged products |
| | Various quality assessment checks during processing and storage |
| Meat and fish | Degree of freshness |
| | Maturity rate |
| | Tenderness and determination of presence of micro-organisms |
| Granular material | Grade quality of grains |
| | Determination of broken and head kennel |
| | Check for the presence of various impurities like stones, insects, or other parasites |

### 17.2.3 STATISTICAL QUALITY CONTROL

Statistical quality control is a technique/method to measure variations in the gathered data to determine any further improvement that is needed in the process line to produce quality safe products for consumers.[10] All processes can have different causes of variation, either common or special. The inherent fault in the process, leading to frequent changes in quality, is known as common cause variations and variation cause which does not occur by itself in process, and are made by outer sources (employees, circumstances, environmental) leads to special cause variations. There are three categories of statistical quality control, which are as follows.

#### 17.2.3.1 DESCRIPTIVE STATISTICS

This type of statistics is used to elaborate various quality characteristics and parameters of food. They are central tendency (mean, median, and mode), variability (standard deviation and range), distribution, and dispersion of the observed data. These are numbers used to summarize and describe data and its relationships. For example, this statistics type may include mean value, standard deviation, the range and a measure of distribution data of food processed, packaged and other characteristics related to color, quality, and safety.

**Measures of Central Tendency** are one of the basic informative descriptions of an average or center value of set data. It is a point about where a number of values clusters: these clusters elaborate mean or an average number of the population of interest. Central tendency is measured through three methods: *mean* is calculated by dividing summation of variable amount with the no. of observations, the *median is* variable mid-value and the most be fallen value is called *mode.*

**Measures of Dispersion** give details about the spread of a variable's values. Range, variance, standard deviation, and skew are four different methods to calculate dispersion.

**Range** is a difference of the highest and lowest value of the calculated data.

**Variance** is the most commonly used method for dispersion. It is computed by mean of average of the differences squared between every single value and the mean.

$$\sigma^2 = \frac{\sum_{i=0}^{n}\left(X_i - \bar{X}\right)^2}{n-1} \tag{17.2}$$

**Standard deviation** is the square root of the variance.

**Skew** measures whether some values of a variable are extremely different from the majority of the values.

**Measures of Association** are the relation among two variables measured through association measurement. *Chi-Square* assessment is for nominal data (e.g., data differentiate into two classes: food secure by gender [Male, Female] and food insecure with [hunger, mild hunger, and severe hunger]) to observe their association. If there is an association between two variables, it is called as significant. On the other hand, if there is not an association, it is called as non-significant. The goodness of fit is measured by the chi-square test. *Correlation* assessment evaluates the potency of association among numeric variables (e.g., density and volume). The details are further explained in this chapter.

#### *17.2.3.2 STATISTICAL PROCESS CONTROL (SPC)*

A statistical method comprises monitoring of specified quality characteristics by detecting a random sample of the output from a process. SPC tool measures changes occurring in a process way, which affects the product quality. The control is done by determining the in or out-of-control limit of the process. For keeping the process within limits, control charts have been used that have upper and lower control range of variation.[9] However, a plot of data falling outside the preset condition limit is known as control chart (Figure 17.2).

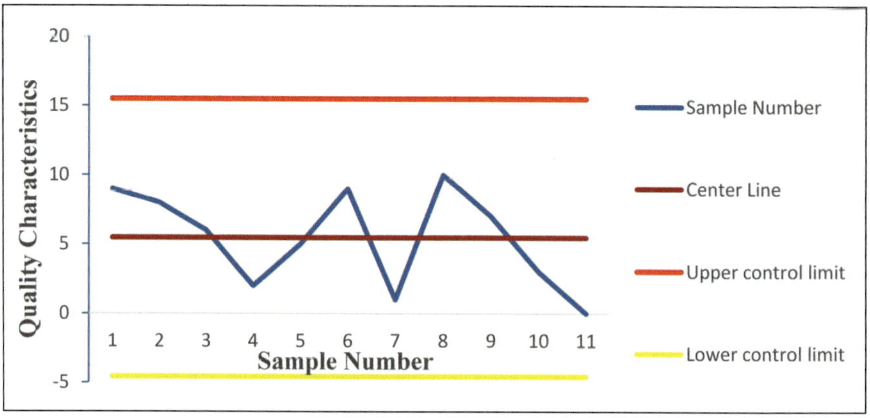


**FIGURE 17.2** Components of "Control charts."[12]

**Control charts** are of two major types: attributes and variables. The average value is denoted by center line (CL) of the control chart. In order to decide the centrality of the process, it describes the values of a parameter, that is, arithmetic mean or any other measurement of central tendency. As shown in Figures 17.3 and 17.4, the upper control limit (UCL) is the highest tolerable variation and the lower control limit (LCL) is lowest modest variation. The following formula is used in the control chart to calculate control line (limits):

$$\text{(Average Process Value)} \pm \{3\times \text{(Standard Deviation)}\} \quad (17.3)$$

**Attributes** are measured as "good" or "bad," therefore, have only discrete, binary, or integer values. These are measured with p-chart and C-chart as follows:

- **p Charts** are used for measurements of the fraction of defects in a sample. The center line denotes the average defective proportion in the products "p" to maintain quality. These p-charts are used only, after the calculation of total size and total number of defects. This chart will plot defective sample proportion (p) movement across the time.[9]

Procedure: Take random sample from process line and look over each item. For sample, defect proportion divides number of defective sample by sample size. Now plot the defects of sample proportion on the control chart with UCL and LCL range and determine whether the process is out of control or in control. This is binomial distribution, which may be computed by normal distribution method.

$$\sigma\rho = \sqrt{\frac{\rho^{-}(1-\bar{\rho})}{n}} \tag{17.4}$$

Where, ρ = defective fraction; $\bar{p}$ = mean; and n = size of sample.

$$\text{Control limits: UCL} = \bar{\rho} + z\sigma_{\rho}, \text{ LCL} = \bar{\rho} - z\sigma_{\rho} \tag{17.5}$$

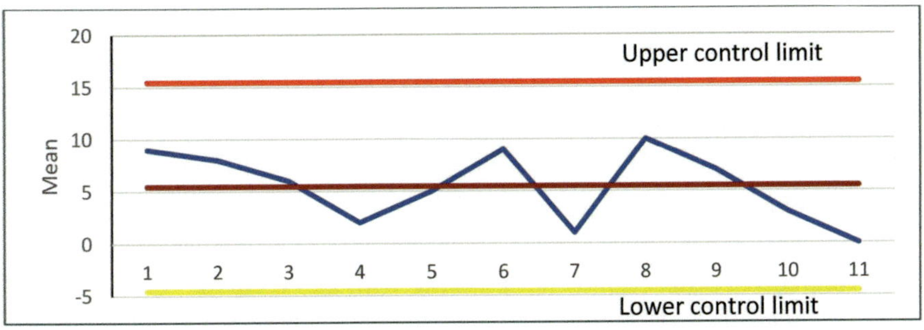


**FIGURE 17.3** Process within limits "in control."[12]

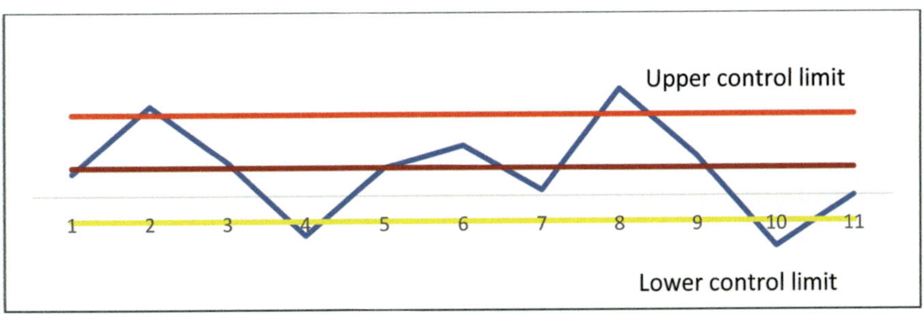


**FIGURE 17.4** Process above limits "out of control."[12]

Where, z is standard deviation from mean. z = 1, the sample limits can exceed 32%. However, mean can also change with limits variations (Figure 17.5). For z = 2, the limits will be increased by 4.5 (or 95.5% confidence) of changed mean value. For z = 3, the limits will be increased by 0.03 (or 99.7% confidence).

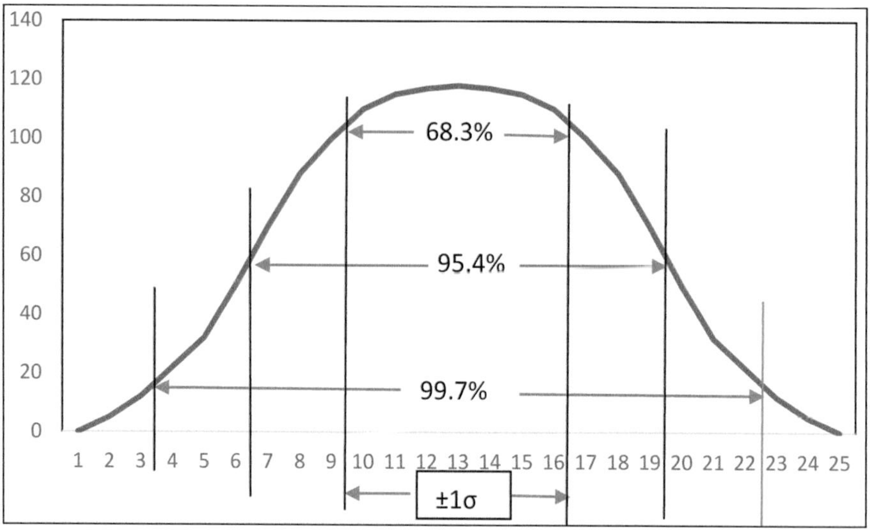


**FIGURE 17.5** Standard deviation variation chart.[9]

- **C chart** is applied for measuring the number of defects per unit. A c-chart can estimate the defects number only. However, evaluation of defective proportion is not possible with this method.[9]

  **Procedure:** To monitor defects in units, choose one lot randomly and calculate the number of defects in that lot. Now, plot that defect on a chart, whether they are in lower control or out-of-upper control limits. This sampling distribution is Poisson distribution, while it can be calculated through normal distribution: c is total assessed defects per unit; and Mean is denoted with $\bar{c}$, **then:**

$$\sigma = \sqrt{\bar{c}} \tag{17.6}$$

Control Limits will be:

$$\text{UCL} = \bar{c} + z\sqrt{\bar{c}}$$
$$\text{LCL} = \bar{c} - z\sqrt{\bar{c}} \tag{17.7}$$

**Variables** are measured on a continuous scale (e.g., weight or volume). These are measured with two types: X-chart and Range (R)-chart. Control charts for variables are: R- chart also known as Range chart are applied to measure process variability; and X-bar chart (average chart) is for average process measurement. These two charts must be in limit for in control process. Furthermore, process flexibility should also be in limits for the development of X bar chart.

- **R-chart** is used for quantitative and continuous data such as dimension or time. The range chart monitors the dispersion or variability between observations in further subgroups (average of series) of data over time.

$$CL = \overline{r} = R$$
$$UCL = D_4 (R)$$
$$LCL = D_3 (R) \quad (17.8)$$

Where, R is past values average and depending upon sample size; $D_3$, and $D_4$ are constants.

- **X chart** is also known as the mean control chart and is used to compute the deviation in the mean process. It is based on average of series called, subgroup. The horizontal axis denotes sample number whereas vertical axis indicates the sample mean for series of observations or subgroup.

$$UCL = X_{GA} + A_2R_A$$
$$LCL = X_{GA} — A_2R_A \quad (17.9)$$

Where, $X_{GA}$ = Grand average of X bar = X-bar bar; $R_A$ = Grand average of range R = R bar.

**Procedure:**

a. First of all, calculate the n-value (number of a subgroup) within each data point. For example, if we are taking temperature reading per min. Therefore, if three readings are taken, then n will be 3; and for same experiment, if four readings are taken 4 then n will be 4.
b. For every time point, take an average of those n-readings and further take average of these average values to compute grand average (X bar bar).
c. In case of a range, find the range of n-readings for all time points. Now calculate a grand average of range by taking average of these ranges.
d. For average standard deviation calculation: first, calculate the standard deviation of the n- readings for all time points, and then take the average of these standard deviations.

**Process capability** is used for process adjustment until it meets the preset limits. These specific limits are range of standard quality parameters. For acceptance, the product must be capable and in control on chart basis. The capability index is measured as below:

$$C_p = \frac{specification\,width}{process\,width} = \frac{USL - LSL}{6\sigma} \tag{17.10}$$

**If *Cp* = 1:** process variability just meets specifications.

**If *Cp* ≤ 1:** process variability exceeds specification range, which means there is need of process improvement.

**If *Cp* ≥ 1:** process variability is snugger than specifications, which means there is possibility of increasing capability minimally.

### *17.2.3.3 ACCEPTANCE SAMPLING*

Acceptance sampling is the third branch of SQC and is decisive about acceptance and rejection of the entire lot through randomly inspecting a certain number of items from a lot or batch. Acceptance sampling is totally different from statistical process control (SPC) because SPC is performed while accepting samples is practiced either *before* or *after* the process. For supplier material, sampling is done before, whereas selection is done for finished products before delivery to outer assembly.

**Sampling Plans:** The criteria for acceptance or rejection of sampling plans are based on some goals for acceptance. So, it includes some terms such as size of lot (N), sample size (n), and rejection of lot with number of defects more than limit (c). Sampling plan can be single, double, or multiple based upon cost and time involved on passing the items. A sampling plan can measure either variables or attributes but is more commonly used for attributes. The acceptance of the sampling goal is based on the level of confidence we wish to attain.

**Operating Characteristics (OC) Curves:** The probability of acceptance of lot is presented through graphs of the operating curve (OC). The X-axis of this curve indicates the percentage of defective items in a lot (Lot Quality) and the Y-axis indicates a probability for acceptance of a lot. The Figure 17.6 clearly demonstrate that the probability of accepting a lot will be decreased with an increased proportion of defects. For example, if the defect is 5%, then the probability of lot acceptance is 90%; similarly with 24% defects, the chance of acceptance is reduced to 10%.

Figure 17.6 illustrates the minimal percentage of defects that consumers accept willingly by denoting AQL, tolerate limit (upper limit of the percentage) of acceptance by consumer of defective items (LTPD); **Consumer's Risk (α)** is a chance of lot acceptance by consumer with Type II error with number of total defects > LTPD (lot tolerance percent defective) limit; and **Producer's**

**Risk (β)** is Type I error with probability of lot rejection having an acceptable quality defects level.

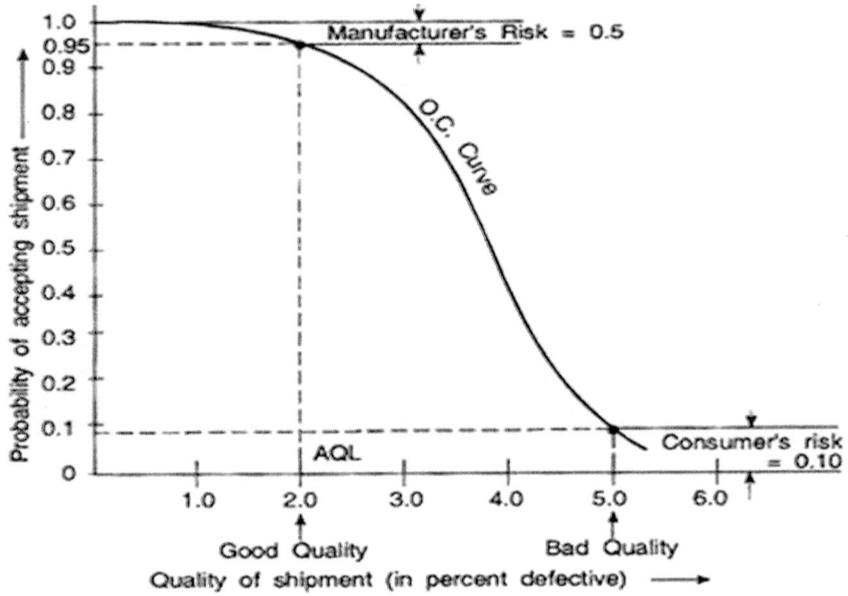


**FIGURE 17.6** "Operating Characteristics (OC)" curve showing sampling risk.

### *17.2.3.4 TOOLS USED IN SQC*

**Histogram:** The numeric data distribution is represented graphically through a histogram. Karl Pearson introduced firstly and estimated probability distribution of quantitative variable. The adjoined rectangles if bar chart with an area equivalent to frequency interval assembled at discrete intervals. The height of the rectangles is measured by dividing the frequency with the width of the range. The number of data represents the total area of the histogram. Different histogram schemes are shown in Figure 17.7.

**Pareto Charts** are a combination of line and bar graphs in which bars exemplified particular values in descending order, whereas, lines represent the cumulative total. Therefore, these charts are used to choose and separate input variables in limited numbers those have the farthest impact on a desirable or undesirable outcome. The basic principle is that proportion of all inputs could not have same impact on outputs. The pareto analysis is also known as the "80/20 rule," which dictates that by addressing 20% of causes of problem, produce 80% of results. A Pareto analysis chart is shown in Figure 17.8.

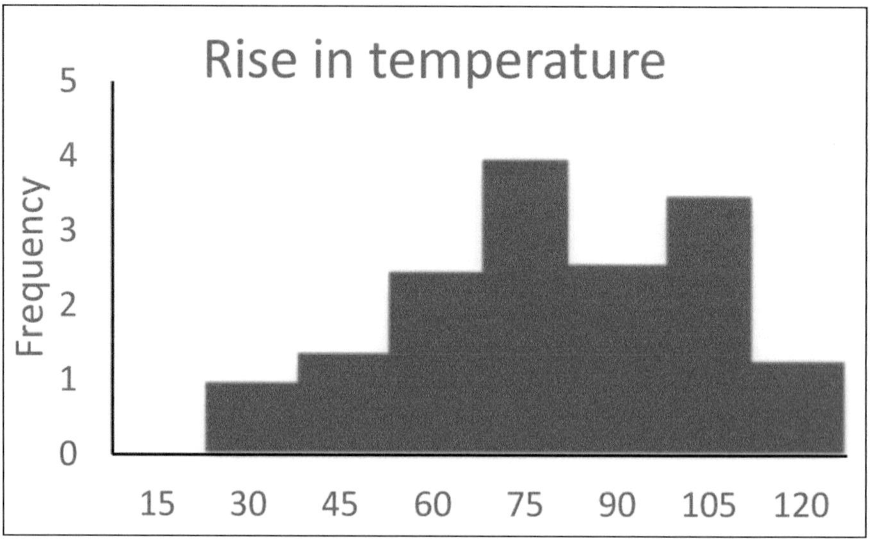


**Time (min)**

**FIGURE 17.7** Histogram chart.

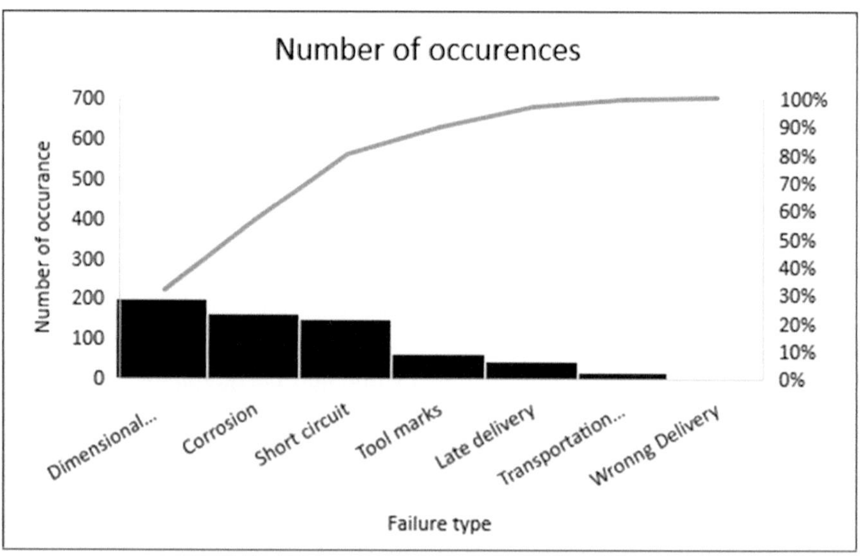


**FIGURE 17.8** Pareto chart.

**Fish-Bone Chart** is useful for organizing ideas and identifying their relationships. The fishbone diagram takes four problem areas into consideration: methods, equipment, materials, and people. This is usually helpful for particular problem, or perhaps a goal (Figure 17.9).

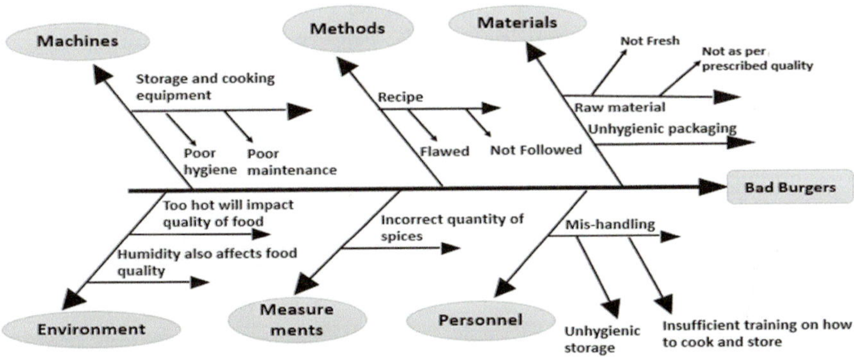


**FIGURE 17.9** Fish Bone chart.

**Defect Concentration Diagram** (Figure 17.10) is also known as problem concentration diagram. This graphical tool is useful in analyzing various defects or causes of the products with location and frequency displayed on Graph.

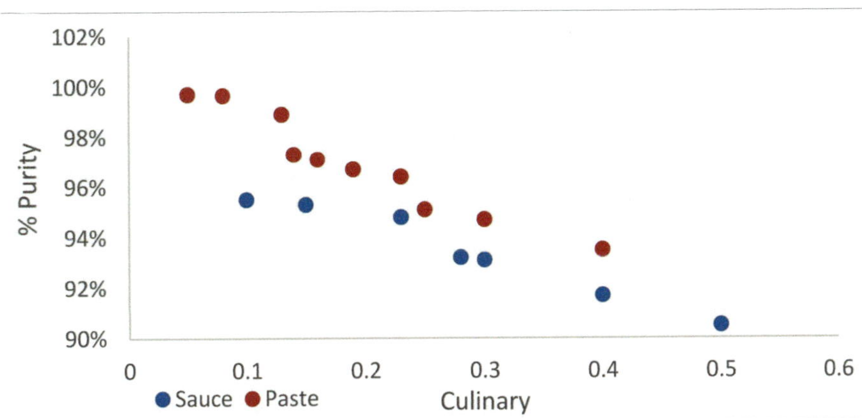


**FIGURE 17.10** Defect concentration diagram.

## 17.3 RELATIONSHIP BETWEEN FOOD QUALITY AND NATURE OF FOOD

Quality of food perceived by consumers is dependent on various characteristics present in the food, which provide basic determinants of food to be chosen. For example, food products with rich in fat content have less satisfaction levels as compared to protein rich foods. Energy density is another quality parameter which is of prime importance and affects the choice of consumers. Foods with more palatable in nature have higher demand. Economic determinants (such as cost and access to the food products) also affect the food choice of people. Food with higher quality grades and standards attract the consumers, thus fetch higher prices in the market. Thus, many factors are linked with nature and type of food, which ultimately decides the consumer preference to the particular food product.

## 17.4 STATISTICAL APPROACH IN FOOD QUALITY

### *17.4.1 STATISTICS AND ITS SIGNIFICANCE*

It is probable to assess the statistical analysis of precise data. Once the data have been accumulated and limited time, a decision may involve error. In these cases, summaries of large statistical information into a single value become convenient for both trainee and expert analyst to depict about statistics. The resulting verdicts and proceedings will now carry out with greater assurance.[6] Furthermore, a reduction in time and funding is possible through analysis. In some cases, judgment is based on the major outcomes of a statistical assessment. The determination of food content, toxins and nutrients is based on reliable chemical analysis methods. Statistical technique plays an essential role in monitoring and communicating these results. This provides certainty about valid results and benefits for consumers because awareness about specific safe foods and dietary regimens can also be planned safely.[4]

### *17.4.2 GENERAL STATISTICAL TECHNIQUE AND DESIGNS*

Researchers make a decision based on statistical results. The analysis would be done based on the nature of data, either qualitative or quantitative. There are several statistical tests; however, the choice has relied upon data structure, distribution, and types of variables (Puri 1989). These testing designs can be divided into correlation, comparison of means, regression, and non-parametric testing. By using sample observations, estimation of point or interval of unknown parameters is done through these tests. For these parameters, testing the validity of certain assumptions is done through hypothesis analysis.

### *17.4.2.1 HYPOTHESIS TESTING*

In this testing, two hypotheses are evaluated. One is a null hypothesis (Ho), called the main hypothesis, and the second is an alternative hypothesis ($H_A$), called the complementary hypothesis. For example, in two production lines, a hypothesis that means filling weight is the same for both lines. So, $\mu_1$ is equal to $\mu_2$, and $H_A$: $\mu_1 \neq \mu_2$. The conclusion is based upon either reject or fail to reject the null hypothesis.

**Correlational and Regression Testing:** For finding linear relationship among two variables, r-value is computed through a correlation co-efficient for n observations of a sample.

$$r = \frac{\sum XY - \frac{(\sum X)(\sum Y)}{n}}{\sqrt{\left(\sum X^2 - \frac{(\sum X)^2}{n}\right)\left(\sum Y^2 - \frac{(\sum Y)^2}{n}\right)}} \tag{17.11}$$

The r value lies in the range -1 to +1. If r is negative, that shows an inverse relationship exists between X and Y. Similarly, a direct relationship is represented with a positive value. However, there is no linear relationship when r is obtained zero. By calculating t value, significance of r is done. Ho: $\rho = 0$.

$$t = \frac{r}{\sqrt{\frac{1-r^2}{n-2}}} \tag{17.12}$$

**Pearson correlation test** is used for two continuous variables association. The ordinal variables that do not rely on normal distribution assumption are analyzed through Spearman correlation, whereas the chi-square test is used for categorical variables. **Regression tests** evaluate one of the variable changes to predict the change in another variable. It can be simple or multiple regression tests based on predicted variables.

**Mean comparison:** The variable means the difference is measured by several tests, such as z-, Paired t-test, Independent t-test, and Mean-variance of the analysis given in Table 17.6.

**TABLE 17.6** Statistical Tests for Comparison of Means.

| Type of test | Use |
|---|---|
| ANOVA | For categorical outcome data and chi-square test for more than two tables, compare proportion between different groups (e.g., controlling for quality, cost, or types of machinery). The tested hypothesis in ANOVA is<br>Ho: $\mu_1 = \mu_2 = \mu_3 = \mu_4 \ldots = \mu_k$<br>Ha: mean of at least one pair is significantly different<br>Calculate F value:<br>**F = ((SSE1 - SSE2)/m)/ SSE2/n-k,**<br>Where, SSE is the residual sum of squares, m is no. of restrictions, and k is no. of independent variables. |
| Independent t-Test | Tests different populations similar variable difference (e.g., comparing boys to girls) |
| Paired t-Test | Tests for same population two different variable difference<br>$\mathbf{t = (x_1 — x_2) / (\sigma / \sqrt{n1} + \sigma / \sqrt{n2})}$,<br>Where $x_1$ and $x_2$ are mean of samples 1 and 2, respectively. |
| z-test | For large sample size, used to calculate differences in two population means with known variance. The parameters used are mean and standard deviation.<br>Ho: $x = \mu$ (Null hypothesis)<br>Ha: $x \neq \mu$ (Alternate hypothesis)<br>$\mathbf{Z = (x - \mu) / (\sigma / \sqrt{n})}$,<br>Where x = sample mean, $\mu$ = population mean, $\sigma/\sqrt{n}$ = standard deviation of a population.<br>If z is less than a critical value, the null hypothesis is accepted; otherwise, reject. |

**Non-parametric testing:** when given data of observation in food quality measurements couldn't just meet critical assumptions for parametric tests; in that case, non-parametric testing is done. The following test used for this type of analysis: (a) **Wilcoxon Rank-Sum test** for two independent variables difference by considering magnitude and direction of difference; (b) **Wilcoxon Sign-Rank test** for the difference among two associated variables; by considering magnitude and direction of difference; (c) **The**

**sign test** is used for computing difference in two related variables; consider the only direction of difference while ignoring the magnitude of change.

### *17.4.2.2 STATISTICAL ANALYSIS FOR QUALITY OPTIMIZATION*

In food processing companies, the numbers of experimental designs are applied; to test ingredients, produce new products or reformulations, and optimize conditions for optimal processing. In order to achieve a final objective, certain designs, such as full factorial, fractional factorial, saturated, central composite, and mixture designs are used depending on the purpose,[3] as presented in Table 17.7.

**TABLE 17.7** Different Design Techniques for Quality Optimization.

| | |
|---|---|
| Fractional factorial design | When various factors are to be studied with less time and raw material, it reduces the number of trails. Researchers should consider which independent variables or factors and their interaction have a statistically significant effect on response variables. **Box-Behnken ($3^{k-p}$)** is a widely used fraction factorial method. It provides a more precise analysis due to more $d_f$ and experimental points.[3] |
| Full-factorial design | Practiced to determine important independent variables with the range of these factors in the study. **Two-level factorial designs ($2^k$)**; a mathematical model which describes linear relationships between factors and response variables.<br>$Y = \beta_o + \beta_1 x_1 + \beta_2 x_2 + \ldots\ldots + \beta_k x_k + \varepsilon$<br>The common type is used to find the optimum value. **The central composite design** is used for this study in response to surface methodology. Several studies related to central composite design using $2^2$, $2^3$, and $2^4$ factorial designs are analyzed by response surface methodology.[3]<br>**Three-level design ($3^k$)** in central composite design is inefficient for quadratic models, owing to lack of significant interactions. |
| Mixture design | They were used for new product development. This design gives an ideal composition of each ingredient for the best product development features (texture, odor, taste, appearance, etc.). Mixture design is used to develop several fruits-based functional products: juices, desserts, conc., and smoothies. |
| Saturated design | Plackett–Burman shows the best results in this case when there is not enough degree of freedom to compute interactions or when the high number of independent variables (>10) is tested. Thus, it allows understanding the effect of various physicochemical, biochemical, and sensory variables with the minimum number of trails. |

Irrespective of design type, there are relevant terms, such as qualitative and quantitative variables and block variables. Initial experimentation should be done to define the level of independent variables. Block variables have an indirect influence on response variables. Many environmental conditions changed day by day, so without block variables, analysis becomes statistically insignificant. For excellent results, CV should be below 10% and in agriculture, biology and microbiology, due to large dispersion in data CV is kept below 30%; otherwise, experiments would need repetition.

Afterward, selecting a mathematical model, significance is determined by ANOVA, that is, variance analysis. For that, the standard deviation of main and interaction factors is considered. If the standard deviation is lower than the mean value, then the mathematical model is significant. Another parameter, regression coefficient ($R^2$), is also one criterion for the goodness of fit. The mathematical models for $R^2$ below <70% are not good, while for >90%, other than the regression coefficient parameter, are used. But it is important to note that $R^2$ above 70% is also good, for example, in the sensory case depending on analysis type.

The last step is an optimization of process factors. RSM's main objective is to achieve an optimum value of control variables that affect maximum or minimum responses to achieve the desired quality. The desirability function is denoted by D and considered as the geometric mean of an individual.

$$D = (d_1 \times d_2 \times \ldots\ldots\ldots\ldots \times d_m)^{1/m} \quad (17.13)$$

Where, m is the number of response variables.

### 17.4.3 MULTIVARIATE STATISTICAL METHODS

Multivariate statistical methods are used to analyze the combined effect of more than one variable and finding the most affecting factor. Therefore, multivariate statistical techniques are utilized to measure the buried attributes and their impact on quality. The main goal of this method is to develop a model for calibration to correlate a set of measurements information to the desired property. It's a powerful tool for computing a large amount of data with multiple numbers of variables. There is a wide range of multivariate techniques available, such as matrix plot, partial least squares regression (PLSR), principal component regression (PCR), modified partial least squares regression (MPLSR), multiple variable correlation analysis, and matrix plot. Overfitting and underfitting of data should be avoided while selecting techniques. The optimum number of components is determined to prevent overfitting.

On the contrary, underfitting caused by few components makes models enable to capture variability in data. Therefore, these methods can depict the most influential element from data. This is the latest technique and is very beneficial for the production line to produce quality products.

### *17.4.4 APPLICATION OF STATISTICAL TECHNIQUES ON VARIOUS QUALITY PARAMETERS OF FOOD*

Statistical analysis of data is of prime importance to analyze the quality of data. The statistical techniques applied for it are descriptive, parametric, and non-parametric. These are highly helpful in extracting the relevant information from the results obtained from actual experimentation.

#### *17.4.4.1 CASE STUDY 1: APPLICATION FOR COMPUTING BASIC STATISTICAL VALUES*

To ascertain clay content, analysis the size of particle, 20 mL a semi-automatic pipette is utilized. This level of volume and opening and closing of taps is estimated. Calibration of pipette for both the accuracy (trueness) and precision is done. The amount of volume (mL) was taken in 10 sets of data given below:

19.41, 19.811, 19.830, 19.829, 19.741, 19.796, 19.938,
19.848, 19.884, 19.805

Therefore: mean = 19.841 mL, $\sigma = 0.0628$ mL.

$$t_{tab} = 2.26 \text{ for } n = 10 \textit{ with } df = 9:$$

$$\text{Pipette volume} = 19.841 \pm 2.26\,(0.0628/\sqrt{10}) = 19.84 \pm 0.04 \text{ mL}$$

(Results indicate that in the volume,
there is a systematic deviation from 20 mL).

#### *17.4.4.2 CASE STUDY 2: CALCULATION OF PROPAGATION OF ERRORS*

To calculate propagation errors, we use derivatives of charged ions of soil, such as Ca, K, Mg, Na, and acids.

The standard derivatives of these values are 0.40, 0.25, 0.15, 0.15, and 0.60 $cmol_c$/kg, respectively.

For precision, $S_{ECEC}$ is calculated:

$$S_{ECEC} = \sqrt{0.4^2 + 0.21^2 + 0.16^2 + 0.14^2 + 0.7^2} = 0.86 \text{ cmol}_c\text{/kg } 0.7393$$

**Interpretation of Results:** precision $S_{ECEC}$ (0.86 $cmol_c$/kg) is more significant than the maximum σ (0.70 $cmol_c$/kg) of an individual value; while below than their summation. Therefore, by reducing maximum individual impact, the total standard deviation can be reduced. Furthermore, exchangeable acidity is found in this case.

#### *17.4.4.3 CASE STUDY 3: MULTIPLICATION CALCULATION*

The Kjeldahl-nitrogen is calculated as follows:

$$\% N = \frac{a-b}{s} \times M \times 1.4 \times mcf \quad (17.14)$$

where, $a$ = requirement of HCl for sample titration; $b$ = titration of blank with HCl; $s$ = weight of dry sample in gram; $M$ = HCl molarity. The atomic weight of N = 14 normality is equal to 1.4, that is, 14 × $10^{-3}$ × 100%, the moisture correction factor is denoted with $mc_f$

Firstly, titration standard deviation *(a-b)* is evaluated, and then it is converted into the relative standard deviation (RSD). Further, the RSD of all individuals was calculated experimentally. **RSD** was 0.8% for distillation, 0.5% (titration), 0.2% (molarity), 0.2% (sample weight), with 0.2% $mc_f$

The total precision is:

$$RSD\,kjeldahl = \sqrt{0.8^2 + 0.5^2 + 0.2^2 + 0.2^2 + 0.2^2} = 1.0\%$$

The RSD of distillation was highest than individuals and dominated the total precision. Sample heterogeneity caused a random error of 1.5% (2.5–1.0% = 1.5%). This means proper mixing of samples should be done for improved results. The moisture correction factor also showed heterogeneity.

#### *17.4.4.4 CASE STUDY 4: DIFFERENCE BETWEEN TWO SET OF VARIABLES FROM THE SAME POPULATION*

Table 17.8 gives information about cation exchange capacity datasets analyzed by the two analysts of a control sample. The $s_1$ and $s_2$ denote the set of data.

$$F = \frac{s_1^2}{s_2^2} \tag{17.15}$$

**TABLE 17.8** CEC Values ($cmol_c$/kg) of a Control Sample Determined by Two Analysts.

| Serial Number | X1 | X2 |
|---|---|---|
| 1 | 10.2 | 9.7 |
| 2 | 10.7 | 9.0 |
| 3 | 10.5 | 10.2 |
| 4 | 9.9 | 10.3 |
| 5 | 9.0 | 10.8 |
| 6 | 11.2 | 11.1 |
| 7 | 11.5 | 9.4 |
| 8 | 10.9 | 9.2 |
| 9 | 8.9 | 9.8 |
| 10 | 10.6 | 10.2 |
| $\bar{x}$ | **10.34** | **9.97** |
| $s$ | **0.819** | **0.644** |
| $n$ | **10** | **10** |
| $\mathbf{F_{cal} = 1.62}$ | $\mathbf{F_{tab} = 4.03}$ | |
| $t_{cal} = 1.12$ | $t_{tab} = 2.10$ | |

The calculated F value is 1.62 and applicable $F_{tab} = 4.03$ with 9 degrees of freedom for both sets. Therefore, $F_{tab}$ is greater than $F_{cal}$, so the null hypothesis is accepted. It is concluded through 95% confidence with no significant difference in precision between Analyst 1 and 2.

### 17.4.4.5 *CASE STUDY 5: EXAMPLE OF ANOVA TECHNIQUE*

**TABLE 17.9** Evaluating ANOVA Technique.

| Source | SS | DF | MS | F |
|---|---|---|---|---|
| **Treatments** | 27.896 | 02 | 13.948 | 9.58 |
| **Error** | 17.451 | 12 | 1.455 | – |
| **Total (corrected)** | 45.348 | 14 | – | – |
| **Correction Factor** | 779.042 | 01 | – | – |

**Interpretation of the values in ANOVA test (Table 17.9):** The test statistic $F_{0.05} = 3.89$ is more significant than the $F_{0.05} = 2.12$ = critical value. Therefore, this implies that the null hypothesis is rejected. Also, it implies that mean of a population is significantly different. The *p*-value is 0.00324 for 9.58; therefore, the test statistic is significant at that level.

## 17.5 SUMMARY

Nowadays, consumer purchasing pattern toward quality food products is changing. They are becoming more health-conscious and concerned about product quality. That is why production companies needs proper quality assessment through statistical methods. To maintain the quality and nature of food, there are several techniques such as discipline analysis, statistical process control, acceptance sampling, and optimization of conditions through the surface methodology. These statistical techniques summarize the data and give one or two conditions under which the best quality products can be developed. So, the application of these statistical techniques ensures and validates the experimental data, thus eliminating results in quality control.

## KEYWORDS

- **statistical process control**
- **statistical quality control**
- **food quality**
- **statistical analysis**
- **food safety**

## REFERENCES

1. Atkins, P.; Bowler, I. *Food in Society: Economy, Culture, Geography*, 1st ed.; Hodder Education Arnold Press: New York, 2001; p 325.
2. Dogan, H. and Kokini J. L. Rheological Properties of Foods. In *Handbook of Food Engineering*; Heldman, D. R.; Lund, D. B.; Sabliov, C., Eds., Chapter 1, Vol. 3; CRC-Press: Boca Raton, 2006; pp 1–124.
3. Granato, D. *Mathematical & Statistical Methods in Food Science & Technology*; Wiley Blackwell: IIT Press: London, 2014; p 511.
4. Granato, D.; Calado, V.; Oliveira, C.; Ares, G. Statistical Approaches to Assess the Association between Phenolic Compounds and the In Vitro Antioxidant Activity of

*Camellia sinensis* and *Ilex paraguariensis* Teas. *Crit. Rev. Food Sci. Nutr.* **2013,** *55* (10), 1456—1473.

5. Iwaoka, W.; Britten, P.; Dong, F. The Changing Face of Food Science Education. *Trends Food Sci. Technol.* **1996,** *7*, 105–112.
6. Keenan, D.; Brunton, N.; Mitchell, M.; Gormley, R.; Butler, F. Flavor Profiling of Fresh and Processed Fruit Smoothies by Instrumental and Sensory Analysis. *Food Res. Int.* **2012,** *45*, 17–25.
7. Kravchuk, O.; Elliott, A.; Bhandari, B. Laboratory Experiment, Based on the Maillard Reaction, Conducted as a Project in Introductory Statistics. *J. Food Sci. Educ.* **2006,** *4*, 70–75.
8. Lasztity, R.; Petro, T. M.; Foldesi, T. History of Food Quality Standards. In *Food Quality and Standards*; Lasztity, R., Ed.; EOLSS Publisher, Inc.: London, 2004; pp 10–21.
9. Montgomery C. D. *Introduction to Statistical Quality Control.* 7th ed.; John Wiley & Sons: New York, 2013; p 749.
10. Puri C. S. *Statistical Methods for Food Quality Management*; Hall Press: New York, 1989; p 167.
11. Rajput, H.; Rehal, J.; Goswami, D.; Mandge, H. M. Methods for Food Analysis and Quality Control. In *State-of-the-Art Technologies in Food Science: Human Health, Emerging Issues and Specialty Topics*; Meghwal, M., Goyal, M. R., Eds., Chapter 10; Apple Academic Press Inc.: New Jersey, 2019; pp 300–318.
12. Shah, S.; Shridhar, P.; Gohil, D. Control Chart: A Statistical Process Control Tool—In Pharmacy. *Asian J. Pharma.* **2010,** *2010*, 184–192.
13. Wilhelm, L. R.; Suter, D. A.; Brusewitz, H. B. Physical Properties of Food Materials. In *Food & Process Engineering Technology*; Wilhelm, L. R., Suter, D. A., Brusewitz, H. B., Eds., Chapter 2; American Society of Agricultural Engineers. Inc.: St. Joseph, MI, 2004; pp 23–52.

# INDEX

## D



## E



## F

## T



## U



## W



## Y